国际材料前沿丛书
International Materials Frontier Series

ELSEVIER

Fundamentals
of Creep in Metals
and Alloys

B
H

M. E. Kassner

（第3版）

金属与合金蠕变的基本原理

Fundamentals of Creep in Metals and Alloys
(Third Edition)

影印版

中南大学出版社
www.csupress.com.cn

·长沙·

图字：18 - 2017 - 162 号

Fundamentals of Creep in Metals and Alloys(Third Edition)

M. E. Kassner

ISBN：9780080994277

Elsevier (Singapore) Pte Ltd.

3 Killiney Road

#08 - 01 Winsland House I

Singapore 239519

Tel：(65) 6349 - 0200

Fax：(65) 6733 - 1817

First Published < 2017 >

< 2017 > 年初版

内容简介

本书主要介绍了金属与合金蠕变的基本原理，评述和分析了与塑性蠕变有关的一系列领域的研究成果。主要内容包括材料蠕变的基本原理、五次幂蠕变、扩散蠕变、Harper – Dorn 蠕变、三次幂黏滞滑移蠕变、超塑性、再结晶、颗粒强化合金的蠕变行为、金属间化合物的蠕变、蠕变断裂、γ/γ´镍基超合金、非晶态金属的蠕变以及低温蠕变性能。

本书在前两个版本的基础上，不仅扩充了其他领域的塑性蠕变现象，增加了金属和合金中蠕变的最新进展，还扩充了陶瓷、共价固体、矿物和聚合物的蠕变行为，着重讨论了各种材料的蠕变基础理论。

本书可供材料、冶金、工业设计、航空航天等领域的科研人员、工程技术人员使用，同时也可作为高等院校材料、冶金、工业设计、航空航天等相关专业学生的参考书。

作者简介

M. E. Kassner 美国洛杉矶南加州大学航空与机械工程系教授。1972 年获得西北大学学士学位，1979 年和 1981 年分别获得美国斯坦福大学材料科学与工程专业硕士、博士学位。主要从事金属塑性理论、蠕变、断裂、相图、疲劳和半固态成型等研究。出版教材、专著 2 部，发表论文 200 多篇。目前为 ELSEVIER《国际可塑性杂志》编委会成员、美国 ASM 国际学会会士。

序

　　这本关于蠕变塑性基本原理的著作，评述和分析了与塑性蠕变有关的一系列领域的研究成果。这些领域包括五次幂蠕变（在金属、合金与陶瓷中有时被称作位错攀移控制蠕变）、合金中的黏滞滑移或三次幂蠕变、扩散蠕变、Harper – Dorn 蠕变、超塑性、第二相强化、蠕变成穴与断裂。在本著作之前，有许多高水平评述和专著对蠕变基本原理进行了广泛评价，要在此基础上有所提高是一个挑战。我们的优势是在上述早期评论之后有大量的相关研究成果供我们写作参考。我们试图在涵盖早期评论中探讨的基础工作之上，把侧重点放在近期研究进展上。

　　这是本书的第 2 版，改正了第 1 版的错误，包含了第 1 版面世后 5 年来的许多研究进展。Maria – Teresa Perez – Prado 博士是第 1 版的合作者。她没有参加第 2 版、第 3 版的撰写，但书中的第 5、6、9 章仍然主要是她的贡献，并在书中作了相关说明。

目　录

Fundamentals of
CREEP IN METALS
AND ALLOYS

Fundamentals of
CREEP IN METALS AND ALLOYS

THIRD EDITION

M.E. KASSNER

Departments of Aerospace and Mechanical Engineering,
Chemical Engineering, and Materials Science
University of Southern California
Los Angeles, CA
USA

AMSTERDAM • BOSTON • HEIDELBERG • LONDON
NEW YORK • OXFORD • PARIS • SAN DIEGO
SAN FRANCISCO • SINGAPORE • SYDNEY • TOKYO

Butterworth-Heinemann is an imprint of Elsevier

ELSEVIER

Butterworth Heinemann is an imprint of Elsevier
225 Wyman Street, Waltham, MA 02451, USA
The Boulevard, Langford Lane, Kidlington, Oxford, OX5 1GB, UK

Notice
Knowledge and best practice in this field are constantly changing. As new research and
experience broaden our understanding, changes in research methods, professional
practices, or medical treatment may become necessary.

Practitioners and researchers must always rely on their own experience and knowledge in
evaluating and using any information, methods, compounds, or experiments described
herein. In using such information or methods they should be mindful of their own safety
and the safety of others, including parties for whom they have a professional
responsibility.

To the fullest extent of the law, neither the Publisher nor the authors, contributors, or
editors, assume any liability for any injury and/or damage to persons or property as a
matter of products liability, negligence or otherwise, or from any use or operation of any
methods, products, instructions, or ideas contained in the material herein.

ISBN: 978-0-08-099427-7

British Library Cataloguing in Publication Data
A catalogue record for this book is available from the British Library

Library of Congress Cataloging-in-Publication Data
A catalog record for this book is available from the Library of Congress

For information on all Butterworth Heinemann publications
visit our web site at http://store.elsevier.com

Working together
to grow libraries in
developing countries

ELSEVIER | Book Aid International

www.elsevier.com • www.bookaid.org

CONTENTS

PREFACE

This book on the fundamentals of creep plasticity is a review and analysis of investigations in a variety of areas relevant to creep plasticity. These areas include five-power-law creep, which is sometimes referred to as dislocation climb-controlled creep (in metals, alloys, and ceramics), viscous glide or three-power-law creep (in alloys), diffusional creep, Harper–Dorn creep, superplasticity, second-phase strengthening, and creep cavitation and fracture. Many quality reviews and books precede this attempt to write an extensive review of creep fundamentals and the improvement was a challenge. One advantage with this attempt is the ability to describe the substantial work published subsequent to these earlier reviews. An attempt was made to cover the basic work discussed in these earlier reviews but especially to emphasize more recent developments.

This is the second edition of this book and one aspect of this recent edition is correcting errors in the first edition, also, many advances occurred over the five years since the first edition and theses are also incorporated. Dr Maria-Teresa Perez-Prado was a co-author of the first edition. While she did not participate in the formulation of the second and third editions, Chapters 5, 6, and 9 remain largely a contribution by Dr Perez-Prado, and her co-authorship is indicated on these chapters.

LIST OF SYMBOLS AND ABBREVIATIONS

a	Cavity radius
a_0	Lattice parameter
$A' - A''''$	Constants
A	
A_F	
A_{C-J}	Solute dislocation interaction parameters
A_{APB}	
A_{SN}	
A_{CR}	
A_{gb}	Grain boundary area
A_{HD}	Harper–Dorn equation constant
A_{PL}	Constants
A_v	Projected area of void
A_0–A_{12}	Constant
APB	Antiphase boundary
b	Burgers vector
B	Constant
BMG	Bulk metallic glass
c	Concentration of vacancies
c^*	Crack growth rate
c_j	Concentration of jogs
c_p	Concentration of vacancies in the vicinity of a jog
c_p^*	Steady-state vacancy concentration near a jog
c_v	Equilibrium vacancy concentration
c_v^D	Vacancy concentration near a node or dislocation
c_0	Initial crack length
c_{1-2}	Constants
C	Concentration of solute atoms
C^*	Integral for fracture mechanics of time-dependent plastic materials
C_{1-2}	Constant
C_{LM}	Larson–Miller constant
CBED	Convergent beam electron diffraction
CGBS	Cooperative grain boundary sliding
CS	Crystallographic slip
CSF	Complex stacking fault
CSL	Coincident site lattice
C_0^*	Constant
C_0–C_5	Constants
d	Average spacing of dislocations that comprise a subgrain boundary
D	General diffusion coefficient or constant
D'	Constant

D_c	Diffusion coefficient for climb
D_{eff}	Effective diffusion coefficient
D_g	Diffusion coefficient for glide
D_{gb}	Diffusion coefficient along grain boundaries
D_i	Interfacial diffusion
D_s	Surface diffusion coefficient
D_{sd}	Lattice self-diffusion coefficient
D_v	Diffusion coefficient for vacancies
D_0	Diffusion constant
DRX	Discontinuous dynamic recrystallization
$\tilde{D}$	Diffusion coefficient for the solute atoms
e	Solute–solvent size difference or misfit parameter
E	Young's modulus or constant
E_j	Formation energy for a jog
EBSP	Electron backscatter patterns
f	Fraction
f_m	Fraction of mobile dislocations
f_p	Chemical dragging force on a jog
f_{sub}	Fraction of material occupied by subgrains
F	Total force per unit length on a dislocation
FEM	Finite element method
g	Average grain size (diameter)
g'	Constant
G	Shear modulus
GBS	Grain boundary sliding
GDX	Geometric dynamic recrystallization
GNB	Geometrically necessary boundaries
h_r	Hardening rate
$\bar{h}_m$	Average separation between slip planes within a subgrain with gliding dislocations
h	Dipole height or strain-hardening coefficient
HAB	High angle boundary
HVEM	High voltage transmission electron microscopy
j	Jog spacing
J	Integral for fracture mechanics of plastic material
J_{gb}	Vacancy flux along a grain boundary
k	Boltzmann constant
$k'-k'''$	Constants
k_y	Hall–Petch constant
k_{MG}	Monkman–Grant constant
k_R	Relaxation factor
k_1-k_{10}	Constants
K	Strength parameter or constant
K_I	Stress intensity factor
K_0-K_7	Constants
ℓ	Link length of a Frank dislocation network
ℓ_c	Critical link length to unstably bow a pinned dislocation

ℓ_m	Maximum link length
l	Migration distance for a dislocation in Harper–Dorn creep
L	Particle separation distance
LAB	Low angle boundary
LM	Larson–Miller parameter
LRIS	Long-range internal stress
m	Strain-rate sensitivity exponent ($=1/N$)
m'	Transient creep time exponent
m''	Strain-rate exponent in the Monkman–Grant equation
m_c	Constant
$\overline{M}$	Average Taylor factor for a polycrystal
M_ρ	Dislocation multiplication constant
n	Steady-state creep exponent or strain-hardening exponent
n^*	Equilibrium concentration of critical sized nuclei
n_m	Steady-state stress exponent of the matrix in a multi-phase material
N	Constant structure stress exponent and dislocation link length per unit volume
$\dot{N}$	Nucleation rate and rate of release of dislocation loops
p	Steady-state dislocation density stress exponent
p'	Inverse grain size stress exponent for superplasticity
PLB	Power law breakdown
POM	Polarized light optical microscopy
PSB	Persistent slip band
q	Dislocation spacing, d, stress exponent
Q_c	Activation energy for creep (with E or G compensation)
Q'_c	Apparent activation energy for creep (no E or G compensation)
Q_p	Activation energy for dislocation pipe diffusion
Q_{sd}	Activation energy for lattice self-diffusion
Q_v	Formation energy for a vacancy
Q^*	Effective activation energies in composites where load transfer occurs
r_r	Recovery rate
R_o	Diffusion distance
R_s	Radius of solvent atoms
s	Structure
SAED	Selected area electron diffraction
SESF	Superlattice extrinsic stacking fault
SISF	Superlattice intrinsic stacking fault
STZ	Shear transformation zone
t	Time
t_c	Time for cavity coalescence on a grain boundary facet
t_f	Time to fracture (rupture)
t_s	Time to the onset of steady-state
T	Temperature
T_d	Dislocation line tension
T_g	Glass transition temperature
T_m	Melting temperature
T_p	Temperature of the peak yield strength

T_x	Onset crystallization temperature
TEM	Transmission electron microscopy
$T-T-T$	Time–temperature–transformation diagram
v	Dislocation glide velocity
v_c	Dislocation climb velocity
v_{cr}	Critical dislocation velocity at breakaway
v_D	Debye frequency
v_p	Jog climb velocity
$\bar{v}$	Average dislocation velocity
$\bar{v}_\ell$	Climb velocity of dislocation links of a Frank network
V	Activation volume
w	Width of a grain boundary
$\bar{x}_c$	Average dislocation climb distance
$\bar{x}_g$	Average dislocation slip length due to glide
XRD	X-ray diffraction
α	Taylor equation constant
α_o	Constant
α'	Climb resistance parameter
α_{1-3}	Constants
β, β_{1-3}	Constants
γ	Shear strain
γ_o	Characteristic strain of an STZ
γ_A	Anelastic unbowing strain
γ_{gb}	Interfacial energy of a grain boundary
γ_m	Surface energy of a metal
$\dot{\gamma}$	Shear creep rate
$\dot{\gamma}_{ss}$	Steady-state shear creep rate
δ	Grain boundary thickness or lattice misfit
Δa	Activation area
ΔG	Gibbs free energy
ΔV_C	Activation volume for creep
ΔV_L	Activation volume for lattice self-diffusion
ϵ	Uniaxial strain
ϵ_o	Instantaneous strain
$\dot{\epsilon}$	Strain rate
$\dot{\epsilon}_{min}$	Minimum creep rate
$\dot{\epsilon}_{ss}$	Steady-state uniaxial strain rate
$\bar{\epsilon}$	Effective uniaxial or von Mises strain
θ	Misorientation angle across high-angle grain boundaries
θ_λ	Misorientation angle across (low-angle) subgrain boundaries
$\theta_{\lambda ave}$	Average misorientation angle across (low-angle) subgrain boundaries
λ	Average subgrain size (usually measured as an average intercept)
λ_s	Cavity spacing
λ_{ss}	Average steady state subgrain size
ν	Poisson's ratio
ν_o	Attempt frequency (often Debye frequency)
ρ	Density of dislocations not associated with subgrain boundaries

ρ_m	Mobile dislocation density
ρ_{ms}	Mobile screw dislocation density
ρ_{ss}	Steady-state dislocation density not associated with subgrain walls
σ	Applied uniaxial stress
σ_i	Internal stress
σ_o	Single crystal yield strength
σ'_o	Annealed polycrystal yield strength
σ''_o	Sintering stress for a cavity
σ_p	Peierls stress
σ_{ss}	Uniaxial steady-state stress
σ_T	transition stress between five-power law and Harper–Dorn creep
σ_{TH_s}	Threshold stress for superplastic deformation
$\sigma_y\big|_{T,\dot{\varepsilon}}$	Yield or flow stress at a reference temperature and strain rate
$\sigma_y^{0.002}$	0.2% offset yield stress
$\bar{\sigma}$	Effective uniaxial, or von Mises, stress
τ	Shear stress
τ_b	Breakaway stress of the dislocations from solute atmospheres
τ_c	Critical stress for climb over a second phase particle
τ_d	Detachment stress from a second phase particle
τ_j	Stress to move screw dislocations with jogs
τ_{or}	Orowan bowing stress
τ_B	Shear stress necessary to eject dislocation from a subgrain boundary
τ_{BD}	Maximum stress from a simple tilt boundary
τ_L	Stress to move a dislocation through a boundary resulting from jog creation
τ_N	Shear strength of a Frank Network
$(\tau/G)_t$	Normalized transition stress
$\phi(P)$	Frank network frequency distribution function
χ	Stacking fault energy
χ'	Primary creep constant
ψ	Angle between cavity surface and projected grain boundary surface
ω_m	Maximum interaction energy between a solute atom and an edge dislocation
Ω	Atomic volume
ω	Fraction of grain boundary area cavitated

CHAPTER 1

Fundamentals of Creep in Materials

Contents

1. INTRODUCTION

1.1 Description of Creep

Creep of materials is classically associated with time-dependent plasticity under a fixed stress at an elevated temperature, often greater than roughly $0.5\ T_m$, where T_m is the absolute melting temperature. The plasticity under these conditions is described in Figure 1 for constant stress (a) and constant strain rate (b) conditions. Several aspects of the curve in Figure 1 require explanation. First, three regions are delineated: Stage I, or primary creep, which denotes that portion where (in (a)) the creep rate (plastic strain rate), $\dot{\varepsilon} = d\varepsilon/dt$ is changing with increasing plastic strain or time. In Figure 1(a), the primary creep rate decreases with increasing strain, but with some types of creep, such as solute drag with "3-power creep," an "inverted" primary occurs where the strain rate increases with strain. Analogously, in (b), under constant strain rate conditions, the metal hardens, resulting in increasing flow stresses. Often, in pure metals, the strain rate decreases or the stress increases to a value that is constant over a range of strain. The phenomenon is termed Stage II, secondary, or steady-state (SS) creep. Eventually, cavitation and/or cracking increases the apparent strain rate or decreases the flow stress. This regime is termed Stage III, or tertiary, creep and leads to fracture. Sometimes, Stage I leads directly to Stage III and an "inflection" is observed. Thus, care must sometimes be exercised in concluding a mechanical SS.

The term "creep" as applied to plasticity of materials likely arose from the observation that at modest and constant stress, at or even below the macroscopic yield stress of the metal (at a "conventional" strain rate), plastic deformation occurs over time as described in Figure 1(a). This is in contrast

Fundamentals of Creep in Metals and Alloys
ISBN 978-0-08-099427-7
http://dx.doi.org/10.1016/B978-0-08-099427-7.00001-3

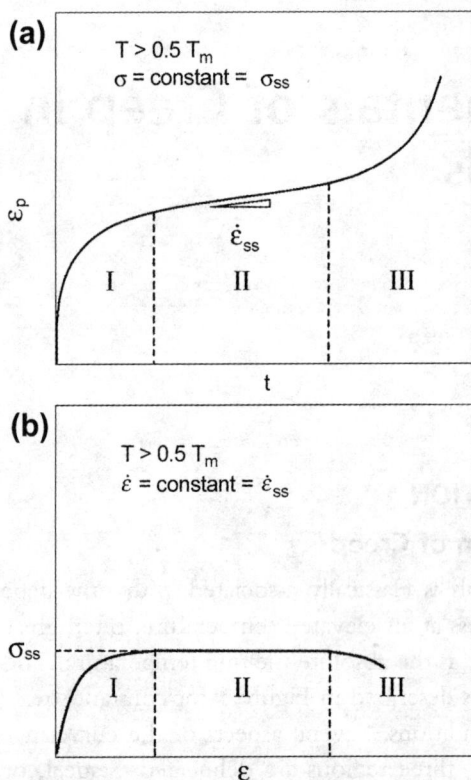

Figure 1 Constant true stress and constant strain rate creep behavior in pure and Class M (or Class I) metals.

with the *general* observation, such as at ambient temperature, where a material deformed at, for example, 0.1–0.3 T_m, shows very little plasticity under constant stress at or below the yield stress, again, at "conventional" or typical tensile testing strain rates (e.g., 10^{-4}–10^{-3} s^{-1}). (The latter observation is not always true as it has been observed that some primary creep is observed (e.g., a few percent strain, or so) over relatively short periods of time at stresses less than the yield stress (e.g., [1,2])).

We observe in Figure 2 that at the "typical" testing strain rate of about 10^{-4} s^{-1}, the yield stress is σ_{y1}. However, if we decrease the testing strain rate to, for example, 10^{-7} s^{-1}, the yield stress decreases significantly, as will be shown is common for metals and alloys at high temperatures. To a "first approximation," we might consider the microstructure (created by dislocation microstructure evolution with plasticity) at just 0.002 plastic strain to be independent of $\dot{\varepsilon}$. In this case, we might describe the change in yield

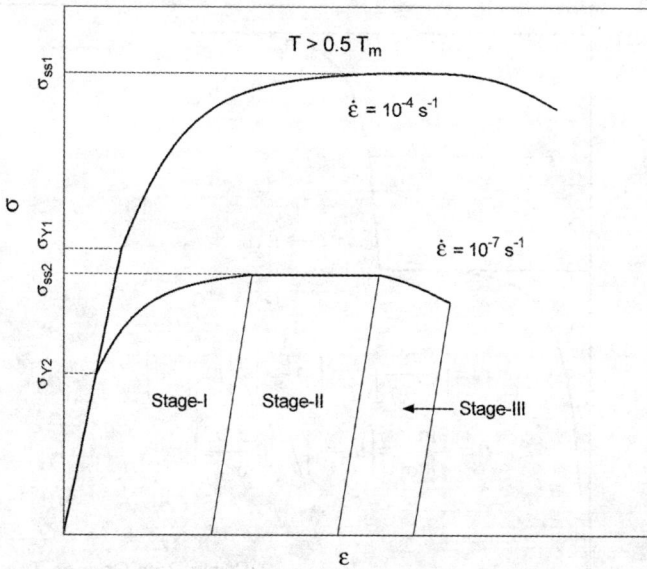

Figure 2 Creep behavior at two different constant strain rates.

stress to be the sole result of the $\dot{\varepsilon}$ change and predicted by the "constant structure" stress-sensitivity exponent, N, defined by.

$$N = [\partial \ln \dot{\varepsilon} / \partial \ln \sigma]_{T,s} \qquad (1)$$

where T and s refer to temperature and the substructural features, respectively. Sometimes, the sensitivity of the creep rate to changes in stress is described by a constant structure strain-rate sensitivity exponent, $m = 1/N$. Generally, N is relatively high at lower temperatures [3] which implies that significant changes in the strain rate do not dramatically affect the flow stress. In pure fcc metals, N is typically between 50 and 250 [3]. At higher temperatures, the values may approach 10, or so [3–10]. N is graphically described in Figure 3. The trends of N versus temperature for nickel are illustrated in Figure 4.

Another feature of the hypothetical behaviors in Figure 2 is that (at the identical temperature) not only is the yield stress at a strain rate of 10^{-7} s^{-1} lower than it is at 10^{-4} s^{-1}, but also the peak stress or, perhaps, SS stress, which is maintained over a substantial strain range, is *less* than the yield stress at a strain rate of 10^{-4} s^{-1}. (Whether SS occurs at, for example, ambient temperature has not been fully settled, as large strains are not easily achievable. Stage IV and/or recrystallization may preclude this SS [11–13].)

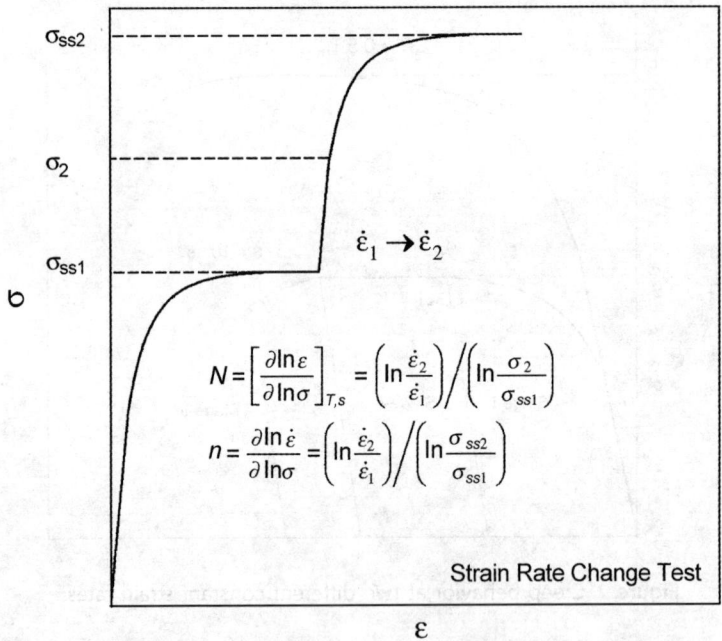

Figure 3 A graphic description of the constant structure strain rate sensitivity experiment, N ($1/m$) and the steady-state stress exponent, n.

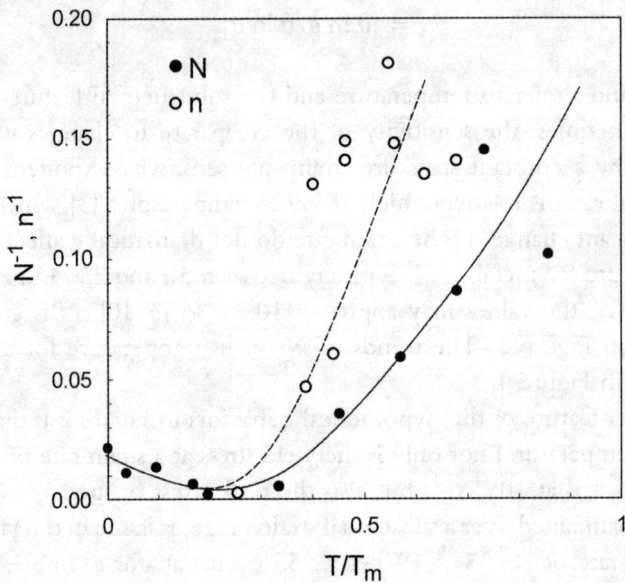

Figure 4 The values of n and N as a function of temperature for nickel. *Data from Ref. [7].*

Thus, if a constant stress σ_{ss_2} is applied to the material, then a substantial strain may be easily achieved at a low strain rate despite the stress being substantially below the "conventional" yield stress at the higher rate of $10^{-4}\,\mathrm{s}^{-1}$. Thus, creep is, basically, a result of significant strain rate sensitivity together with low strain hardening. We observe in Figure 4 that N decreases to relatively small values above about 0.5 T_m, while N is relatively high below about this temperature. This implies that we would expect that "creep" would be more pronounced at higher temperatures, and less obvious at lower temperatures, because, as will be shown subsequently, work-hardening generally diminishes with increasing temperature and N also decreases (more strain rate sensitive). The above description/explanation for creep is consistent with earlier descriptions [14]. Again, it should be emphasized that the maximum stress, σ_{ss_2}, in a constant strain rate $(\dot{\varepsilon})$ test, is often referred to as a SS stress (when it is the result of a balance of hardening and dynamic recovery processes, which will be discussed later). The creep rate of $10^{-7}\,\mathrm{s}^{-1}$ that leads to the SS stress (σ_{ss_2}) is the *same* creep rate that would be achieved in a constant stress test at σ_{ss_2}. Hence, at σ_{ss_2}, $10^{-7}\,\mathrm{s}^{-1}$ is the SS creep rate. The variation in the SS creep rate with the applied stress is often described by the SS stress exponent, n, defined by.

$$n = [\delta \ln \dot{\varepsilon}_{ss}/\delta \ln \sigma_{ss}]_T \qquad (2)$$

This exponent is described in Figure 3. Of course, with hardening, n is expected to be less than N. This is illustrated in Figures 3 and 4. As just mentioned, generally, the lower the strain rate, or higher the temperature, the less pronounced is the strain hardening. This is illustrated in Figure 5, reproduced from [15], where the stress-versus-strain behavior of high-purity aluminum is illustrated over a wide range of temperatures and strain rates. All these tests use a constant strain rate. The figure shows that with increasing temperature, the yield stress decreases, as expected. Also, for a given temperature, increases in strain rate are associated with increases in the yield stress of the annealed aluminum. That is, increases in temperature and strain rate tend to oppose each other with respect to flow stress. This can be rationalized by considering plasticity to be a thermally activated process. Figure 5 also illustrates that hardening is more dramatic at lower temperatures (and higher strain rates). The general trend that the strain to achieve SS increases with increasing stress (decreasing temperature and/or increasing strain rate) is also illustrated. This explanation describes constant-stress behavior in terms of constant strain rate stress-versus-strain behavior and cannot be considered a fundamental explanation of creep.

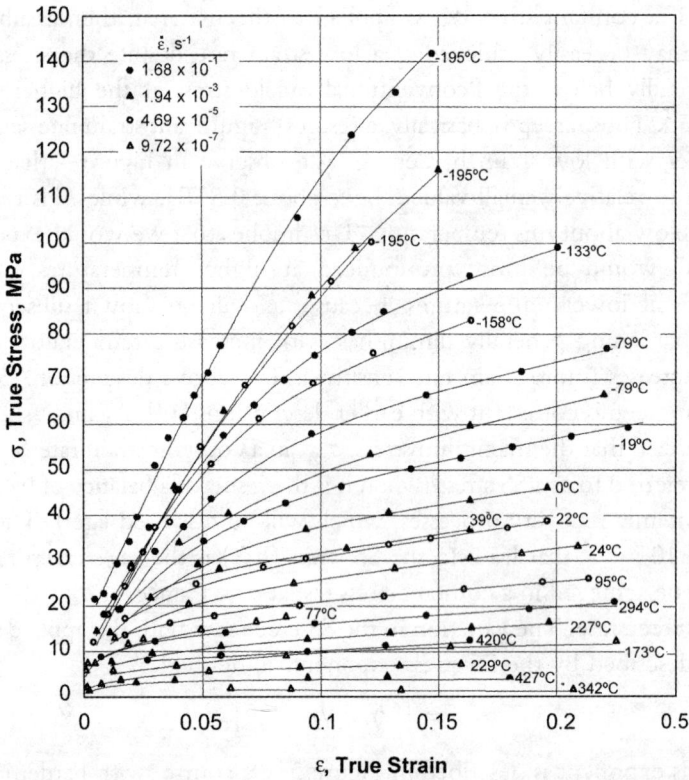

Figure 5 The stress-versus-strain behavior of high-purity aluminum. *Data from Ref. [15].*

1.2 Objectives

There have been other, often short, reviews of creep—notably, Sherby and Burke [16], Takeuchi and Argon [17], Argon [18], Orlova and Cadek [19], Cadek [20], Mukherjee, [21], Blum [22], Nabarro and de Villiers [23], Weertman [24,25], Evans and Wilshire [28], Kassner and Pérez-Prado [29], and others [30–32]. These, however, often do not include some important recent work and have sometimes been relatively brief (and, as a result, are not always very comprehensive). Thus, it was believed important to provide a new description of creep that is extensive, current, and balanced. Creep is discussed in the context of traditional 5-power-law creep, Nabarro-Herring, Coble, diffusional creep, Harper-Dorn, low-temperature creep (power-law-breakdown), as well as with 3-power viscous glide creep, particle-strengthening super-plasticity, low-temperature creep, creep of amorphous metals and alloys, and creep fracture. Each will be discussed separately. Deformation maps have been formulated for a variety of metals [33].

CHAPTER 2

Five-Power-Law Creep

Contents

Fundamentals of Creep in Metals and Alloys
ISBN 978-0-08-099427-7
http://dx.doi.org/10.1016/B978-0-08-099427-7.00002-5

1. MACROSCOPIC RELATIONSHIPS

1.1 Activation Energy and Stress Exponents

In pure metals and Class M alloys (similar creep behavior to pure metals), there is an early, established, largely phenomenological relationship between the steady-state strain rate, $\dot{\varepsilon}_{ss}$ (or creep rate), and stress, σ_{ss}, for steady-state five-power-law (PL) creep:

$$\dot{\varepsilon}_{ss} = A_0 \exp[-Q_c/kT](\sigma_{ss}/E)^n \tag{3}$$

where A_0 is a constant, k is Boltzmann's constant, and E is Young's modulus (although, as will be discussed subsequently, the shear modulus, G, can also be used). This is consistent with Norton's Law [34]. The activation energy for creep, Q_c, has been found to often be about that of lattice self-diffusion, Q_{sd}. The exponent n is constant and is about 5 (4–7) for pure metals, ceramics, and many alloys over a relatively wide range of temperatures and strain rates (hence "five-power-law" behavior) until the temperature decreases below roughly 0.5–0.6 T_m, where power-law breakdown (PLB) occurs, and n increases and Q_c generally decreases. Steady-state creep is often emphasized over primary or tertiary creep due to the relatively large fraction of creep life within this regime. The importance of steady-state creep is evidenced by the empirical Monkman-Grant relationship [35]:

$$\dot{\varepsilon}_{ss}^{m''} t_f = k_{MG} \tag{4}$$

where t_f is the time to rupture and m'' and k_{MG} are constants.

A hyperbolic sine (sinh) function is often used to describe the transition from PL to PLB:

$$\dot{\varepsilon}_{ss} = A_1 \exp[-Q_c/kT][\sinh \alpha_1 (\sigma_{ss}/E)]^5 \tag{5}$$

(although some have suggested that there is a transition from five- to seven-power-law behavior prior to PLB [25,36], and this will be discussed later). Equations (3) and (5) will be discussed in detail subsequently. The discussion of five-power-law creep will be accompanied by a significant discussion of the lower-temperature companion, PLB.

As discussed earlier, time-dependent plasticity or creep is traditionally described as a permanent or plastic extension of the material under fixed applied stress. This is usually illustrated for pure metals or Class M alloys (again, similar behavior to pure metals) by the constant stress curve of Figure 1, which also illustrates, of course, that creep plasticity can occur under constant strain-rate conditions as well. Stage I, or primary creep, occurs when the material experiences hardening through changes in the dislocation substructure. Eventually Stage II, or secondary or steady-state creep, is observed. In this region, hardening is balanced by dynamic recovery (e.g., dislocation annihilation). The consequence of this is that the creep rate or plastic strain rate is constant under constant true von Mises stress (tension, compression, or torsion). In a constant strain-rate test, the flow stress is independent of plastic strain except for changes in texture (e.g., changes in the average Taylor factor of a polycrystal), often evident in larger strain experiments (such as $\varepsilon > 1$) [37–39]. It will be illustrated that a genuine mechanical steady state is achievable. As mentioned earlier, this stage is particularly important as large strains can accumulate during steady state at low, constant stresses that can lead to failure.

Since Stage II or steady-state creep is important, the creep behavior of a material is often described by the early steady-state creep plots such as in Figure 6 for high-purity aluminum [16]. The tests were conducted over a range of temperatures from near the melting temperature to as low as $0.57\ T_m$. Data have been considered unreliable below about $0.3\ T_m$, as it has recently been shown that dynamic recovery is not the exclusive restoration mechanism [11], since dynamic recrystallization in 99.999% pure Al has been confirmed. Dynamic recrystallization becomes an additional restoration mechanism that can preclude a constant flow stress (for a constant strain rate) or a "genuine" mechanical steady state, defined here as a balance between dynamic recovery and hardening. The plots in Figure 6 are important for several reasons. First, the steady-state data are in sets at fixed temperatures and it is not necessary for the stress to be modulus-compensated to illustrate stress dependence, e.g., Eqn (3). Thus, the power-law behavior is clearly evident for each of the four temperature sets of high-purity aluminum data without any ambiguity (from modulus compensation). The stress exponent is about 4.5 for aluminum. Although this is not precisely 5, it is constant over a range of temperature, stress, and strain rate, and falls within the range of 4–7 observed in pure metals and class M alloys (as will be shown later, ceramics may have exponents less than this range). This range has been conveniently termed "five power." Some

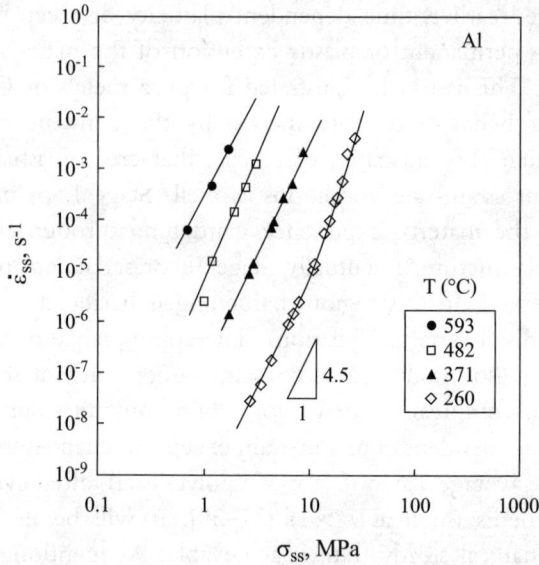

Figure 6 The steady-state stress versus strain rate for high-purity aluminum at four temperatures. *From Ref. [16].*

have referred to five-power-law creep as "dislocation climb-controlled creep," but this term may be misleading as climb control appears to occur in other regimes such as Harper-Dorn, superplasticity, power-law breakdown (PLB), etc. We note from Figure 6 the slope increases with increasing stress and the slope is no longer constant with changes in the stress at higher stresses (often associated with lower temperatures). Again, this is PLB and will be discussed later. The activation energy for steady-state creep, Q_c, calculations have been based on plots similar to Figure 6. The activation energy here simply describes the change in (steady-state) creep rate for a given substructure (strength), at a fixed applied "stress" with changes in temperature. It will be discussed in detail later, for at least steady state, that the microstructures of specimens tested at different temperatures appear *approximately* identical, by common microstructural measures, for a fixed *modulus-compensated* stress, σ_{ss}/E or σ_{ss}/G. (Modulus compensation (modest correction) will be further discussed later.) For a given substructure, s, and relevant "stress," σ_{ss}/E (again, it is often assumed that a constant σ_{ss}/E or σ_{ss}/G implies constant structure, s), the activation energy for creep, Q_c, can be defined by:

$$Q_c = -k[\delta(\ln/\dot{\varepsilon}_{ss})/\delta(1/T)]_{\sigma_{ss}/E,s} \qquad (6)$$

It has been very frequently observed that Q_c seems to be essentially equal to the activation energy for lattice self-diffusion Q_{sd} for a large number of materials. This is illustrated in Figure 7, where over 20 (bcc, fcc, hcp, and other crystal structures) metals show excellent correlation between Q_c and Q_{sd} (although it is not certain that this figure includes the (small) modulus compensation). Another aspect of Figure 7 that is strongly supportive of the activation energy for five-power-law creep being equal to Q_{sd} is based on activation volume (ΔV) analysis by Sherby and Weertman [5]. That is, the effect of (high) pressure on the creep rate $(\partial \dot{\varepsilon}_{ss}/\partial P)_{T,\sigma_{ss}/E(\text{or } G)} = \Delta V_c$ is the same as the known dependence of self-diffusion on the pressure $(\partial D_{sd}/\partial P)_{T,\sigma_{ss}/E(\text{or } G)}$. Other more recent experiments by Campbell, Tao, and Turnbull on lead have shown that additions of solute that affect self-diffusion also appear to identically affect the creep rate [40]. Some modern superalloys have their creep resistance improved by solute additions that decrease D_{sd} [600].

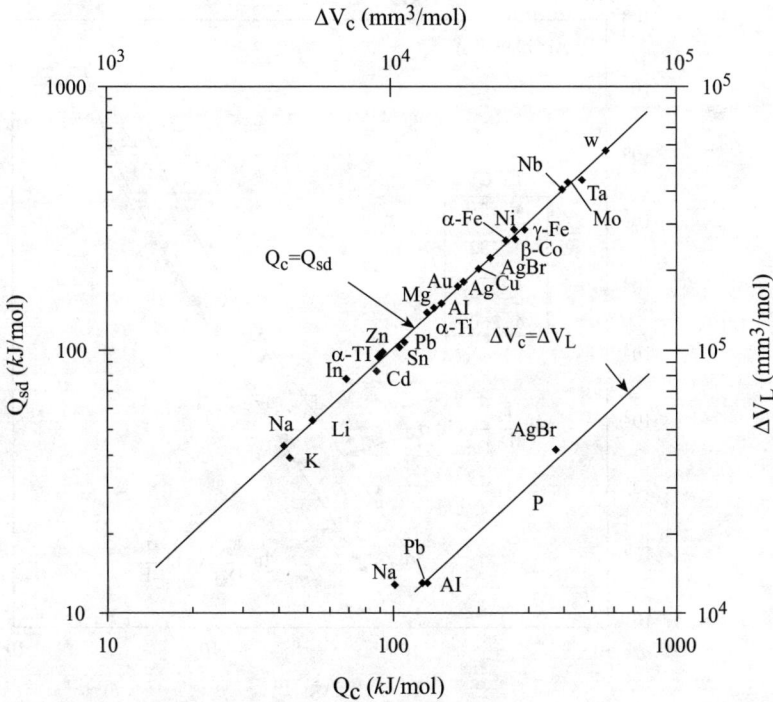

Figure 7 The activation energy and volume for lattice self-diffusion versus the activation energy and volume for creep. *Data from Ref. [26].*

Figure 8 describes the data of Figure 9 on what appears as a nearly single line by compensating the steady-state creep rates ($\dot{\varepsilon}_{ss}$) at several temperatures by the lattice self- diffusion coefficient, D_{sd}. At higher stresses PLB is evident, where n continually increases. The above suggests for power-law creep, typically above $0.6\ T_m$ (depending on the creep rate):

$$\dot{\varepsilon}_{ss} = A_2 \exp[-Q_{sd}/kT](\sigma_{ss})^{n(\cong 5)} \tag{7}$$

where A_2 is a constant, and varies significantly among metals. For aluminum, as mentioned earlier, $n = 4.5$, although for most metals and class M alloys $n \cong 5$, hence "five-power" (steady-state) creep. Figure 7 also shows that, phenomenologically, the description of the data may be improved by normalizing the steady-state stress by the elastic (Young's in this case) modulus. This will be discussed more later. (The correlation between Q_c and Q_{sd} (the former calculated from Eqn (6)) utilized modulus

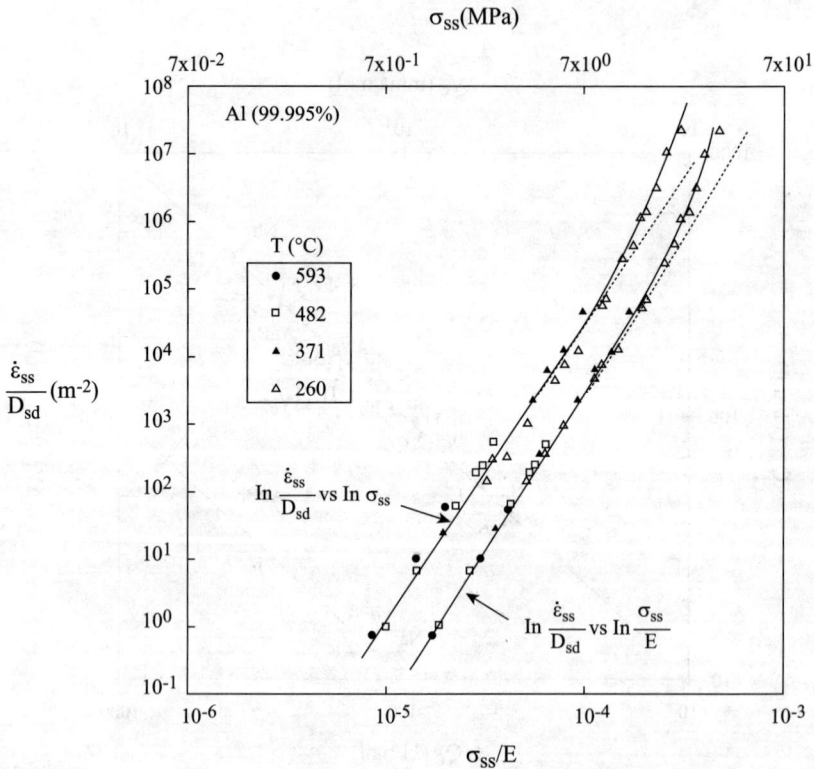

Figure 8 The lattice self-diffusion coefficient compensated steady-state strain rate versus the Young's modulus compensated steady-state stress. *From Ref. [16].*

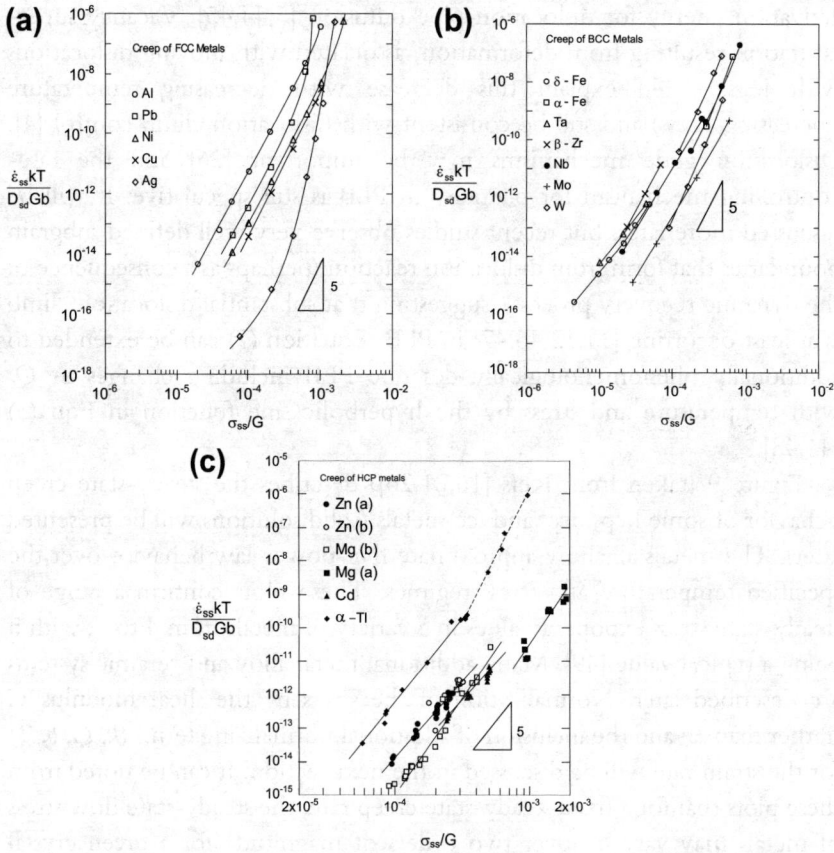

Figure 9 (a) The compensated steady-state strain rate versus modulus compensated steady-state stress. *(Based on Ref. [26] for selected FCC metals.)* (b) The compensated steady-state strain rate versus modulus compensated steady-state stress. *(Based on Ref. [26] for selected BCC metals.)* (c) The compensated steady-state strain rate versus modulus compensated steady-state stress. *(Based on Ref. [21] for selected HCP metals.)*

compensation of the stress. Hence, Eqn (7) actually implies some modulus compensation.)

It is now widely accepted that the activation energy for five-power-law creep closely corresponds to that of lattice self-diffusion, D_{sd}, or $Q_c \cong Q_{sd}$, although this is not a consensus judgment [41–43]. Thus, most have suggested that the mechanism of five-power-law creep is associated with dislocation climb.

Although within PLB, Q_c generally decreases as n increases, some still suggest that creep is dislocation climb controlled, but Q_c corresponds to the activation energy for dislocation-pipe diffusion [5,44,45]. Vacancy super-saturation resulting from deformation, associated with moving dislocations with jogs, could explain this decrease with decreasing temperature (increasing stress) and still be consistent with dislocation climb control [4]. Dislocation glide mechanisms may be important [26] and the rate-controlling mechanism for plasticity in PLB is still speculative. It will be discussed more later, but recent studies observe very well defined subgrain boundaries that form from dislocation reaction (perhaps as a consequence of the dynamic recovery process), suggesting that substantial dislocation climb is at least occurring [11,12,46,47] in PLB. Equation (7) can be extended to additionally phenomenologically describe PLB including changes in Q_c with temperature and stress by the hyperbolic sine function in Eqn (5) [44,48].

Figure 9 (taken from Refs [16,21,26]) describes the steady-state creep behavior of some hcp, bcc, and fcc metals (solid solutions will be presented later). The metals all show approximate five-power-law behavior over the specified temperature and stress regimes. These plots confirm a range of steady-state stress exponent values in a variety of metals from 4 to 7, with 5 being a typical value [49]. Many additional metal alloy and ceramic systems are described later. Normalization of the stress by the shear modulus G (rather than E) and the inclusion of additional normalizing terms (k, G, b, T) for the strain rate will be discussed in the next section. It can be noted from these plots that for a fixed steady-state creep rate, the steady-state flow stress of metals may vary by over two orders of magnitude for a given crystal structure. The reasons for this will also be discussed later. A decreasing slope (exponent) at lower stresses has often been suggested to be due to diffusional creep or Harper-Dorn creep [50]. Diffusional creep includes Nabarro-Herring [51] and Coble [52] creep. These will be discussed more later, but briefly, Nabarro-Herring creep consists of volume diffusion induced strains in polycrystals while Coble creep consists of mass transport under a stress by vacancy migration via short circuit diffusion along grain boundaries. Harper-Dorn creep is not fully understood [53–55] and appears to involve dislocations within the grain interiors. There has been some recent controversy as to the existence of diffusional creep [56–61] as well as Harper-Dorn creep [55,949].

1.2 Influence of the Elastic Modulus

Figure 10 plots the steady-state stress versus the Young's modulus at a *fixed* lattice self-diffusion-coefficient-compensated steady-state creep rate.

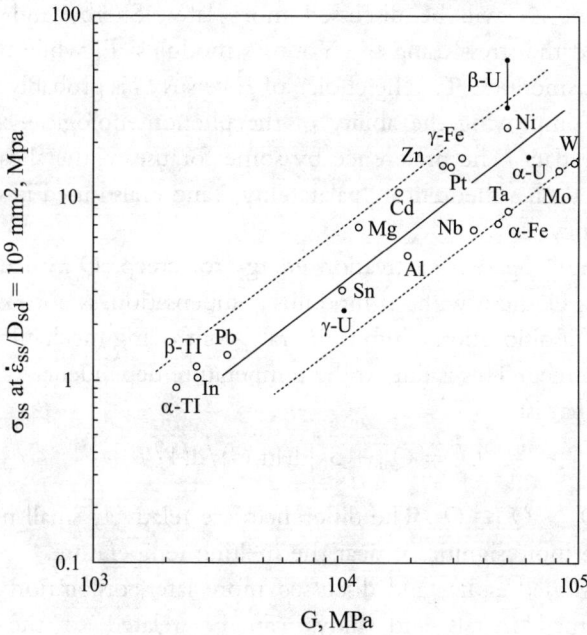

Figure 10 The influence of the shear modulus on the steady-state flow stress for a fixed self-diffusion-coefficient-compensated steady-state strainrate, for selected metals. *Based on Ref. [26].*

Clearly, there is an associated increase in creep strength with Young's modulus, and the flow stress can be described by:

$$\sigma_{ss}|_{\dot{\varepsilon}_{ss}/D_{sd}} = K_0 G \tag{8}$$

where K_0 is a constant. This, together with Eqn (7), can be shown to imply that five-power-law creep is described by the equation utilizing modulus-compensation of the stress, such as with Eqn (3):

$$\dot{\varepsilon}_{ss} = A_3 \exp[-Q_{sd}/kT](\sigma_{ss}/G)^5 \tag{9}$$

where A_3 is a constant. Utilizing modulus compensation produces less variability of the constant A_3 among metals, as compared to A_2, in Eqn (7). It was shown earlier that the aluminum data of Figure 7 could, in fact, be more accurately described by a simple power law if the stress is modulus compensated (Q_{sd} used). The modulus compensation of Eqn (9) may also be sensible for a given material, as the dislocation substructure is better related to the modulus–compensated stress rather than just the applied stress.

The constant A_3 will be discussed more later. Sherby and coworkers compensated the stress using the Young's modulus, E, while most others use the shear modulus, G. The choice of E versus G is probably not critical in terms of improving the ability of the phenomenological equation to describe the data. The preference by some for use of the shear modulus may be based on a theoretical "palatability," and is also used in this review for consistency.

Thus, the "apparent" activation energy for creep, Q'_c, calculated from plots such as Figure 6 without modulus compensation, is not exactly equal to Q_{sd} even if dislocation climb is the rate-controlling mechanism for five-power-law creep. This is due to the temperature dependence of the elastic modulus. That is:

$$Q'_c = Q_{sd} + 5k[d(\ln G)/d(1/T)] \qquad (10)$$

Thus, $Q'_c > Q_{sd} \cong Q_c$. The differences are relatively small near 0.5 T_m but become more significant near the melting temperature.

As mentioned earlier and discussed more later, dislocation features in creep-deformed metals and alloys can be related to the modulus-compensated stress. Thus, the s in Eqn (6), denoting constant structure, can be omitted if constant modulus compensated stress is indicated, since for steady-state structures in the power law regime, a constant σ_{ss}/E (or σ_{ss}/G) will imply, at least approximately, a fixed structure. Figure 11 [6] illustrates some of the Figure 6 data, as well as additional PLB data on a strain rate versus modulus-compensated stress plot. This allows a direct determination of the activation energy for steady-state creep, Q_c, since changes in $\dot{\varepsilon}_{ss}$ can be associated with changes in T for a fixed structure (or σ_{ss}/G). Of course, Konig and Blum [62] showed that with a change in temperature at a constant applied stress, the substructure changes, due at least largely to a change in σ/G in association with a change in temperature. We observe in Figure 11 that activation energies are comparable to that of lattice self-diffusion of aluminum (e.g., 123 kJ mol^{-1} [63]). Again, below about 0.6 T_m or so, depending on the strain rate, the activation energy for creep Q_c begins to significantly decrease *below* Q_{sd}. This occurs at about PLB where $n > 5$ (>4.5 for Al). Figure 12 plots steady-state aluminum data along with steady-state silver activation energies from references [12] and [44]. Other descriptions of Q_c versus T/T_m for Al [64] and Ni [65] are available that utilize temperature-change tests in which constant structure is assumed (but not assured) and σ/G is not constant. The trends observed are nonetheless consistent with those of Figure 12. The question as to whether

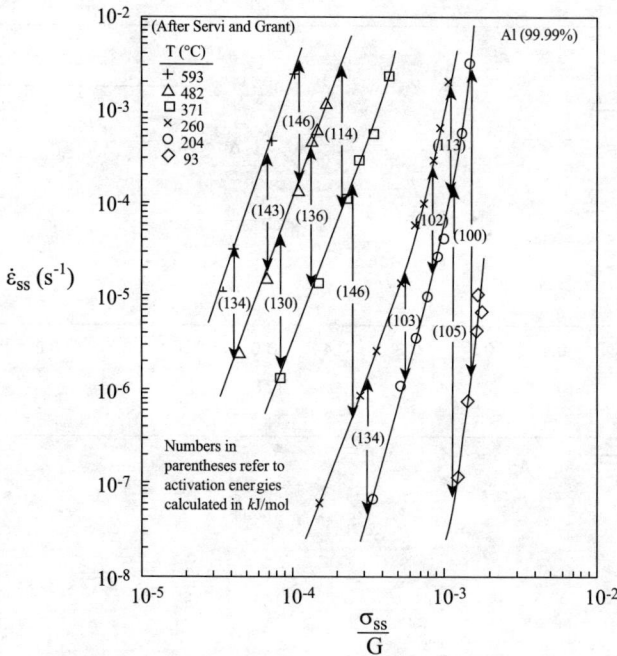

Figure 11 The steady-state strain rate versus the modulus-compensated stress for six temperatures. This plot illustrates the effect of temperature on the strain rate for a fixed modulus-compensated steady-state stress (constant structure) leading to the calculation for activation energies for creep, Q_C. Based on Ref. [6].

the activation energy for steady-state and primary (or transient, i.e., from one steady-state to another) creep are identical does not appear established, and this is an important question. However, some [66,67,978] have suggested that the activation energy from primary to steady state does not change substantially. Luthy, Miller, and Sherby, [44] and Sherby and Miller [68], present a Q_c versus T/T_m plot for steady-state deformation of W, NaCl, Sn, and Cu that is frequently referenced. This plot suggests two activation energies, one regime where $Q_c \cong Q_{sd}$ (as Figure 12 shows for Ag and Al) from 0.60 to 1.0 T/T_m. They additionally suggest that, with PLB, Q_c is approximately equal to that of vacancy diffusion through dislocation pipes, Q_P (two "plateaus" of Q_c). That is, it was suggested that the rate-controlling mechanism for steady-state creep in PLB is still dislocation climb, but facilitated by short-circuit diffusion of vacancies via the elevated density of dislocations associated with increased stress between 0.3 and about 0.6 T/T_m. (The interpretation of the NaCl results are ambiguous and

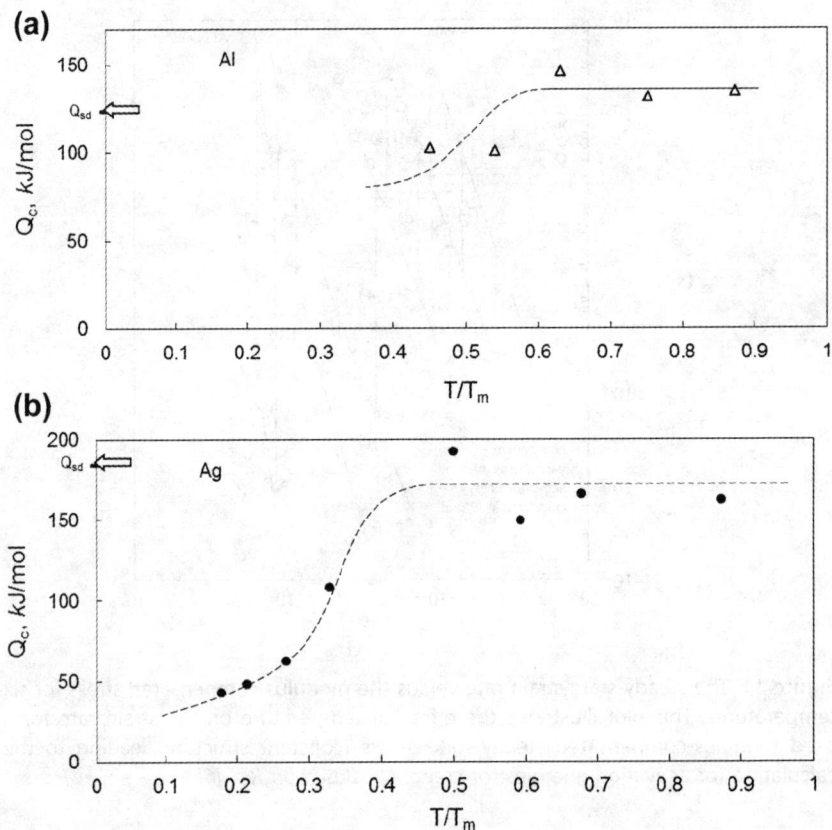

Figure 12 The variation of the activation energy for creep versus fraction of the melting temperature for (a) Al *(based on Ref. [44])* and (b) Ag *(based on Ref. [12])*.

may actually be more consistent with Figure 12 if the original NaCl data is reviewed [16]). The situation for Cu is ambiguous. Raj and Langdon [70] reviewed activation energy data and it appears that Q_c may decrease continuously below at least 0.7 T_m from Q_{sd}, in contrast to earlier work on Cu that suggested $Q_c = Q_{sd}$ above about 0.7 T_m and "suddenly" decreases to Q_P. As mentioned earlier, the steady-state torsion creep data of Luthy, Miller, and Sherby, on which the lower temperature activation energy plateau calculations were based, are probably unreliable. Dynamic recrystallization is certainly occurring in their high-purity aluminum along with probable (perhaps 20%) textural softening (decrease in the average Taylor factor, $\overline{M}$) along with adiabatic heating. Use of solid specimens also complicates the interpretation of steady state as outer portions may soften while inner portions are hardening. Lower-purity specimens could be used to avoid dynamic recrystallization, but Stage IV hardening [11,13] may occur and may preclude (although sometimes just postpone) a mechanical steady state. Thus, steady state, as defined here, as a balance between dislocation hardening and, exclusively, dynamic recovery, is not relevant. Weertman [25] suggested that the Sn results may show an activation energy transition to a plateau value of Q_p over a range of elevated temperatures. This transition occurs already at about 0.8 T_m and Q_c values at temperatures less than 0.6 T_m do not appear available. Thus, the values of activation energy, between 0.3 and 0.6 T_m (PLB), and the question as to whether these can be related to the activation energy of dislocation pipe diffusion, are probably unsettled. Quality activation energy measurements over a wide range of temperatures both for steady-state and primary creep for a variety of pure metals are surprisingly unavailable.

Sherby and Burke have suggested that vacancy supersaturation may occur at lower temperatures where PLB occurs (as have others [71]). Thus, vacancy diffusion may still be associated with the rate-controlling process despite a low, activation energy. Also, as suggested by others [9,26,41–43], cross-slip or the cutting of forest dislocations (glide) may be the rate-controlling dislocation mechanisms rather than dislocation climb.

1.3 Stacking Fault Energy and Summary

In the above, the steady-state creep rate for five-power-law creep was described by:

$$\dot{\varepsilon}_{ss} = A_4 D_{sd}(\sigma_{ss}/G)^5 \tag{11}$$

where:

$$D_{sd} = D_o \exp(-Q_{sd}/kT). \tag{12}$$

Many investigators [4,21,26,72] have attempted to decompose A_4 into easily identified constants. Mukherjee et al. [72] proposed that:

$$\dot{\varepsilon}_{ss} = A_5 (D_{sd}Gb/kT)(\sigma_{ss}/G)^5 \tag{13}$$

This review will utilize the form of Eqn (13) since this form has been more widely accepted than Eqn (11). Equation (13) allows the expression of the power law on a logarithmic plot more conveniently than Eqn (11), due to dimensional considerations.

The constants A_0 through A_5 depend on stacking fault energy, in at least fcc metals, as illustrated in Figure 13. The way by which the stacking fault

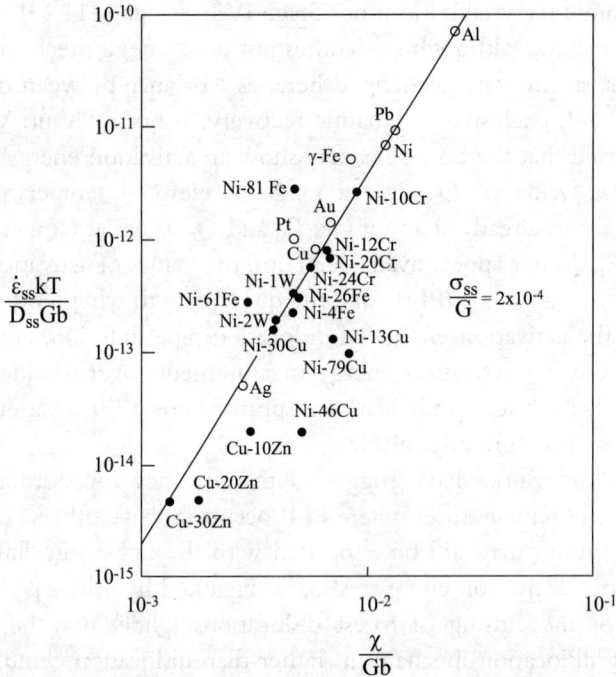

Figure 13 The effect of stacking-fault energy on the (compensated) steady-state strain rate for a variety of metals and Class M alloys. *Based on Ref. [73].*

energy affects the creep rate is unclear. For example, it does not appear known whether the climb rate of dislocations is affected (independent of the microstructure) and/or whether the dislocation substructure, which affects creep rate, is affected (or both). In any case, for fcc metals, Mohamed and Langdon [73] phenomenologically suggested:

$$\dot{\varepsilon}_{ss} = A_6 (\chi/Gb)^3 (D_{sd} Gb/kT)(\sigma_{ss}/G)^5 \qquad (14)$$

where χ is the stacking fault energy.

Thus, in summary it appears that, over five-power creep, the activation energy for steady-state creep is approximately that of lattice self-diffusion. (Exceptions with pure metals above 0.5 T/T_m have been suggested. One example is Zr, where a glide-control mechanism [74] has been suggested to be rate controlling, but self-diffusion may still be viable [75], just obscured by impurity effects.) This suggests that dislocation climb is associated with the rate-controlling process for five-power-law creep. The activation energy decreases below about 0.5 T_m, depending, of course, on the strain rate. There is a paucity of reliable steady-state activation energies for creep at these temperatures, and it is difficult to associate these energies with specific mechanisms. The classic plot of effective diffusion coefficient D_{eff}-compensated strain rate versus modulus-compensated stress for aluminum by Luthy, Miller, and Sherby may be the most expansive in terms of the ranges of stress and temperature. It appears in other creep reviews [23,24] and may have some critical flaws. They modified Eqn (11) to a Garofalo (hyperbolic sine) [48] equation to include PLB:

$$\dot{\varepsilon}_{ss} = BD_{eff}[\sinh \alpha_1 (\sigma_{ss}/E)]^5 \qquad (15)$$

where α_1 and B are constants. Here, again, D_{eff} reflects the increased contribution of dislocation pipe diffusion with decreasing temperature. D_{eff}-compensated strain rate utilizes a composite strain rate controlled by lattice and dislocation pipe diffusion. The contributions of each of these to D_{eff} depend on both the temperature and the dislocation density (which at steady-state is non-homogeneous, as will be discussed). Equation (15), above, was later modified by Wu and Sherby [53] for aluminum to account for internal stresses, although a dramatic improvement in the modeling of the data of the five-power law and PLB was not obvious. The subject of internal stresses will be discussed later. Diffusion is not a clearly established mechanism for plastic flow in PLB, and D_{eff} is not precisely known. For this reason, this text will avoid the use of D_{eff} in compensating strain rate.

Just as PLB bounds the high-stress regime of five-power-law creep, a diffusional creep mechanism or Harper-Dorn creep may bound the low-stress portion of five-power-law creep (for alloys, superplasticity (2-power) or viscous glide (3-power) may also be observed, as will be discussed later). For pure aluminum, Harper-Dorn creep is generally considered to describe the low-stress regime and is illustrated in Figure 14. The stress exponent for Harper-Dorn is 1 with an activation energy often of Q_{sd}, and Harper-Dorn is grain-size independent. The precise mechanism for Harper-Dorn creep is not understood [53,54] and some have suggested that it may not exist [55,949]. Figure 14, from Blum and Straub [76,77], is a compilation of high-quality steady-state creep in pure aluminum and describes the temperature range from about 0.5 T_m to near the melting temperature, apparently showing three separate creep regimes. It appears that the range of steady-state data for Al may be more complete than for any other metal. The aluminum data presented in earlier figures are

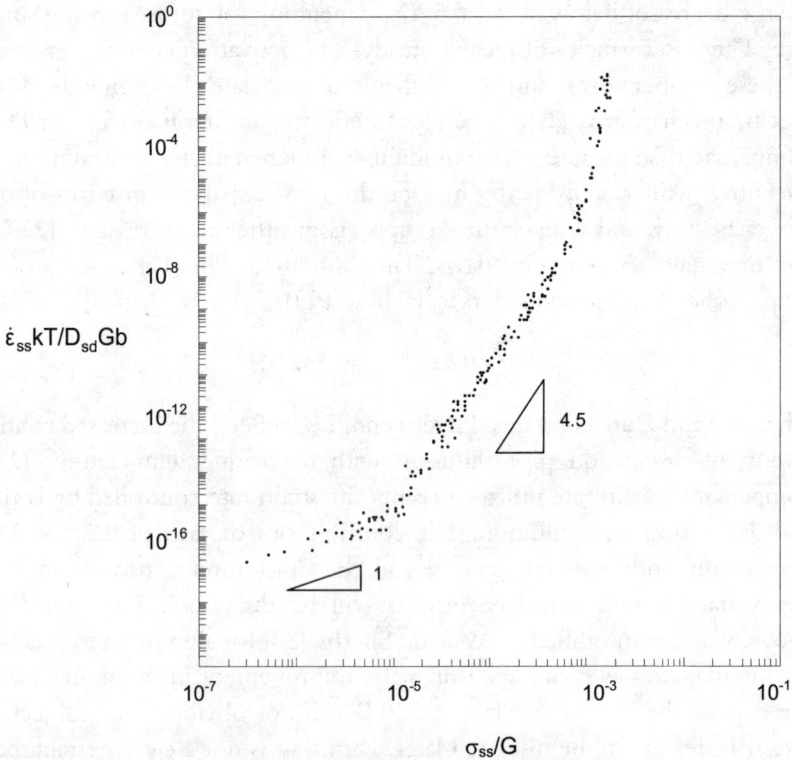

Figure 14 The compensated steady-state strain rate versus the modulus-compensated steady-state stress for 99.999 pure Al. *Based on Refs [76,77].*

consistent with the data plotted by Blum and Straub. It is intended that this data not include the temperature/stress regime where Stage IV and recrystallization may obfuscate recovery-controlled steady state. This plot also (probably not critical to the PLB transition) uses the same activation energy, Q_{sd} (142 kJ mol^{-1}) [76], over the entire stress/strain-rate/temperature regime. As discussed earlier, Q_c seems to decrease with decreasing temperature (increasing strain rate) within PLB. The aluminum data shows a curious undulation at $\sigma_{ss}/G = 2 \times 10^{-5}$, that is not understood, although impurities were a proposed explanation [76]. It will be discussed in the Harper-Dorn section, but other more recent Al data in the low-stress regime will be mentioned.

There are other metallic systems for which a relatively large amount of data from several investigators can be summarized. One is copper, which is illustrated in Figure 15. The summary was reported recently in references

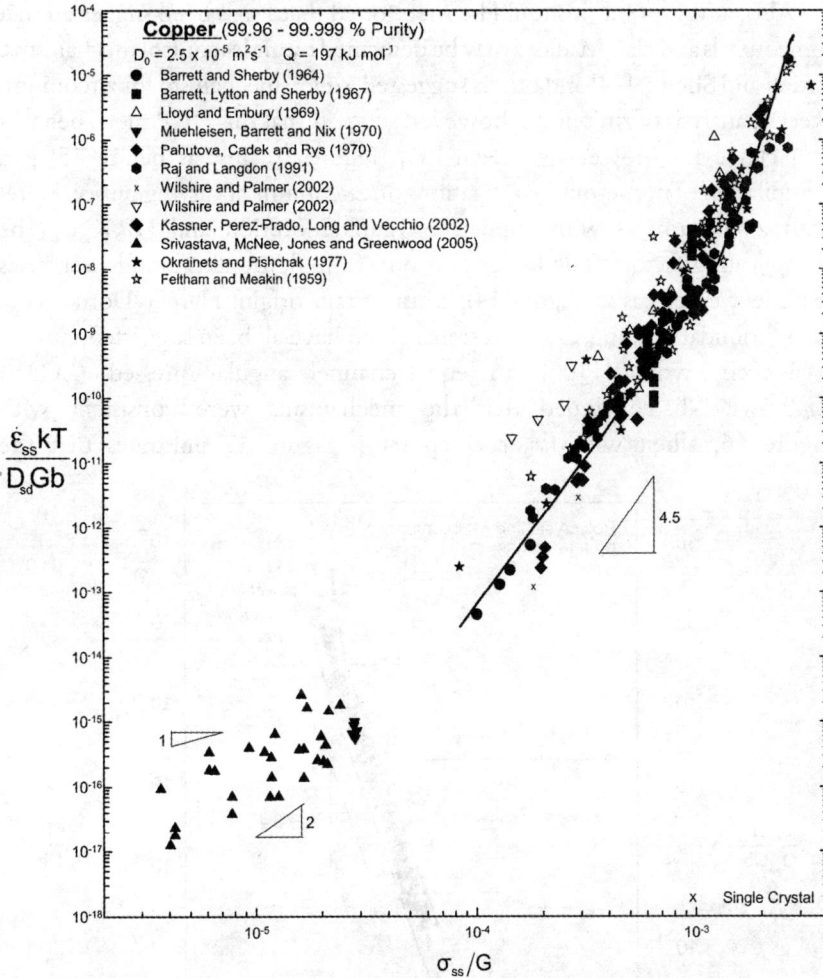

Figure 15 Summary of the diffusion-coefficient compensated steady-state strain rate versus the modulus-compensated steady-state stress for copper of various high purities from various investigations. *From Ref. [78].*

[78] and [949]. Again, a well-defined five-power-law regime is evident. Again, this data is consistent with the (smaller quantity of reported) data of Figure 8. Greater scatter is evident here as compared with Figure 14, as the results of numerous investigators were used and the purity varied. Copper is a challenging experimental metal as oxygen absorption and discontinuous dynamic recrystallization can obfuscate steady-state behavior in five-power-law creep, which is a balance between dislocation hardening and dynamic recovery.

Also, at this point, it should be mentioned that it has been suggested that some metals and class M alloys may be deformed by glide-control mechanisms. Ardell and Sherby [74] and others suggested a glide mechanism for zirconium. Recent analysis of zirconium, however, suggests that this HCP metal behaves as a classic five-power-law metal [80]. Figure 16, just as Figure 15, is a compilation of numerous investigations on zirconium of varying purity. Here, with zirconium, as with copper, oxygen absorption and DRX can be complicating factors. The lower portion of the figure illustrates lower stress exponent creep (as in Figure 14), of uncertain origin. Harper-Dorn creep, grain boundary sliding, and diffusional creep have all been suggested.

Recent work [1100] on equal-channel angular pressed (ECAP) Zr-2.5wt%Nb concluded that the mechanisms were consistent with Figure 16, albeit with faster creep rates. Figure 17 illustrates that the

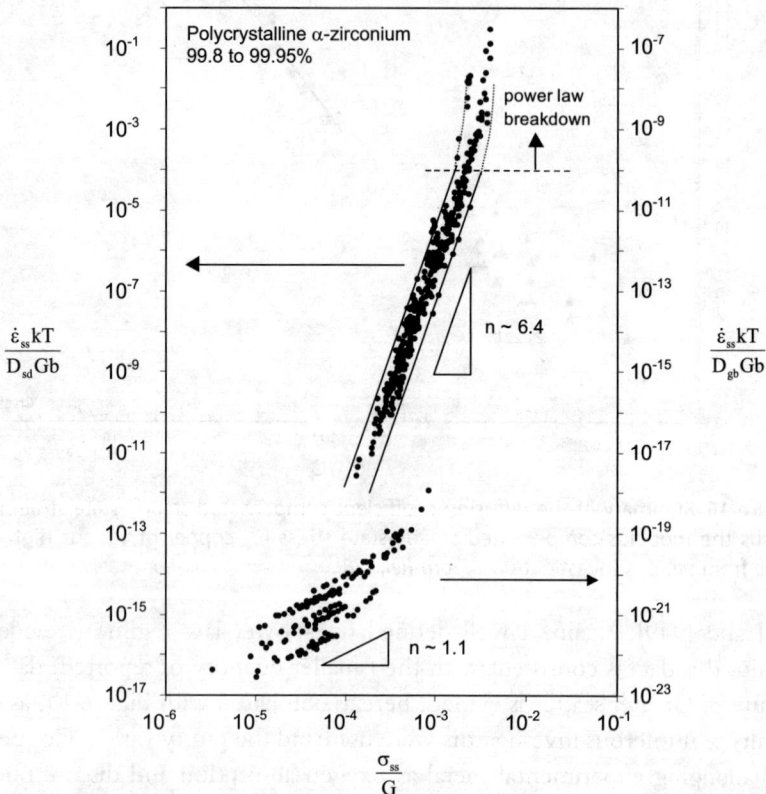

Figure 16 The diffusion-coefficient-compensated steady-state strain rate versus modulus-compensated steady-state stress for polycrystalline zirconium of various purities from various investigations. *From Ref. [80].*

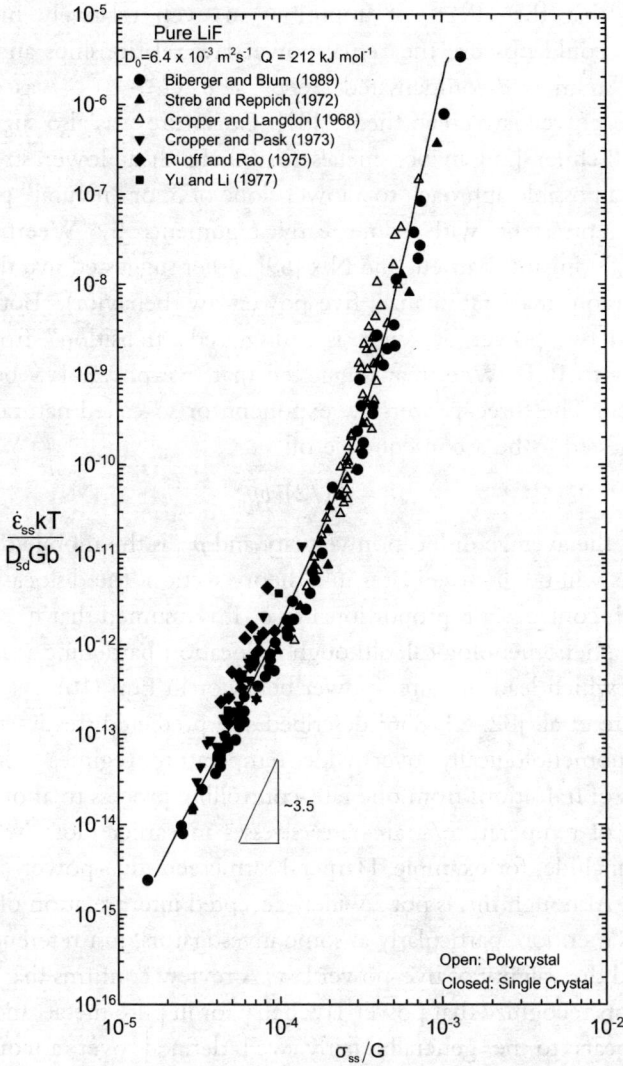

Figure 17 Plot of steady-state creep rate of LiF.

five-power-law regime in ceramics is associated with somewhat lower stress exponents than metals. In LiF, an exponent of 3.5 is observed, as is typical for other ceramics described in the Harper-Dorn chapter.

1.4 Natural Three-Power Law

It is probably important to note here that Blum [22] suggests the possibility of some curvature in the five-power-law regime in Figure 14. Blum cautioned that the effects of impurities in even relatively high-purity aluminum could obscure the actual power-law relationships and that the value of "strain rate-compensated creep" at $\sigma/G \cong 10^{-5}$ was consistent with three-power-law creep theory [81]. Curvature was also suggested by Nix and Ilschner [26] in bcc metals (in Figure 8) at lower stresses, and suggested a possible approach to a lower slope of 3, or "natural" power law exponent consistent with some early arguments by Weertman [25] (although Springarn, Barnett, and Nix [69] earlier suggested that dislocation core diffusion may rationalize five-power-law behavior). Both groups interpreted five-power behavior as a disguised "transition" from three-power-law to PLB. Weertman suggested that five-power-law behavior is unexpected. The three-power-law exponent, or so-called natural law, has been suggested to be a consequence of:

$$\dot{\varepsilon} = (1/2)\bar{v}b\rho_{\mathrm{m}} \qquad (16)$$

where $\bar{v}$ is the average dislocation velocity and ρ_{m} is the mobile dislocation density. As will be discussed later in a theory section, the dislocation climb rate, which controls $\bar{v}$, is proportional to σ. It is assumed that $\sigma^2 \propto \rho_{\mathrm{m}}$ (the relation is phenomenological, although dislocation hardening is *not* always assumed), which leads to three-power behavior in Eqn (16).

Wilshire et al. [82, 83, 856] described and predicted steady-state creep rates phenomenologically over wide temperature regimes without assumptions of transitions from one rate-controlling process to another across the range of temperature/strain-rates/stresses in earlier plots (which suggested to include, for example, Harper-Dorn creep, five-power-law creep, and PLB). Although this is not a widely accepted interpretation of the data, it deserves mention, particularly as some investigators, just referenced, have questioned the validity of five-power law. A review confirms that nearly all investigators recognize that power-law behavior in pure metals and Class M alloys appears to be generally fairly well defined over a considerable range of modulus-compensated steady-state stress (or diffusion-coefficient-compensated steady-state creep rate). Although this value varies, a typical value is 5. Thus, for this review, the designation of five-power-law creep is judged meaningful. The full meaning of three-power-law creep will be addressed in the rate-controlling mechanisms section of this chapter.

1.5 Substitutional Solid Solutions

Two types of substitutional solutions can be considered: cases where a relatively large fraction of solute alloying elements can be dissolved, and those cases with small amounts of either intentional or impurity (sometimes interstitial) elements. The addition of solute can lead to basically two phenomena within the five-power-law regime of the solvent. Hardening or softening while maintaining five-power-law behavior can be observed, or three-power, viscous glide behavior, the latter being discussed in a separate section, may be evident. Figure 18 shows the effects of substitutional solid-solution additions for a few alloy systems for which five-power-law behavior is maintained. This plot was adapted from Mukherjee [21].

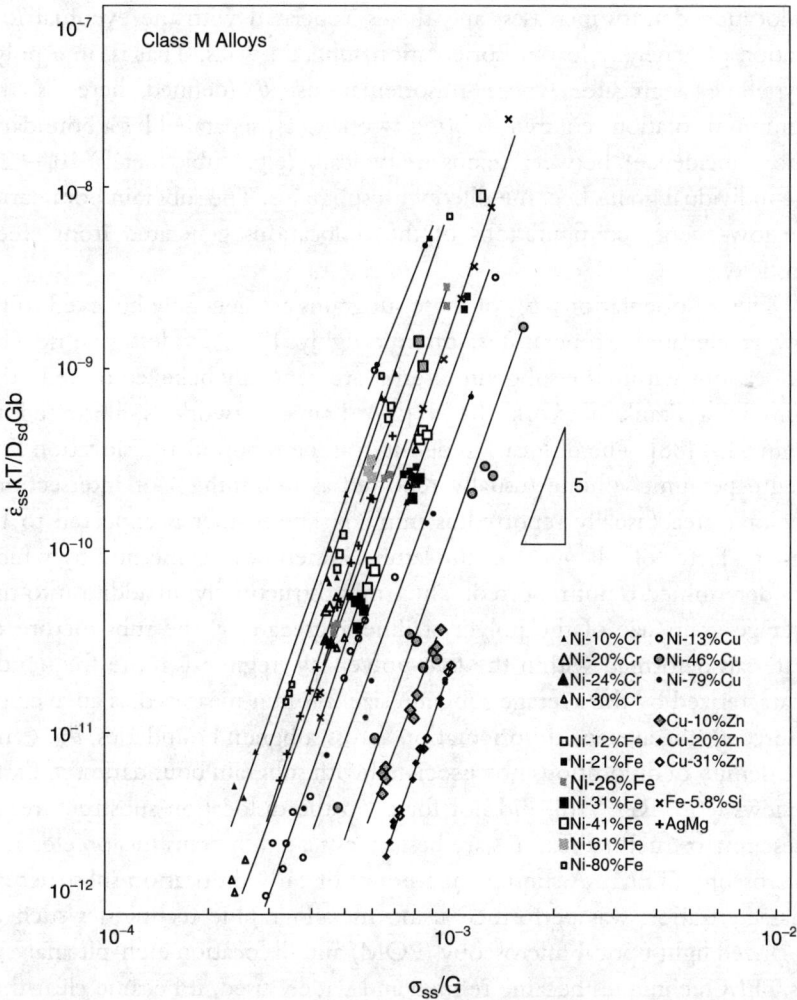

Figure 18 The compensated steady-state strain rate versus modulus-compensated steady-state stress for a variety of Class M (Class I) alloys. *Based on Ref. [21].*

2. MICROSTRUCTURAL OBSERVATIONS

2.1 Subgrain Size, Frank Network Dislocation Density, Subgrain Misorientation Angle, and the Dislocation Separation within the Subgrain Walls in Steady-State Structures

Certain trends in the dislocation substructure evolution have been fairly well established when an annealed metal is deformed at elevated temperature (e.g., under constant stress or strain rate) within the five-power-law regime. Basically, on commencement of plastic deformation, the total dislocation density increases, and this is associated with the eventual formation of generally low-misorientation subgrain walls. That is, in a polycrystalline aggregate, where misorientations, θ (defined here as the minimum rotation required to bring two lattices, separated by a boundary, into coincidence), between grains are typically (e.g., cubic metals) $10°-62°$, the individual grains become filled with subgrains. The subgrain boundaries are low-energy configurations of the dislocations generated from creep plasticity.

The misorientations, θ_λ, of these subgrains are generally believed to be low at elevated temperatures, often roughly $1°$ at modest strains. The dislocations within the subgrain interior are generally believed to be in the form of a Frank network [84–87]. A Frank network is illustrated in Figure 19) [88]. The dislocation density can be reported as dislocation line length per unit volume (usually reported as $mm\,mm^{-3}$) or intersections per unit area (usually reported as mm^{-2}). The former is expected to be about a factor of 2 larger than the latter. Sometimes the method by which ρ is determined is not reported. Thus, microstructurally, in addition to the average grain size of the polycrystalline aggregate, g, the substructure of materials deformed within the five-power-law regime is more frequently characterized by the average subgrain size λ (often measured as an average intercept), the average misorientation across subgrain boundaries, $\theta_{\lambda_{ave}}$, and the density of dislocations not associated with subgrain boundaries, ρ. Early reviews (e.g., Ref. [16]) did not focus on the dislocation substructure, as these microstructural features are best investigated by transmission electron microscopy (TEM). A substantial amount of early dislocation substructure characterization was performed using metallographic techniques such as polarized light optical microscopy (POM) and dislocation etch-pit analysis. As TEM techniques became refined and widely used, it became clear that the optical techniques are frequently unreliable, often, for example, overestimating subgrain size [89] partly due to a lack of ability (particularly

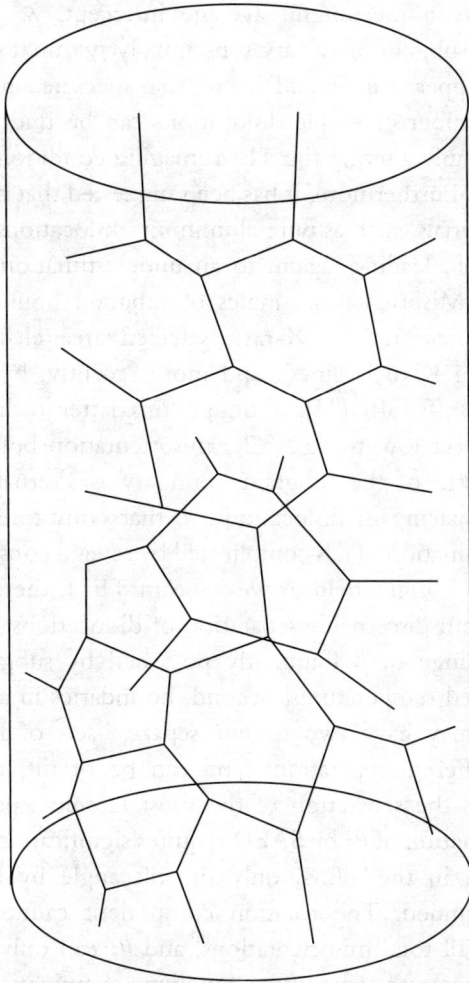

Figure 19 A three-dimensional Frank network of dislocations within a subgrain or grain. *Based on Ref. [88].*

POM) to detect lower-misorientation-angle subgrain boundaries. Etch-pit-based dislocation density values may be unreliable, particularly in materials with relatively high ρ values [90]. Unambiguous TEM charac-terization of the dislocation substructure is not trivial. Although dislocation density measurements in metals may be most reliably performed by TEM, several shortcomings of the technique must be overcome. Although the TEM can detect subgrain boundaries of all misorientations, larger areas of thin (transparent to the electron beam) foil must be examined to statistically ensure a meaningful average intercept, λ. The dislocation density within a subgrain may vary substantially within a given specimen. This variation appears independent of the specimen-preparation procedures. The number of visible dislocations can be underestimated by a factor of 2 by simply altering the TEM imaging conditions depending on the material [91]. Furthermore, it has been suggested that in high stacking-fault energy materials, such as pure aluminum, dislocations may "recover" from the thin foil, leading, again, to an underestimation of the original density [92,93]. Misorientation angles of subgrain boundaries, θ_λ, have generally been measured by X-rays, selected area electron diffraction (SAED, including Kikuchi lines), and more recently, by electron back-scattered patterns (EBSP) [94], although this latter technique, to date, cannot easily detect lower (e.g., $<2°$) misorientation boundaries. Sometimes the character of the subgrain boundary is alternatively described by the average spacing of dislocations, d, that constitute the boundary. A reliable determination of d is complicated by several considerations. First, with conventional bright-field or weak-beam TEM, there are limitations as to the minimum discernable separation of dislocations. This appears to be within the range of d frequently possessed by subgrain boundaries formed at elevated temperatures. Second, boundaries in at least some fcc metals [95–97] may have two to five separate sets of Burgers vectors, perhaps with different separations, and can be of tilt, twist, or mixed character. Often, the separation of the most closely separated set is reported. Determination of θ_λ by SAED requires significant effort, since for a *single* orientation in the TEM, only the tilt angle by Kikuchi shift is accurately determined. The rotation component cannot be accurately measured for small total misorientations, and θ_λ can only be determined using Kikuchi lines with some effort, involving examination of the crystals separated by a boundary using several orientations [42,43,966]. Again, EBSP cannot always detect lower θ_λ boundaries, which may comprise a large fraction of subgrain boundaries.

Figure 20 is a TEM micrograph of 304 austenitic stainless steel, a class M alloy, deformed to steady state within the five-power-law regime. A well-defined subgrain substructure is evident and the subgrain walls are "tilted" to expose the sets of dislocations that comprise the walls. The dislocations not associated with subgrain walls are also evident. Because of the finite thickness of the foil ($\cong 100$ nm), the Frank network has been disrupted. The heterogeneous nature of the dislocation substructure is evident. A high magnification TEM micrograph of a hexagonal array of screw dislocations

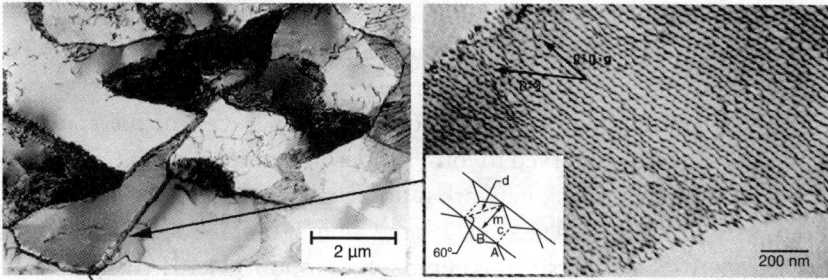

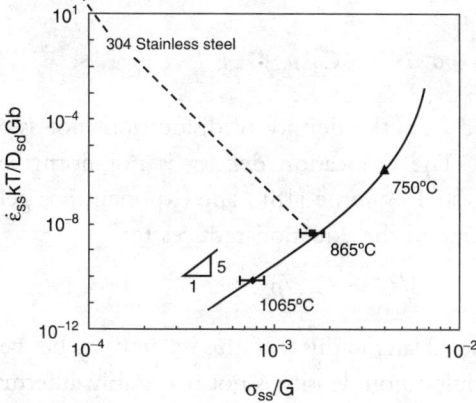

Figure 20 TEM micrographs illustrating the dislocation microstructure of 304 stainless steel deformed at the indicated conditions on the compensated steady-state strain rate versus modulus- compensated steady-state stress plot. The micrograph on the right is a high magnification image of the subgrain boundaries such as illustrated on the left. *(Based on [98,99].)* (Modulus based on 316 values.)

comprising a subgrain boundary with one set (of Burgers vectors) satisfying invisibility [98,99] is also in Figure 20.

It has long been observed, phenomenologically, that there is an approximate relationship between the subgrain size and the steady state flow stress:

$$\sigma_{ss}/G = C_1(\lambda_{ss})^{-1} \tag{17}$$

It should be emphasized that this relationship between the *steady-state* stress and the steady-state subgrain size is not for a fixed temperature and strain rate. Hence, it is not of a same type of equation as, for example, the Hall–Petch relationship, which relates the strength to the grain size, *g*, at a reference T and $\dot{\varepsilon}$. It will be discussed later that the variation of the stress with subgrain size at a fixed T and $\dot{\varepsilon}$ is different than Eqn (17). Some (e.g., Refs [89,100]) have normalized the subgrain size in Eqn (17) by the Burgers vector, although this is not a common practice.

The subgrains contain a dislocation density in excess of the annealed values, and are often believed to form a three-dimensional, Frank, network. The conclusion of a Frank network is not firmly established for five-power creep, but indirect evidence of a large number of nodes in thin foils [54,85,101–103] supports this common contention [84,104–109]. Analogous to Eqn (17), there appears to be a relationship between the density of dislocations not associated with subgrain boundaries [17] and the steady-state-stress:

$$\sigma_{ss}/G = C_2(\rho_{ss})^p \qquad (18)$$

where C_2 is a constant and ρ_{ss} is the density of dislocations not associated with subgrain boundaries. The dislocation density is not normalized by the Burgers vector as suggested by some [18]. The exponent p is generally considered to be about 0.5, and the equation reduces to:

$$\sigma_{ss}/G = C_3\sqrt{\rho_{ss}} \qquad (19)$$

As will be discussed in detail later in this text, this relationship between the steady-state stress and the dislocation density is not necessarily athermal, i.e., independent of temperature and strain rate. Hence, it is not necessarily of a same type of microstructure-strength equation as the classic Taylor relationship, which *is* generally presumed athermal. That is, this equation tells us the dislocation density not associated with subgrain walls that can be expected for a given steady-state stress that varies with the temperature and strain rate, analogous to Eqn (17). However, the flow stress associated with this density could depend on $\dot{\varepsilon}$ and T. We will further discuss this equation later.

As mentioned in the discussion of the "natural stress exponent," ρ_{ss} is sometimes presumed equal, or at least proportional, to the mobile dislocation density, ρ_m. This appears unlikely for a Frank network and the fraction mobile over some time interval, dt, is unknown, and may vary with stress. In Al, at steady state, ρ_m may only be about $1/3\,\rho$ or less [107], although this fraction was not firmly established.

Proponents of dislocation hardening also see a resemblance of Eqn (19) to:

$$\tau = \frac{Gb}{\ell_c} \qquad (20)$$

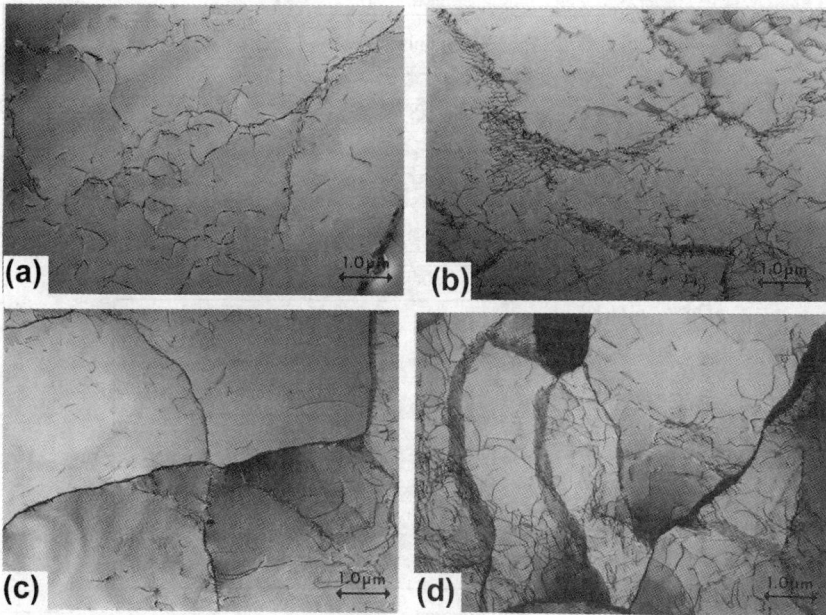

Figure 21 TEM micrographs illustrating the evolution of the dislocation substructure during primary creep of AISI 304 stainless steel torsionally deformed at 865 °C at $\dot{\bar{\varepsilon}} = 3.2 \times 10^{-5}\text{s}^{-1}$, to strains of 0.027 (a), 0.15 (b), 0.30 (c), and 0.38 (d).

leading to:

$$\tau \approx Gb\sqrt{\rho_{ss}}$$

where τ is the stress necessary to activate a Frank Read source of critical link length, ℓ_c. Again, the above equation must be regarded as athermal or valid only at a specific temperature and strain rate. It does not consider the substantial solute *and* impurity strengthening evident in (even 99.999% pure) metals and alloys, as shown in Figures 21 and 22, which is very temperature dependent. Weertman [25] appears to justify Eqn (19) by the dislocation (all mobile) density necessarily scaling with the stress in this manner, although he does not regard these dislocations as the basis for the strength. This seems contradictory; a dislocation relationship to stress would appear to suggest dislocation hardening, particularly with $\rho^{1/2}$. Consequently, Al–5at%Zn alloy, which has basically identical creep behavior, but possible precipitation pinning of dislocations during cooling, was used to determine the ρ versus σ_{ss} trends in Al instead. These data (Al and stainless steel) were used for Figures 23 and 24, as particular reliability is assigned to these investigations.

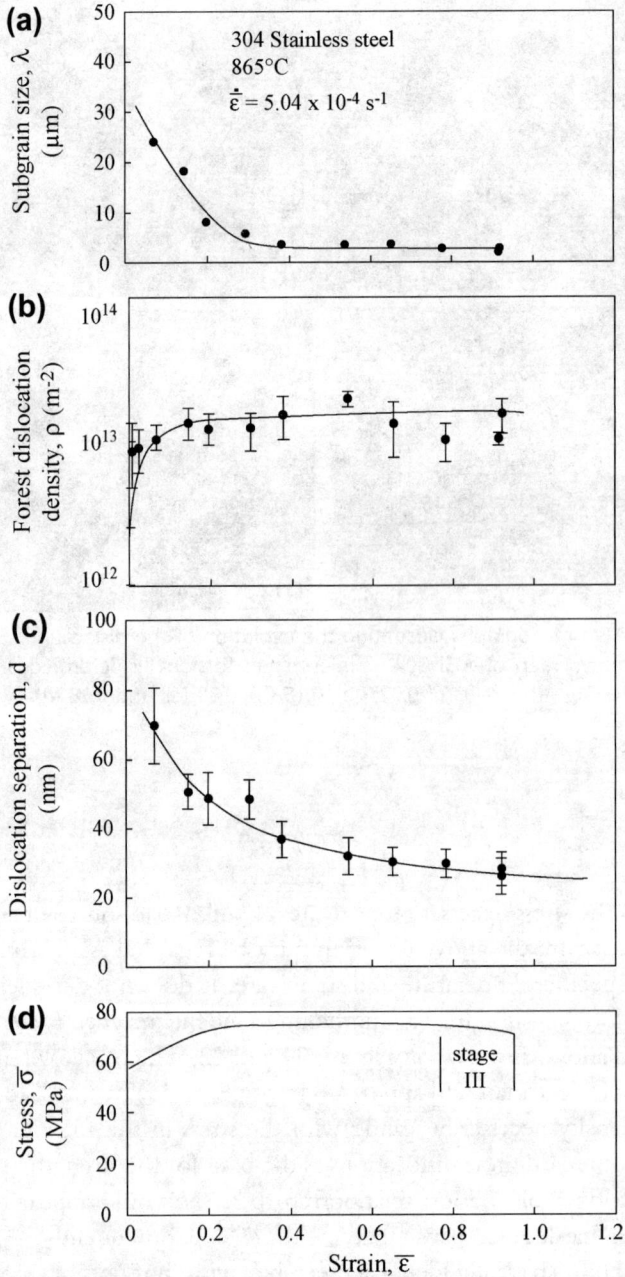

Figure 22 Work-hardening at a constant strain-rate primary creep transient in AISI 304 stainless steel, illustrating the changes in (a) λ, (b) ρ, and (c) d and (d) stress with strain. *Based on Ref. [98].*

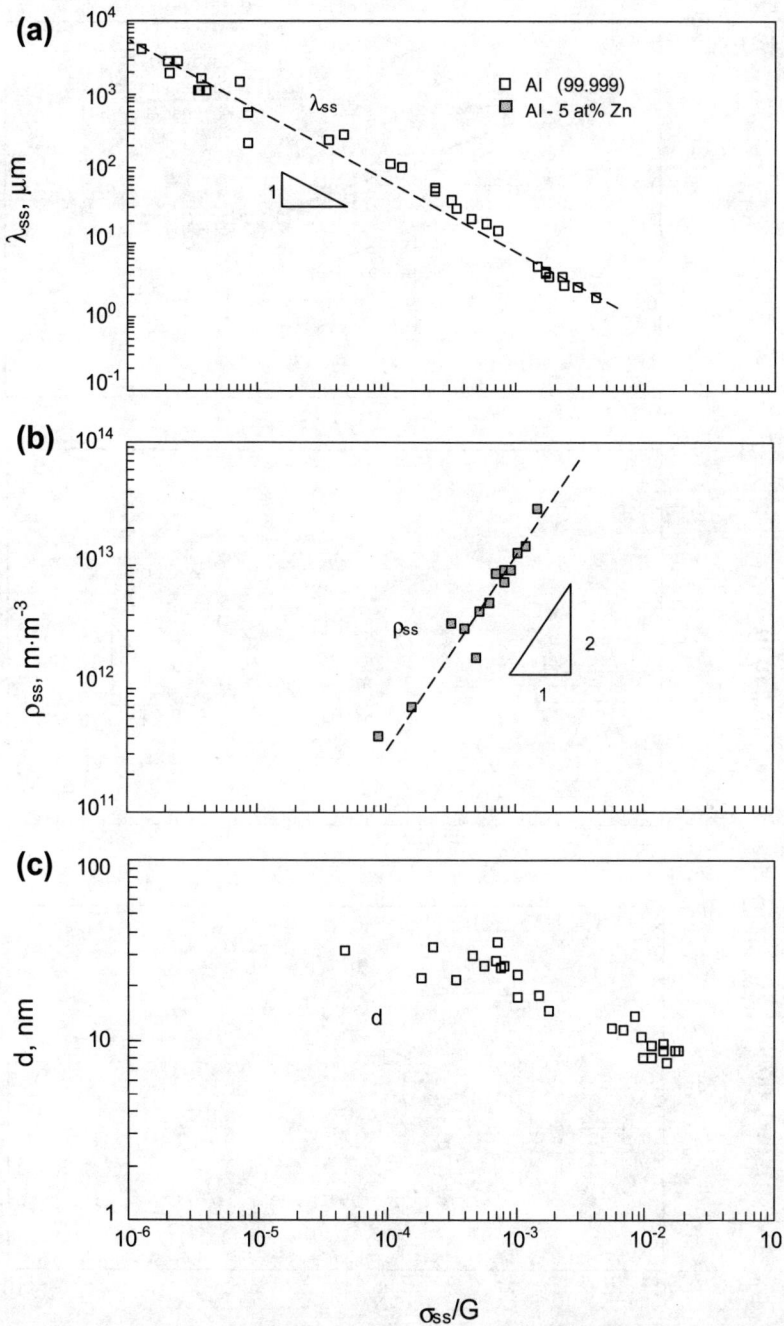

Figure 23 The average steady-state subgrain intercept, (a) λ, density of dislocations not associated with subgrain walls, (b) ρ, and the average separation of dislocations that comprise the subgrain boundaries, (c) d, for Al (and Al-5at%Zn that behaves, mechanically, essentially identical to Al, but is suggested to allow for a more accurate determination of ρ by TEM). *Based on Refs [22,90]*.

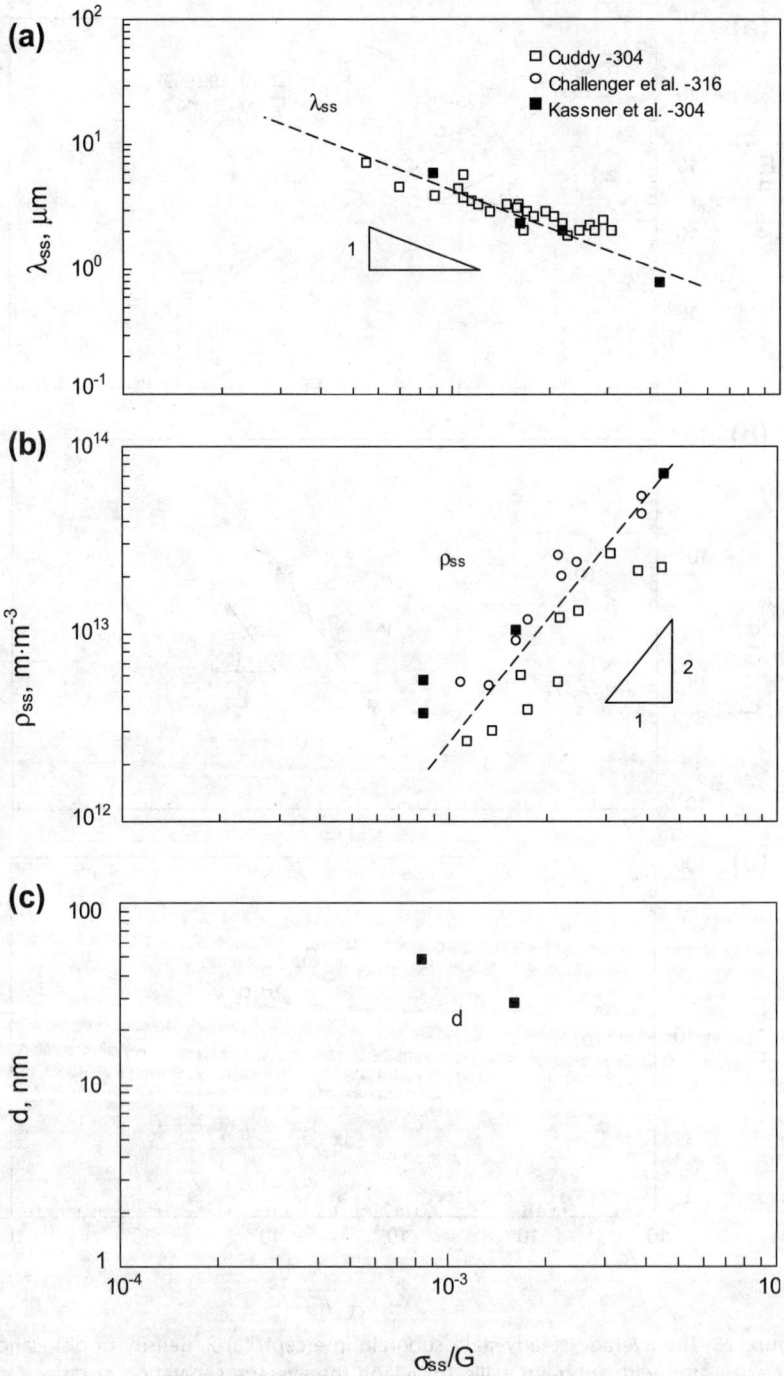

Figure 24 The average steady-state subgrain intercept, (a) λ, density of dislocations not associated with subgrain boundaries, (b) ρ, and average separation of dislocations that comprise the subgrain boundaries, (c) d, for 304 stainless steel. *Data from Refs [66,98,103,110,111].*

They are reflective of the general observations of the community, and are also supportive of Eqn (17) with an exponent of -1. Both sets of data are consistent with Eqn (18) with $p \cong 0.5$. It should, however, be mentioned that there appears to be some variability in the observed exponent, p, in Eqn (18). For example, some TEM work on αFe [112] suggests 1 rather than 0.5. Hofmann and Blum [113] more recently suggested 0.63 for Al modeling. Figure 25(a) and (b) illustrate the NaCl trends and suggests $\rho = 0.5$–1.0. Equation (18) may not uniquely relate ρ_{ss} to σ_{ss}/G. It may only be approximately valid. The above two equations, of course, mandate a relationship between the steady-state subgrain size and the density of dislocations not associated with subgrain boundaries:

$$\lambda_{ss} = C_4 (\rho_{ss})^{-p'} \tag{21}$$

rendering it difficult, simply by microstructural inspection of steady-state substructures, to determine which feature is associated with the rate-controlling process for steady-state creep or elevated-temperature strength.

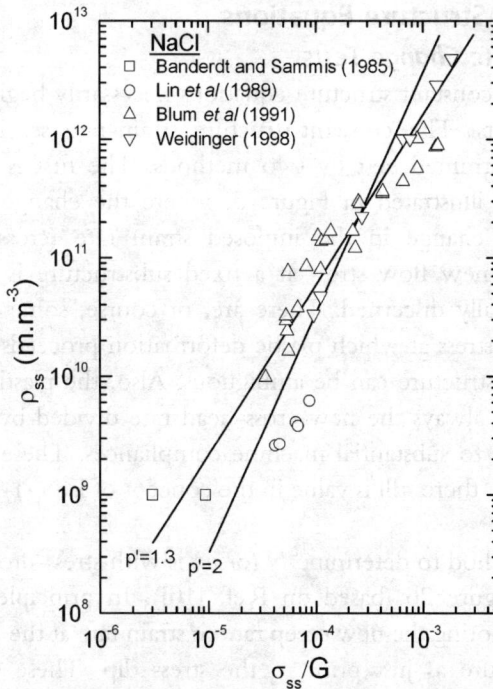

Figure 25 Modulus-compensated steady-state stress versus dislocation density. Illustrates NaCl trends and suggests $\rho = 0.5$–1.0.

Figures 23 and 24 additionally report the spacing, d, of dislocations that constitute the subgrain walls. The relationship between d and σ_{ss} is not firmly established, but Figures 23 and 24 suggest that:

$$\frac{\sigma_{ss}}{G} \cong d^{-q} \qquad (22)$$

where q may be between 2 and 4. Other [90] work on Fe- and Ni-based alloys suggests that q may be closer to values near 4. One possible reason for the variability may be that d (and $\theta_{\lambda_{ave}}$) may vary during steady state, as will be discussed in a later section. Figure 24 relies on d data well into steady state using torsion tests ($\bar{\varepsilon} \cong 1.0$). Had d been selected based on the onset of steady state, a q value of about 4 would also have been obtained. It should, of course, be mentioned that there is probably a relationship between d and θ_{λ}. Straub and Blum [90] suggested that:

$$\theta_{\lambda} \cong 2\ \arcsin(b/2d) \qquad (23)$$

2.2 Constant Structure Equations

2.2.1 Strain-Rate Change Tests

A discussion of constant structure equations necessarily begins with strain-rate change tests. The constant-structure strain-rate sensitivity, N, can perhaps be determined best by two methods. The first is the strain-rate increase test, as illustrated in Figure 3, where the change in flow stress with a sudden change in the imposed strain rate (cross-head rate) is measured. The new flow stress at a fixed substructure is that at which plasticity is initially discerned. There are, of course, some complications. One is that the stress at which plastic deformation proceeds at a "constant dislocation" substructure can be ambiguous. Also, the plastic strain rate at this stress is not always the new cross-head rate divided by the specimen gage length due to substantial machine compliances. These complications notwithstanding, there still is value in the concept of Eqn (1) and strain-rate increase tests.

Another method to determine N (or m) is with stress-drop (or dip) tests illustrated in Figure 26 (based on Ref. [10]). In principle, N could be determined by noting the new creep rate or strain rate at the lower stress for the same structure as just prior to the stress dip. These stress dip tests originated about 50 years ago by Gibbs [114], and the interpretation is still ambiguous [10]. Biberger and Gibeling [10] published an overview of creep

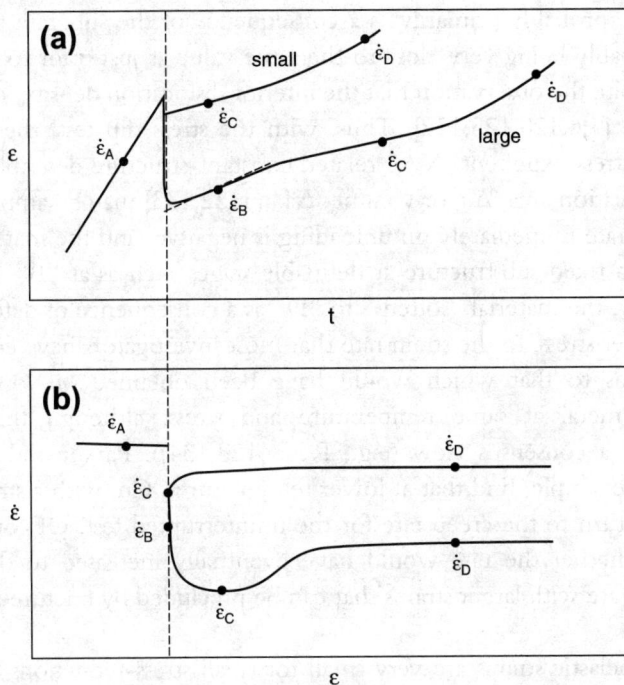

Figure 26 Description of the strain (a) and strain rate (b) versus time and strain for stress-dip (drop) tests associated with relatively small and large decreases in the applied stress. *From Ref. [10].*

transitions in pure metals following stress drops, emphasizing work by Gibeling and coworkers [10,87], Nix and coworkers [27], and Blum and coworkers [22,77, 92,121–125], all of whom have long studied this area, as well as several others [66,104,114,126–131]. The following discussion on the stress dip test relies on this overview.

With relatively large stress drops, there are quick contractions that may occur as a result of an initial, rapid, anelastic component, in addition, of course, to elastic contractions, followed by slower anelastic backflow [119]. Researchers in this area tend to report a "first maximum" creep rate, which occurs at "B" in Figure 26. It has been argued that the plastic strain preceding "B" is small and that the dislocation microstructure at "B" is essentially identical to that just prior to the stress drop. Some investigators have shown, however, that the interior or network density, ρ, may be different [131]. Also, since the creep rate *decreases* further to a minimum value at "C" (in Figure 26), the creep at "B" has been occasionally termed "anomalous." $\dot{\varepsilon}_C$ (point "C") has also been referred to as "constant structure," probably primarily as a consequence of the subgrain size, λ, at "C", probably being very close to the same value as just prior to the stress drop, despite the observation that the interior dislocation density, ρ, appears to change [4,6,120,125,132]. Thus, with the stress dip test, the constant structure stress exponent, N, or related constant-structure descriptors, such as the activation area, Δa, or volume, ΔV_0, [132,133] may be ambiguous as the strain rate immediately on unloading is negative, and the material does not have a fixed substructure at definable stages such as at "B" and "C." Eventually, the material "softens" to "D" as a consequence of deformation at the lower stress, to the strain rate that most investigators have concluded corresponds to that which would have been obtained on loading the annealed metal at same temperature and stress, although this is not necessarily a consensus view (e.g., Refs [128,134]). Parker and Wilshire [134], for example, find that at lower temperatures, Cu, with a stress drop, did not return to the creep rate for the uninterrupted test. Of course, it is unclear whether the rate would have eventually increased to the uninterrupted rate with larger strains that can be precluded by fracture in tensile tests.

The anelastic strains are very small for small stress-reductions and may not be observed. The creep rate cannot be easily defined until $\dot{\varepsilon}_C$, and an "anomalous" creep is generally not observed as with large stress reductions. Again, the material eventually softens to a steady state at "D."

The stress–dip test appears to at least be partially responsible for the introduction of the concept of an internal backstress. That is, the backflow associated with the stress dip, observed in polycrystals and single crystals alike, has been widely presumed to be the result of an internal stress. At certain stress reductions, a zero initial creep rate can result, which would presumably be at an applied stress about equal to the backstress [81,115,126,135]. Blum and coworkers [22,77,122,125,135,136], Nix and coworkers [26,117], Argon and coworkers [18,137], Morris and Martin [42,43], and many others [20,53,138] (to reference a few) have suggested that the backflow or backstress is a result of high local internal stresses that are associated with the heterogeneous dislocation substructure, or subgrain walls. Recent justification for high internal stresses beyond the stress–dip test has included X-ray diffraction (XRD) and convergent beam electron diffraction (CBED) [136]. Gibeling and Nix [117] have performed stress relaxation experiments on aluminum single crystals and found large anelastic "backstrains," which they believed were substantially in excess of that which would be expected from a homogeneous stress state [26,117].

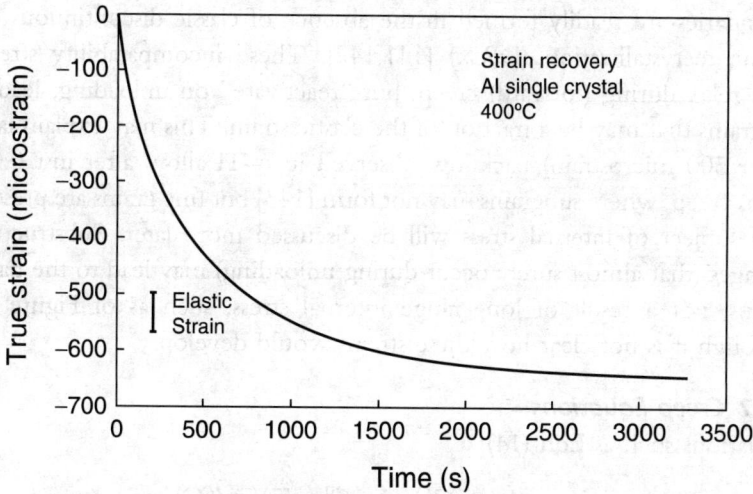

Figure 27 The backstrain associated with unloading an aluminum single crystal from 4 to 0.4 MPa at 400 °C. *From Ref. [117].*

This experiment is illustrated in Figure 27 for a stress drop from 4 to 0.4 MPa at 400 °C. If a sample of cubic structure is assumed with only one active slip system, and an orthogonal arrangement of dislocations, with a density ρ, and all segments are bowed to a critical radius, then the anelastic unbowing strain is about:

$$\gamma_A = \cong \frac{\pi b \sqrt{\rho}}{8\sqrt{3}} \qquad (24)$$

or about 10^{-4}, much smaller than that suggested by Figure 27. Because the average stress is the applied stress, single crystals should only suggest backstrains comparable with the elastic unloading strains, so the results of Figure 27 are perplexing. Equation (22) only assumes one-third of ρ are bowing as two-thirds are on planes without a resolved shear stress. In reality, slip on {111} and {110} [130] may lead to a higher fraction of bowed dislocations. Furthermore, it is known that the subgrain boundaries are mobile. The motion may well involve (conservative) glide, leading to line tension and the potential for substantial backstrain. It is known that substantial elastic incompatabilities are associated with grain boundaries [140]. Although Nix et al. suggest that backstrain from grain boundary sliding (GBS) is not a consideration for single crystals, it has been later demonstrated that for single crystals of Al, in creep, such as Figure 27, high angle boundaries are readily formed in the absence of classic discontinuous dynamic recrystallization (DRX) [141,142]. These incompatability stresses may relax during (forward) creep, but "reactivate" on unloading, leading to strains that may be a fraction of the elastic strain. This may explain large (over 500 microstrain) backflow observed in α-Ti alloys after just 0.002 strain creep, where subgrains may not form [143] but fine grains are present. The subject of internal stress will be discussed more later. Substructural changes (that almost surely occur during unloading) may lead to the back-strains, not a result of long-range internal stress, such as of Figure 27, although it is not clear how these strains would develop.

2.2.2 Creep Equations

Equations such as Eqn (14):

$$\dot{\varepsilon}_{ss} = A_6(\chi/Gb)^3(D_{sd}Gb/kT)(\sigma_{ss}/G)^5$$

are capable of relating, at a fixed temperature, the creep rate to the steady-state flow stress. However, in associating different steady-state creep rates with (steady-state) flow stresses, it must be remembered that the dislocation structures are different. This equation does not relate different stresses and substructures at a fixed temperature *and* strain rate as, for example, the Hall–Petch equation, at ambient temperature and a "conventional" strain rate.

Sherby and coworkers reasoned that relating the flow stress to the (e.g., steady-state) substructure at a fixed strain rate and temperature may be performed with knowledge of N (or m) in Eqn (1):

$$N = \left(\frac{\partial \ln \dot{\varepsilon}}{\partial \ln \sigma}\right)_{T,s}$$

Sherby and coworkers suggested that the flow stress at a fixed elevated temperature and strain rate is predictable through [4,6]:

$$\dot{\varepsilon} = A_7(\lambda^3)\exp[-Q_{sd}/kT](\sigma/E)^N \tag{25}$$

for substructures resulting from steady-state creep deformation in the five-power regime. It was suggested that $N \cong 8$. Steady-state (ss) subscripts are notably absent in this equation. This equation is important in that the flow stress can be directly related to the microstructure at any fixed (e.g., reference) elevated-temperature and strain rate. Sherby and coworkers, at least at the time that this equation was formulated, believed that subgrain boundaries were responsible for elevated-temperature strength. Sherby and coworkers believed that particular value was inherent in this equation, since if Eqn (17):

$$\frac{\sigma_{ss}}{G} = C_1 \lambda_{ss}^{-1}$$

were substituted into Eqn (25), then the well-established five-power-law equation (Eqn (9)) results:

$$\dot{\varepsilon}_{ss} = A_3 \exp[-Q_{sd}/kT]\left(\frac{\sigma_{ss}}{G}\right)^5$$

Equation (25) suggests that at a fixed temperature and strain rate:

$$\frac{\sigma}{E}\bigg|_{\dot{\varepsilon},T} = C_5(\lambda)^{-3/8} \qquad (26)$$

Sherby normalized stress with the Young's modulus although the shear modulus could have been used. This is very different from Eqn (17). This equation, of course, does not preclude the importance of the interior dislocation network over the heterogeneous dislocation substructure (or subgrain walls) for steady-state substructures, on which Eqn (25) was based. This is because there is an approximately fixed relationship between the steady-state subgrain size and the steady-state interior dislocation density. Equation (26) could be reformulated, without a loss in accuracy, as:

$$\frac{\sigma}{G}\bigg|_{\dot{\varepsilon},T} = k_2(\rho)^{-3p/8 \cong -3/16} \qquad (27)$$

2.2.3 Dislocation Density- and Subgrain-Based Constant–Structure Equations

Equation (26) for subgrain strengthening does not have a strong resemblance to the well-established Hall–Petch equation for high-angle grain-boundary strengthening:

$$\sigma_y\big|_{\dot{\varepsilon},T} = \sigma_o + k_y g^{-1/2} \qquad (28)$$

where $\sigma_y\big|_{\dot{\varepsilon},T}$ is the yield or flow stress (at a reference or fixed temperature and strain rate), k_y is a constant, g is the average grain diameter, and σ_o is the single crystal strength and can include solute strengthening as well as dislocation hardening. Of course, subgrain boundaries may be the microstructural feature associated with elevated temperature strength and the rate-controlling process for creep, without obedience to Eqn (28). Nor does Eqn (27) resemble the classic dislocation hardening equation [144]:

$$\sigma_y\big|_{\dot{\varepsilon},T} = \sigma_0' + \alpha MGb(\rho)^{1/2} \qquad (29)$$

where $\sigma_y\big|_{\dot{\varepsilon},T}$ is the yield or flow stress (at a reference or fixed temperature and strain rate), σ_0' is the near-zero dislocation density strength and can include solute strengthening as well as grain-size strengthening, M is the Taylor factor, 1–3.7, and α is a constant, often about 0.3 at ambient temperature. As Eqn (29) has an athermal hardening component ($\alpha MGb\rho^{1/2}$) (σ_0' varies with T and $\dot{\varepsilon}$), the designation of a constant T and $\dot{\varepsilon}$ of σ_y is not necessary. (This constant will be dependent upon the units of ρ, as line-length per unit volume, or intersections per unit area, the latter being a factor of two lower for identical structure.) Both Eqns (28) and (29) assume that these hardening features can be simply summed to obtain their combined effect. Although this is reasonable, there are other possibilities [145]. Equation (29) can be derived on a variety of bases (e.g., bowing stress, passing stress, or cutting stress in a "forest" of dislocations, etc.), some essentially athermal, and others not, and may not always include a σ_0' term.

Even with high-purity aluminum experiments (99.999% pure), it is evident in constant strain-rate mechanical tests that annealed polycrystal has a yield strength (0.002 plastic strain offset) that is about one-half the steady state flow stress [4,146] that cannot be explained by subgrain (or dislocation) hardening; yet this is not explicitly accounted in the phenomenological equations, e.g., Eqns (26) and (27). When accounted, by assuming that σ_0(or σ_0') $= \sigma_y|_{T,\dot{\varepsilon}}$ for annealed metals, Sherby and coworkers showed that the resulting subgrain-strengthening equation that best describes the data form would not resemble Eqn (28), the classic Hall–Petch equation; the best-fit $(1/\lambda)$ exponent is somewhat high at about 0.7. Kassner and Li [147] also showed that there would be problems with assuming that the creep strength could be related to the subgrain size by a Hall–Petch equation. The constants in Eqn (28), the Hall–Petch equation, were experimentally determined for high-purity annealed aluminum with various (HAB) grain sizes. The predicted (extrapolated) strength (at a fixed elevated temperature and strain rate) of aluminum with grain sizes comparable to those of steady-state subgrain sizes was substantially *lower* than the observed value. Thus, even if low misorientation subgrain walls strengthen in a manner analogous to HABS, then an "extra strength" in steady-state, subgrain-containing structures appears from sources other than that provided by boundaries. Kassner and Li suggested that this extra strength may be due to the steady-state dislocation density not associated with the subgrains, and dislocation hardening was observed. Additional discussion of grain-size effects on the creep properties will be presented later.

The hypothesis of dislocation strengthening was tested using data on high-purity aluminum as well as a Class M alloy, AISI 304 austenitic stainless steel (19Cr–10Ni) [107,148]. It was discovered that the classic dislocation hardening (e.g., Taylor) equation is reasonably obeyed if σ_0' is approximately equal to the annealed yield strength. Furthermore, the constant α in Eqn (29), at 0.29, is comparable to the observed values from ambient temperature studies of dislocation hardening [144,149–151] as will be further discussed later. Figure 28 illustrates the polycrystalline stainless steel results. The λ and ρ values were manipulated by combinations of creep and cold work. Note that the flow stress at a reference temperature and strain rate that corresponds to nearly within five-power-law creep (750 °C in Figure 23) is independent of λ for a fixed ρ. The dislocation-strengthening conclusions are consistent with the experiments and analysis of Ajaja and Ardell [152,153] and Shi and Northwood [154,155] also on austenitic stainless steels.

Henshall, Kassner, and McQueen [156] also performed experiments on Al-5.8at%Mg in the 3-power regime where subgrain boundaries only sluggishly form. Again, the flow stress was completely independent of the subgrain size (although these tests were relevant to 3-power creep). The Al-Mg results are consistent with other experiments by Weckert and Blum [121] and the elevated temperature in situ TEM experiments by Mills [157]. The latter experiments did not appear to show interaction between subgrain walls and gliding dislocations. The experiments of this paragraph will be discussed in greater detail later.

2.3 Primary Creep Microstructures

Previous microstructural trends in this review emphasized steady-state substructures. This section discusses the development of the steady-state substructure during primary creep where hardening is experienced. A good discussion of the phenomenological relationships that describe

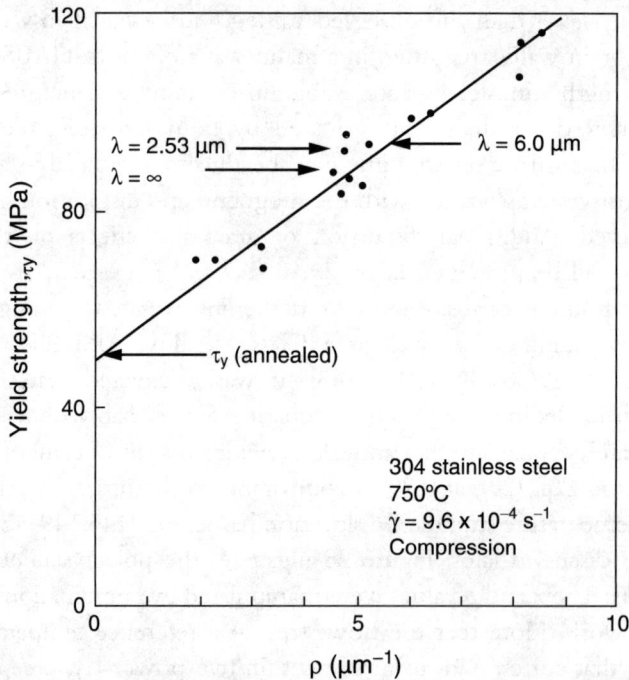

Figure 28 The elevated temperature yield strength of 304 stainless steel as a function of the square root of the dislocation density (not associated with subgrain boundaries) for specimens of a variety of subgrain sizes. (Approximately five-power-law temperature/strain-rate combination.) *Based on Ref. [148].*

primary creep was presented by Evans and Wilshire [28]. Primary creep is often described by the phenomenological equation:

$$\ell = \ell_o\left(1 + \beta t^{1/3}\right)e^{\chi' t} \tag{30}$$

This is the classic Andrade [158] equation. Here, ℓ is the instantaneous gage length of a specimen and ℓ_o is the gage length on loading (apparently including elastic deflection) and β and χ' are constants. This equation leads to equations of the form:

$$\varepsilon = at^{1/3} + ct + dt^{4/3} \tag{31}$$

which is the common phenomenological equation used to describe primary creep. Modifications to this equation include [159]:

$$\varepsilon = at^{1/3} + ct \tag{32}$$

and [160]:

$$\varepsilon = at^{1/3}bt^{2/3} + ct \tag{33}$$

or:

$$\varepsilon = at^b + c^t \tag{34}$$

where [161]:

$$0 < b < 1.$$

These equations cannot be easily justified, fundamentally [23].

For a given steady-state stress and strain rate, the steady-state microstructure appears to be independent as to whether the deformation occurs under constant stress or constant strain-rate conditions. However, there are some differences between the substructural development during a constant stress as compared to constant strain-rate primary-creep. Figure 29 shows Al-5at%Zn at 250 °C at a constant stress of 16 MPa [77]. Again, this is a class M alloy, which mechanically behaves essentially identically to pure Al.

The strain rate continually decreases to a strain of about 0.2, where mechanical steady state is achieved. The density of dislocations not associated with subgrain boundaries is decreasing from a small strain after loading (<0.01) to steady state. This constant-stress trend with the "free" dislocation density is consistent with early etch pit analysis of Fe-3at%Si [162], and the TEM analysis of 304 stainless steel [163], α-Fe [164], and Al [165,166]. Some have suggested that the decrease in dislocation density in association with hardening is evidence that hardening cannot be associated with dislocations and is undisputed proof that subgrains influence the rate of plastic deformation [81]. However, as will be discussed later, this may not be accurate. Basically, Kassner [107,901] suggested that for constant-stress transients, the network dislocations cause hardening but the fraction of mobile dislocations may decrease, leading to strain rate decreases not necessarily associated with subgrain formation. Figure 29 plots the average subgrain size only in areas of grains that contain subgrains. The volume of Al-5at%Zn is not completely filled with subgrains until steady state, at $\varepsilon \cong 0.2$. Thus, the subgrain size averaged over the entire volume would show a more substantial decrease during primary creep. The average spacing, d, of dislocations that comprises subgrain walls decreases *both* during primary, and at least during early steady-state. This trend in d and/or θ_λ was also observed by Suh, Cohen, and Weertman in Sn [167], Morris and Martin [42,43] and Petry et al. [168] in Al-5at%Zn, Orlova, et al. [166] in

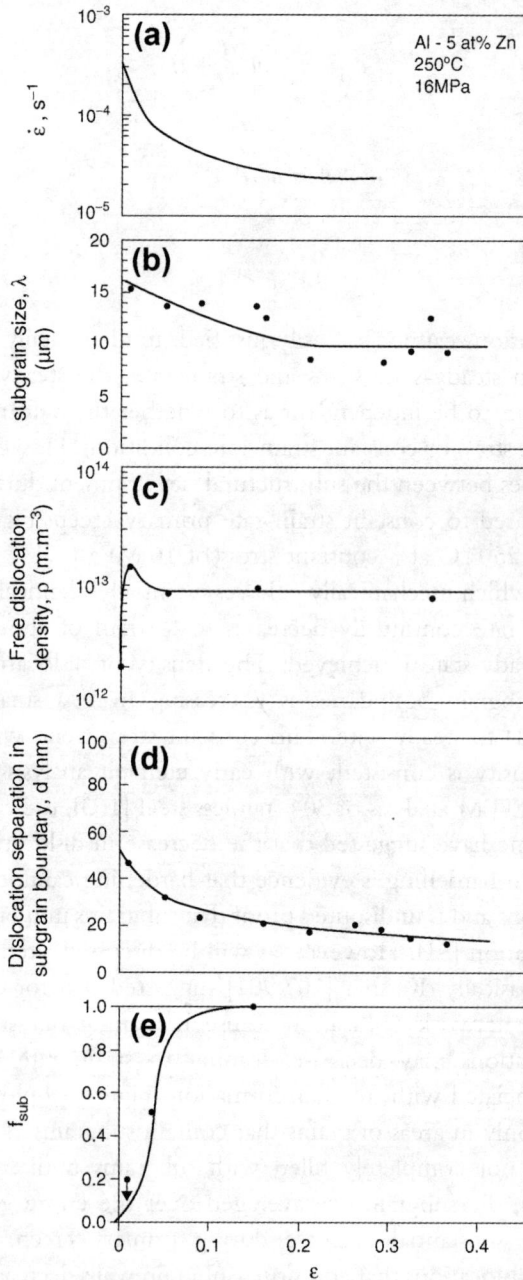

Figure 29 (a) The constant-stress primary creep transient in Al-5at%Zn (essentially identical behavior to pure Al) illustrating the variation of the average subgrain intercept, (b) λ, (in those areas in which subgrains are observed) density of dislocations not associated with subgrain walls, (c) ρ, and the spacing, (d) d, of dislocations that comprise the boundaries. (e) The fraction of material occupied by subgrains is indicated by f_{sub}. *Based on Ref. [77].*

Al, Karashima et al. [112] in αFe, and Kassner et al. in Al [146] and 304 [98] stainless steel. These data are illustrated in Figure 30.

Work hardening for constant strain–rate creep, microstructural trends were examined in detail by Kassner and coworkers [98,107,146] and are

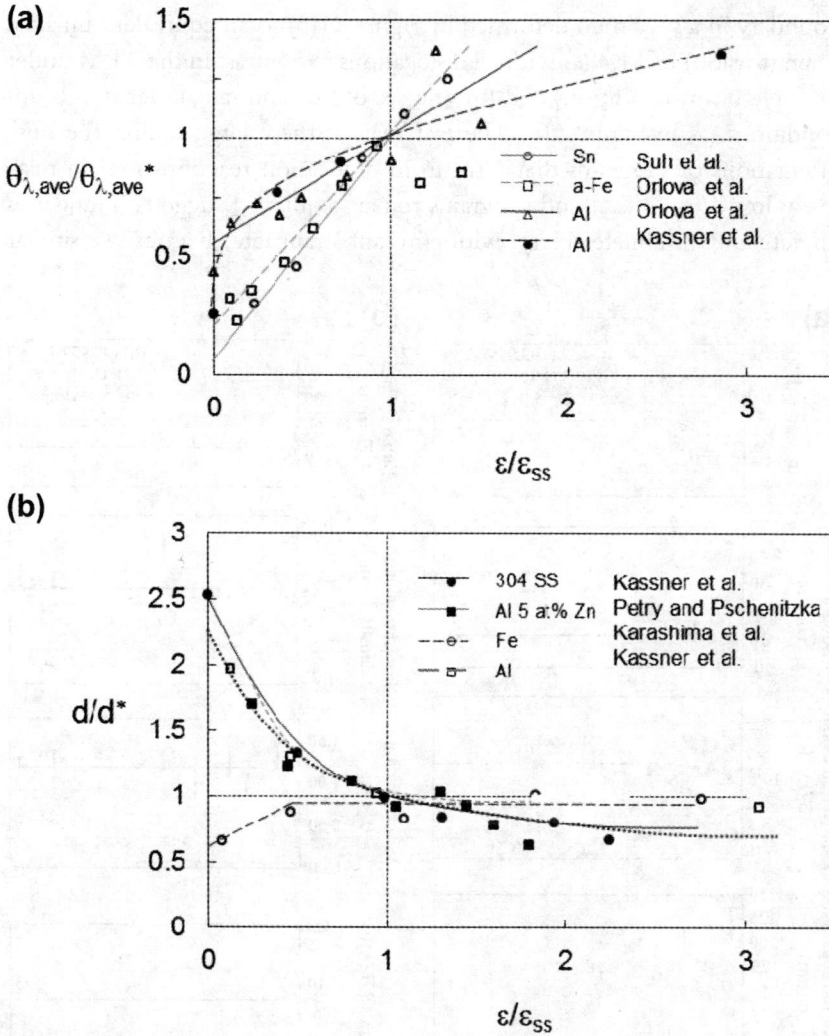

Figure 30 (a) The variation of the average misorientation angle across subgrain walls, $\theta_{\lambda,ave}$, and (b) separation of dislocations comprising subgrain walls with fraction of strain required to achieve steady state, $\varepsilon/\varepsilon_{ss}$ for various metals and alloys. $\theta^*_{\lambda,ave}$ and d^* are values at the onset of steady-state.

illustrated in Figures 21 and 22 for 304 stainless steel and Figure 31 for pure Al. Figure 21 illustrates the dislocation substructure, quantitatively described in Figure 22. Figure 31(a) illustrates the small strain region and that steady state is achieved by $\bar{\varepsilon} = 0.2$. Figure 31(b) considers larger strains achieved using torsion of solid aluminum specimens. Figure 32 illustrates a subgrain boundary in a specimen deformed in Figure 31(b) to an equivalent uniaxial strain (torsion) of 14.3 (a) with all dislocations in contrast in the TEM under multiple beam conditions and (b) one set out of contrast under two beam conditions (as in Figure 20). The fact that, at these large strains, the mis-orientations of subgrains that form from dislocation reactions remain relatively low $(\theta_{\lambda_{ave}} < 2°)$ and subgrains remain equiaxed suggests boundaries migrate and annihilate. Here, with constant strain rate, we observe similar

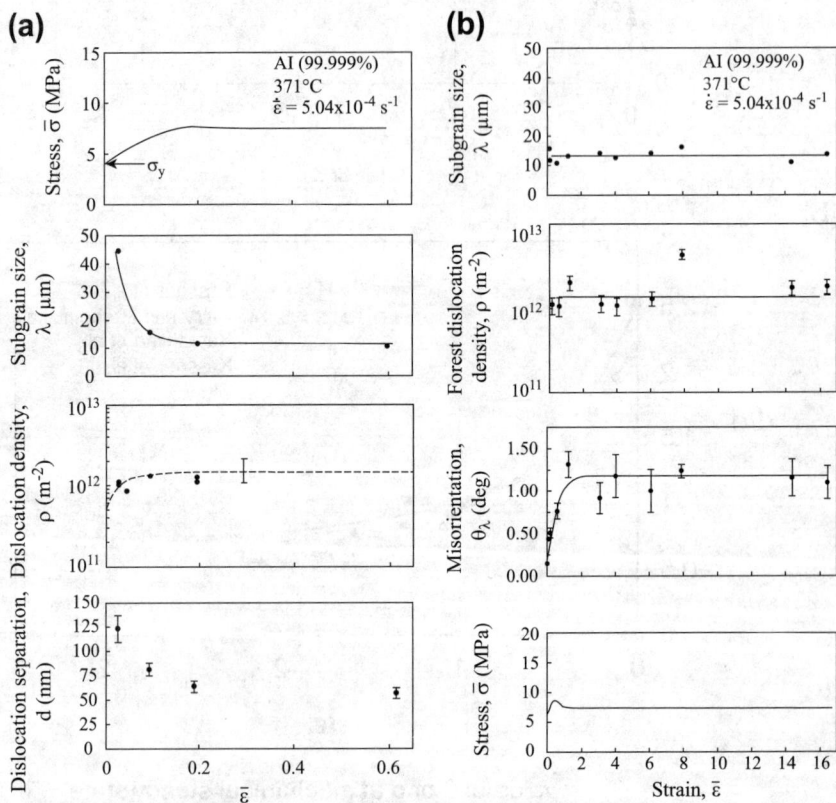

Figure 31 The work hardening during a constant strain-rate creep transient for Al, illustrating the variation of λ, ρ, d, and $\theta_{\lambda_{ave}}$ over primary and secondary creep. The bracket refers to the range of steady-state dislocation density values observed at larger strains; (a) for strains to 0.6 and (b) to very large steady-state strains to over 16.

(a)   (b)

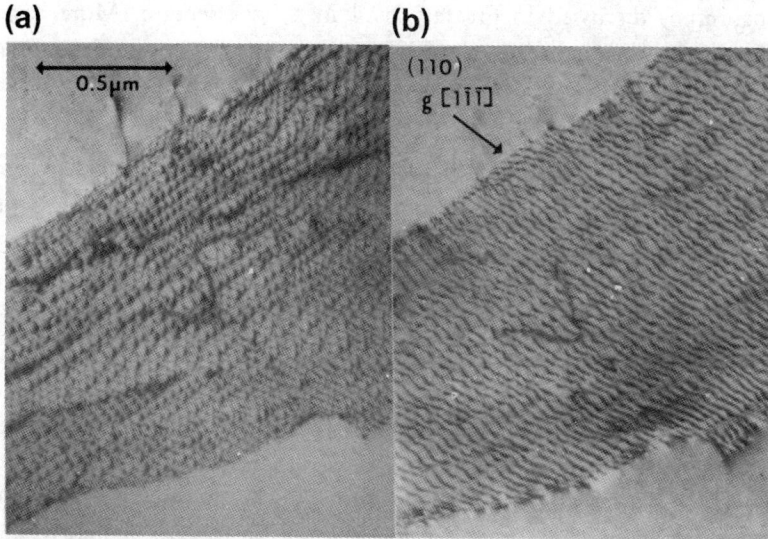

Figure 32 TEM micrographs of a subgrain boundary in Al deformed at 371 °C at $\dot{\bar{\varepsilon}} = 5.04 \times 10^{-4}\text{s}^{-1}$, to steady state under (a) multiple and (b) two-beam diffraction conditions. Three sets of dislocations, of, apparently, nearly screw character. *From Ref. [146].*

subgrain trends to the constant stress trends of Blum in Figure 29 at a similar fraction of the absolute melting temperature. Of course, 304 has a relatively low stacking fault energy while aluminum is relatively high. In both cases, the average subgrain size (considering the *entire* volume) decreases over primary creep. The lower stacking fault energy 304 austenitic stainless steel, however, requires substantially more primary creep strain (0.4 vs 0.2) to achieve steady state at a comparable fraction of the melting temperature. It is possible that the subgrain size in 304 stainless steel continues to decrease during steady state.

Under constant-strain rate conditions, the density of dislocations not associated with subgrain boundaries *monotonically increases* with increased flow stress for both austenitic stainless steel and high-purity aluminum. This is opposite to constant-stress trends. Similar to the constant-stress trends, both pure Al and 304 stainless steel show decreasing d (increasing θ) during primary and "early" steady-state creep. Measurements of d were considered unreliable at strains beyond 0.6 in Al and only misorientation angles are reported in Figure 30(b). It should be mentioned that HABs here form primarily by elongation of the starting grains through geometric dynamic recrystallization, but these are not included in Figure 31(b). This mechanism is discussed in greater detail in a later section. More recent experiments by Sadnabadi et al. [593] on CaF_2 show very similar primary creep trends as just described for 304 stainless steel and Al.

2.4 Creep Transient Experiments

As mentioned earlier, creep transient experiments have been performed by several investigators [127,129,170] on high-purity and commercial-purity aluminum, where a steady state is achieved at a fixed stress/strain rate followed by a change in the stress/strain rate. The strain rate/stress change is followed by a creep "transient," which leads to a new steady state with, presumably, the characteristic dislocation substructure associated with an uninterrupted test at the (new) stress/strain rate. These investigators measured the subgrain size during the transient and subsequent mechanical steady state, particularly following a drop in stress/strain rate. Although Ferriera and Stang [127] found, using less reliable polarized light optical microscopy (POM), that changes in λ in Al correlate with changes in $\dot{\varepsilon}$ following a stress drop, Huang and Humphreys [129] and Langdon et al. [170] found the opposite using TEM; the λ continued to change even once a new mechanical steady state was reached. Huang and Humphreys [129] and Langdon et al. [170] showed that the dislocation microstructure changes with a stress drop, but the dislocation density follows the changes in creep rate more closely than the subgrain size in high-purity aluminum. This led Huang and Humphreys to conclude, as did Evans et al. [171], that the "free" dislocation density was critical in determining the flow properties of high-purity aluminum. Parker and Wilshire [134] made similar conclusions for Cu in the five-power-law regime. Blum [22] and Biberger and Gibeling [10] suggest that interior dislocations can be obstacles to gliding dislocations, based on stress drop experiments leading to aluminum activation area calculations.

2.5 Internal Stress

2.5.1 Introduction

One of the important suggestions within the creep community is that of the internal (or back) stress, which of course has been suggested for plastic deformation, in general. Stress fields in crystals can be either short range or long range, the former occurring on the nanometer scale. Long-range internal stress (LRIS) refers to variations in stress that occur over longer length scales, such as the microstructure (often microns). LRIS appears to be important for a variety of practical reasons possibly including the Bauschinger effect and cyclic deformation (fatigue) and springback in metal forming [27].

Bauschinger Effect

The concept of long-range internal backstresses in materials may have been first discussed in connection with the Bauschinger effect (BE). The material strain hardens, and on reversal of the direction of the straining, the metal plastically flows at a stress less in magnitude than in the forward direction, in contrast to what would be expected based on isotropic hardening. (A Bauschinger effect is illustrated in Figure 33.) Not only is the flow stress lower on reversal but the hardening features are different as well. The BE is

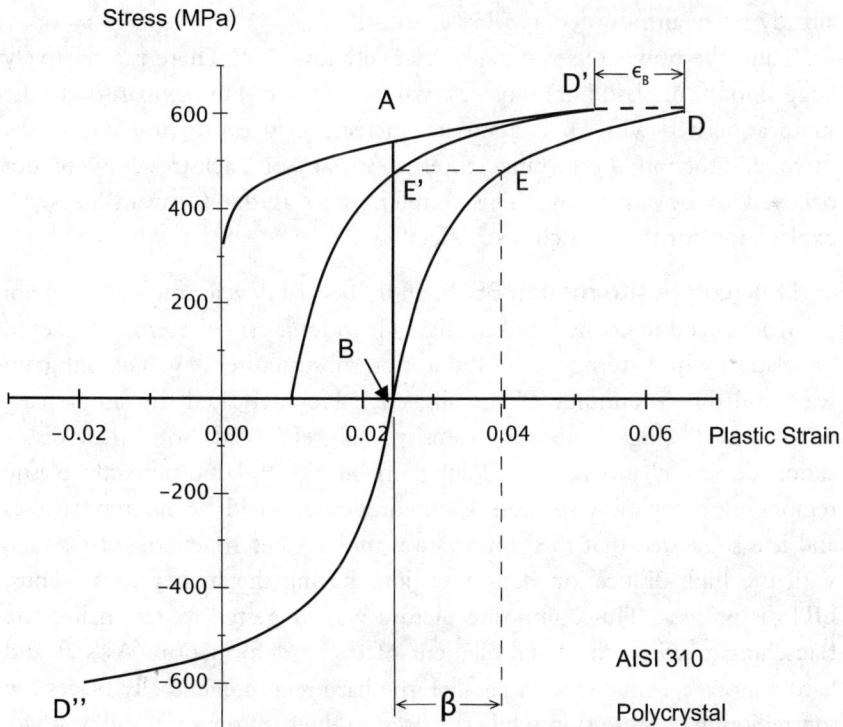

Figure 33 The Bauschinger effect in austenitic stainless steel. The specimen is loaded to A, unloaded and reverse deformation occurs at B. This curve (B–D'') is "reversed" to B–D. Subtracting the Orowan-Sleeswyk strain β, similar to a Bauschinger strain ε_B, brings the reversed curve and is nearly coincident to the forward curve, AD'. *From Ref. [172].*

important as it appears to be the basis for low hardening rates and low saturation stresses (and failure stresses) in cyclic deformation (fatigue).

Non-LRIS explanation for BE Sleeswyk et al. [173] analyzed the hardening features in several metals at ambient temperature and found that the hardening behavior on reversal can be modeled by that of the monotonic case provided a small (e.g., 0.01) "reversible" strain is subtracted from the (early) plastic strain associated with each reversal. This led to the conclusion of an Orowan-type mechanism (no long-range internal stress or backstress) [174] with dislocations easily reversing their motion (across cells). Sleeswyk et al. suggested that gliding dislocations, during work hardening, encounter increasingly effective nonregularly spaced obstacles and the stress necessary to activate further dislocation motion or plasticity continually increases. On reversal of the direction of straining from a "forward" sense, σ, the dislocations will need to move only past those obstacles that have already been surmounted at a lower stress.

Thus, the flow stress is initially relatively low, $<\sigma$. There is a relatively large amount of plastic strain on reversal to $-(\sigma + d\sigma)$ in comparison to the strain associated with an incremental increase in stress to $(\sigma + d\sigma)$ in the forward direction. Long-range internal stresses or backstresses were not believed to be important. This is referred to as the Orowan-Sleeswyk explanation for the Bauschinger effect.

LRIS (composite model) BE In an influential development, Mughrabi [138] advanced the concept of relatively high long-range internal stresses in association with heterogeneous dislocation substructures (e.g., cell/subgrain walls and dipole bundles, PSB walls, etc.). He advocated the simple case where "hard" (high dislocation density walls, etc.) and "soft" (low dislocation density channels, or cell subgrain interiors) elastic perfectly plastic regions are compatibly sheared. Each component yields at different stresses and it is suggested that the composite is under a heterogeneous stress state with the high-dislocation density regions having the higher stress. Thus, LRIS is present. This composite picture was suggested to rationalize the Bauschinger effect, the basic element of cyclic deformation. As soft and hard regions are unloaded in parallel, the hard region eventually places the soft region in compression while the stress in the hard region is still positive. When the total, or average, stress is zero, the stress in the hard region is positive while negative in the soft region. Thus, a BE may be observed where plasticity occurs on reversal at a lower average magnitude of stress than just prior to unloading due to reverse plasticity in the soft region. The composite model for backstress is illustrated in Figure 34. This concept has also been widely embraced for monotonic deformation [954,181] including elevated-temperature creep deformation. The long-range internal stresses are defined by:

$$\tau_w^y = \tau_a + \Delta\tau_w \tag{35}$$

$$\tau_I^y = \tau_a + \Delta\tau_I \tag{36}$$

where τ_a is the applied stress, and $\Delta\tau_w$ and $\Delta\tau_I$ are LRIS in the composite substructure. Simply stated, LRIS is the deviation from the average (or applied) stress in a loaded material.

Experimental History

Bauschinger effect In early work performed by Kassner et al. a random dislocation arrangement in monotonically deformed stainless steel produces nearly the *same* elevated temperature Bauschinger effect (BE) as one with cells and/or subgrains where LRIS should be more substantial from the composite model [111]. Also, it can be noted from the aluminum experiments of [180,955] that a very pronounced BE is evident in the first cycle (1/20% monotonic plastic strain) at 77 K when a cellular substructure is not expected to be evident during the very early Stage I deformation. The BE is comparable to the case where a heterogeneous vein/channel substructure developed after hundreds or thousands of cycles. Perhaps consistent with the Orowan-Sleeswyk explanation, the principal features of the BE may be independent of LRIS.

X-ray peak asymmetry Evidence for internal stresses in the past has also been suggested based on X-ray diffraction (XRD) experiments during both cyclic and monotonic deformation at both high and low temperatures [138,954,956–959]. Basically, X-ray peaks may asymmetrically broaden with plastic deformation. A decomposition can be performed on the asymmetric peak profile into two symmetric peaks, one peak is suggested to represent the material in the vicinity of the high dislocation-density heterogeneities such as dipole bundles, such as, PSB (dipole) walls or cell walls or subgrain boundaries with elevated local stresses. The second peak in cyclically deformed metals is suggested to represent the lower dislocation density material (e.g., cell interiors) where the stresses are smaller in magnitude than the applied stress. This asymmetry is illustrated in Figure 35. Asymmetry analysis varies but often suggests that (under load) stresses in the walls that are about a factor of 1.3–2.8 larger than the applied

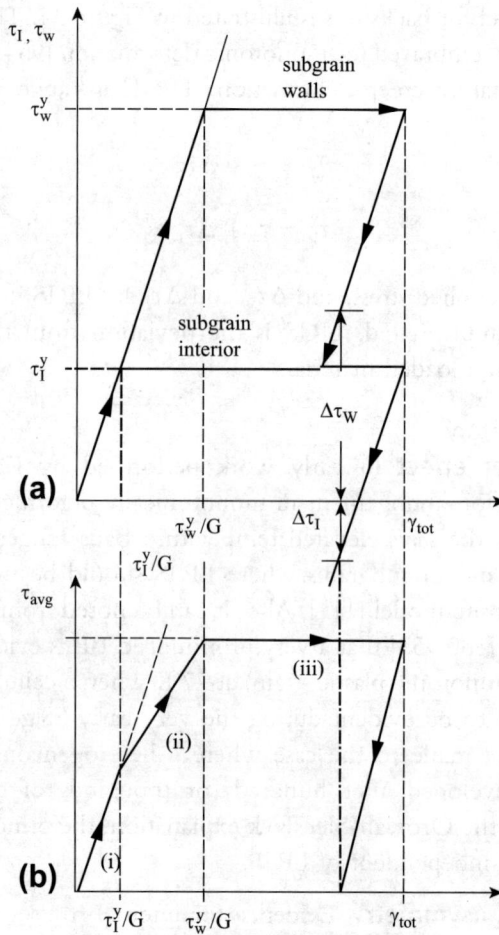

Figure 34 The composite model illustrating the Bauschinger effect. The different stress versus strain behaviors of the (hard) subgrain walls and the (soft) subgrain interiors are illustrated in (a), while the stress versus strain behavior of the composite is illustrated in (b). When the composite is completely unloaded, the subgrain interior is under compressive stress. This leads to a yielding of the softer component in compression at a "macroscopic" stress less than τ_I^y under initial loading. Hence, a Bauschinger effect due to inhomogeneous (or internal) stresses is observed. Note that the individual components are elastic-perfectly plastic.

stress [137,954,961] with PSB walls in cyclically deformed metals having relatively high values. The single crystal Cu compression deformation by Ungar, Mughrabi, et al. suggested the lowest values by this technique at $+0.10\,\sigma_A$ in the cell interiors and $-0.4\,\sigma_A$ in the cell walls [960]. X-ray

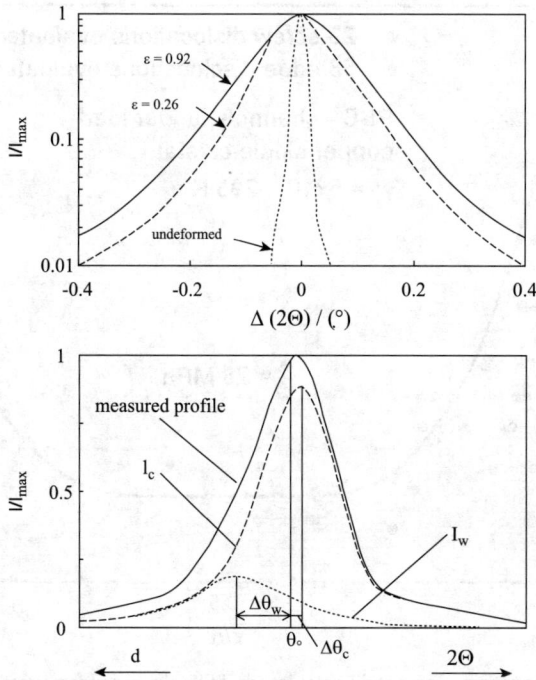

Figure 35 The X-ray diffraction peak in Cu deformed to various strains, showing asymmetric broadening. A decomposition is performed that leads to two symmetric peaks that has been interpreted as a heterogeneous stress state. *Based on Ref. [181].*

peak asymmetry persists *after* unloading in creep–deformed Cu. This suggests that if, in fact, the asymmetry is due to internal stresses, then these stresses persist in the unloaded case [956].

In-situ experiments In situ deformation in the TEM experiments by Lepinoux and Kubin [179] and particularly the neutron irradiation experiments by Mughrabi [138] of Figure 36 have long been cited as early evidence for long-range internal stresses in cyclically deformed metals; the stresses are, roughly, a factor of 3 higher than the applied stress for dipole bundles in Cu PSB walls (i.e., LRIS is about twice the applied stress). These two studies assessed LRIS by measuring dislocation loop radii as a function of position within the heterogeneous microstructure in cyclically deformed single-crystal Cu. More recent work by Mughrabi et al. suggests that these values may be 50% higher than the actual LRIS [962]. One could also interpret [963] the data of Figure 36 to suggest that there is a real variation in LRIS at heterogeneities and an *average* value of LRIS to roughly half the

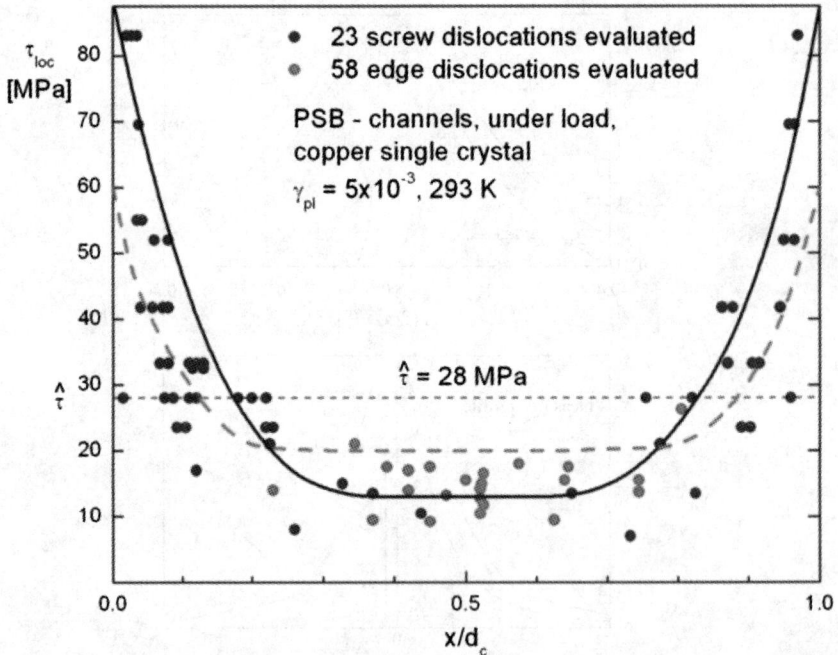

Figure 36 Dislocation loop-radii-calculated LRIS in neutron-irradiated cyclically deformed Cu single crystals (under load) with persistent slip bands (PSBs). *From Ref. [138] with recent modifications [962].*

maximum value. Figure 37 illustrates *very* high local stresses at subgrain boundaries (up to 20 × the applied stress) in Al-5at%Zn creep-deformed alloys based on pinned dislocation loops via a precipitation reaction on cooling under load. Other in situ TEM experiments were inconclusive [180]. In situ X-ray diffraction experiments will be discussed later.

Dipole height measurements Dipole heights may allow the prediction of stresses in cyclically deformed materials. The approximate stress to separate a dipole of height h can be calculated from:

$$\tau_d = \frac{Gb}{8\pi(1-\nu)h} \tag{37}$$

where G is the shear modulus, b is the Burgers vector, and ν is Poisson's ratio.

Cyclic deformation experiments on aluminum and copper single crystals [955] showed that the dipole heights in the presaturation microstructure are

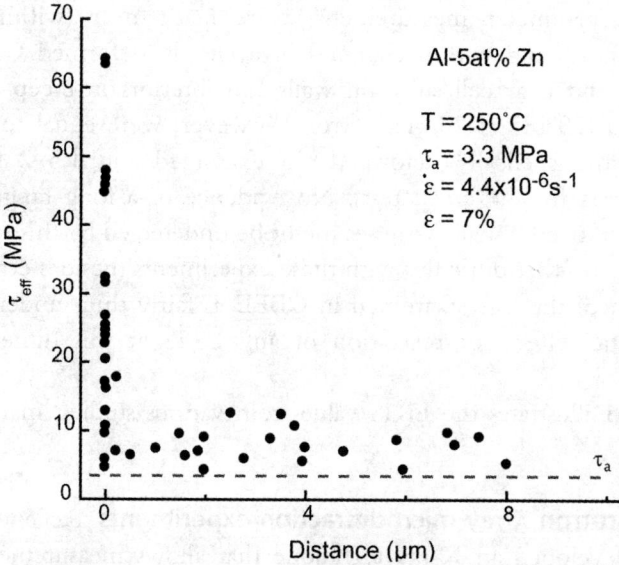

Figure 37 Dislocation loop-radii-calculated LRIS based on precipitate pinning (under load) in creep deformed Al–Zn alloy with subgrains. *(From [42].)* The high stresses are located at the subgrain walls.

also approximately *independent of location*, being equal in the dipole bundles (or veins) and the channels, consistent with other Ni work [964,965]. This could suggest a uniform stress state across the microstructure. Furthermore, the maximum dipole heights are the widest stable dipoles and suggest, through Eqn (37), a maximum stress in the vicinity of the dipole. The maximum dipole height translates to a stress, according to Eqn (37), that is about *equal* to the applied (cyclic) stress for Al. The stress to separate dipoles is within a factor of 2–3 in Cu while for Ni is within a factor of 4 [78,955,964,965].

Convergent beam electron diffraction (CBED) experiments were used to assess internal stresses in unloaded monotonically deformed Cu [181]. CBED, with a 20–100 nm beam size, can probe smaller volumes than the X-ray studies referenced so far, which irradiate over large portions of the specimens. The CBED results by Borbely et al. [181] suggest high local stresses in the vicinity of subgrain boundaries in Cu, based on CBED; however, this may be speculative. The data are sparse and definitive trends were not evident. CBED tests are most easily performed on unloaded thin films.

Lattice parameter measurements were made near (within 80 nm) dipole bundles and within channels in cyclically deformed Cu to pre-saturation and near cell/subgrain walls and interiors in creep-deformed Al and Cu [78,185, 186]. The errors, however, were equal to the flow stress for the cyclically deformed Cu examined and 1.5–2 times the applied stress for the creep tests. No evidence of a long-range internal stress was noticed. Smaller stresses might be undetected by this technique. Of course, another difficulty with these experiments (besides error) is that the region of the foils examined in CBED is fairly thin, under 100 nm. Any "extra" effects of relaxation of any LRIS at this dimension are unknown.

Table 1 illustrates the LRIS value from various studies, many already discussed.

Synchrotron X-ray microdiffraction experiments Recently, Larson and Ice developed an X-ray technique that allows measurement of the X-ray line profile as well as having the ability to calculate the internal stress state in relatively small volumes within the bulk specimen [974]. The submicron ($\cong 0.5$ μm) spot size of this new technique is small enough to allow probing within individual dislocation cells and subgrains within a bulk specimen [974,975]. Local data acquisition is possible on the length scale that is necessary to answer many questions regarding LRIS discussed earlier.

This technique was applied to analyze the strain state along a line that traverses the various high and low dislocation density regions of a deformed substructure in [001] oriented copper single crystals in both tension and compression to strains that led to dislocation cell structures. Thus, a syn-chrotron beam can probe across cells of bulk deformed Cu samples in which the length scale of the microstructure has been characterized by TEM. Broadening of the diffraction profiles confirmed that the mono-tonically deformed Cu samples exhibit the asymmetry expected based on other work [957].

It should be mentioned that the microbeam studies were performed on unloaded specimens. Borbely, Blum, and Ungar made asymmetry mea-surements in situ or under load [954]. The XRD line profiles provided the expected asymmetry under stress at elevated temperature. However, the unloaded specimen retained a majority of the asymmetry. If the asymmetry under load is reflective of LRIS, then unloaded specimens appear to sub-stantially "lock in" these stresses.

Table 1 LRIS values from various studies

Material	References	Deformation mode	Strain	LRIS (σ_A) Walls	Interior	Observation method	Temp.	Notes
Cu	Mughrabi [138] Mughrabi et al. [962]	Cyclic	Saturation w/PSBs	+2.0 +1.3	−0.5 −0.37	In-situ neutron irrad.	RT	Reanalysis of above
	Mughrabi et al. [958,960,967,968]	Tension		+0.4	−0.1	X-ray peak asymm.		Unloaded [001] oriented single crystal
	Lepinoux and Kubin [179]	Cyclic	Saturation w/PSBs	+2.5	−0.5	In-situ TEM	RT[a]	Loaded single crystal
	Kassner [185]	Cyclic	Pre-sat. no PSBs	0	0	CBED Dipole sep.	RT	Unloaded [123] single crystal
	Borbely et al. [954]	Creep	Steady state	+1.0	−0.08	X-ray peak asymm.	527 K	Loaded
	Kassner et al. [78]	Creep	Steady state	—	0	CBED	823 K	Unloaded
	Straub et al. [136]	Compression		−0.3 to −0.6	+0.05−0.08	X-ray peak asymm.	RT−633 K	
		Compression		—	Observed	CBED	RT−633 K	
	Levine et al. [970]			—	+0.29	X-ray microbeam	RT	Unloaded [001] oriented single crystal
		Tension		—	−0.17			

Continued

Table 1 LRIS values from various studies—cont'd

Material	References	Deformation mode	Strain	LRIS (σ_A) Walls	LRIS (σ_A) Interior	Observation method	Temp.	Notes
Ni	Levine et al. [199]	Compression		−0.1	+0.1		RT	Loaded
	Hecker et al. [961]	Cyclic	Saturation w/PSBs Pre-sat. no PSBs	± 1.4–1.8 0	±0.16–0.2 0	X-ray peak asymm.	RT	
Al	Kassner et al. [180,185]	Cyclic	Pre-sat no PSBs	0	0	Dipole separation	77 K	Unloaded [123] oriented single crystal
	Kassner et al. [78]	Creep	Steady state	0	0	CBED	664 K	Unloaded
Al-5at%Zn	Morris and Martin [42]	Creep	Steady state	+25	+1	Disl. loops from precipitation pinning	483–523 K	
—	Sedlacek et al. [971]	Creep		1.5–10.0	0.5–1.0	Theoretical	Creep	
—	Gibeling and Nix [26,118]	Creep		7.7	0.1–0.2	Theoretical	Creep	
Si	Legros et al. [119]	Cyclic	Pre-sat.	0	0	CBED	RT	Unloaded

Figure 38 (a) and (b) illustrate the (006) diffraction peaks for compression (−0.277) and tension (+0.306) deformed [100]-oriented Cu single crystals (also recently reported in Ref. [970]). The deformation-induced change in the lattice parameter in cell interiors reverses sign in tension and compression (negative in tension, positive in compression as measured locally by DAXM). Also, (006) X-ray Bragg reflections (not DAXM), from tension- and compression-deformed samples (Figure 38), were found to contain asymmetry, and the asymmetry "sense" was reversed when comparing tension and compression samples. These two results, when taken together, are the most direct evidence that a long-range internal stress remains after plastic deformation, specifically showing that cell interiors are under compressive stress after tensile deformation and vice versa. Further, these two results are in qualitative agreement with the predictions of the composite model. Preliminary results [199] suggest the presence of a stress with a sign opposite to that of the cell interior in the cell wall (that appears of a similar magnitude) is consistent with the data.

Thus X-ray peak asymmetry here (and perhaps in general) appears reflective of LRIS. The magnitude of the range of average of the LRIS in cell interiors for tension and compression appears to be on the order of 0.16–0.29 the applied stress [199], while cell walls appear of opposite sign and a similar magnitude. Of course, the interpretation of these stresses is the summation of the individual stress fields of the dislocations in each region, unlike multiphase systems where coherency can lead to long-range residual stress [182]. More recent work [115] refined the earlier work and directly measured LRIS in both cell interiors and cell walls as $+0.1\ \sigma_a$ in cell interiors and $-0.1\ \sigma_a$ in cell walls for compression. Recent molecular dynamics work [896] also appears consistent with these findings.

There is reasonable agreement between the Cu microbeam work and the work of Mughrabi and Ungar et al. [958] that was revised in [960]. The values are lower than other X-ray asymmetry measurements and other measurements of LRIS as discussed earlier. Our estimates for the volume fractions of cell wall and cell interiors at −0.248 strain in [001] oriented Cu is $f_w \cong 0.44$–0.58 based on a fairly simple image analysis of TEM micrographs that partitions high and low dislocation density regions. High and low were qualitatively based on image analysis as well as by specifically designating regions with about $10 \times$ dislocation density as cell interiors. The cell wall estimates of LRIS are in rough agreement with the Mughrabi

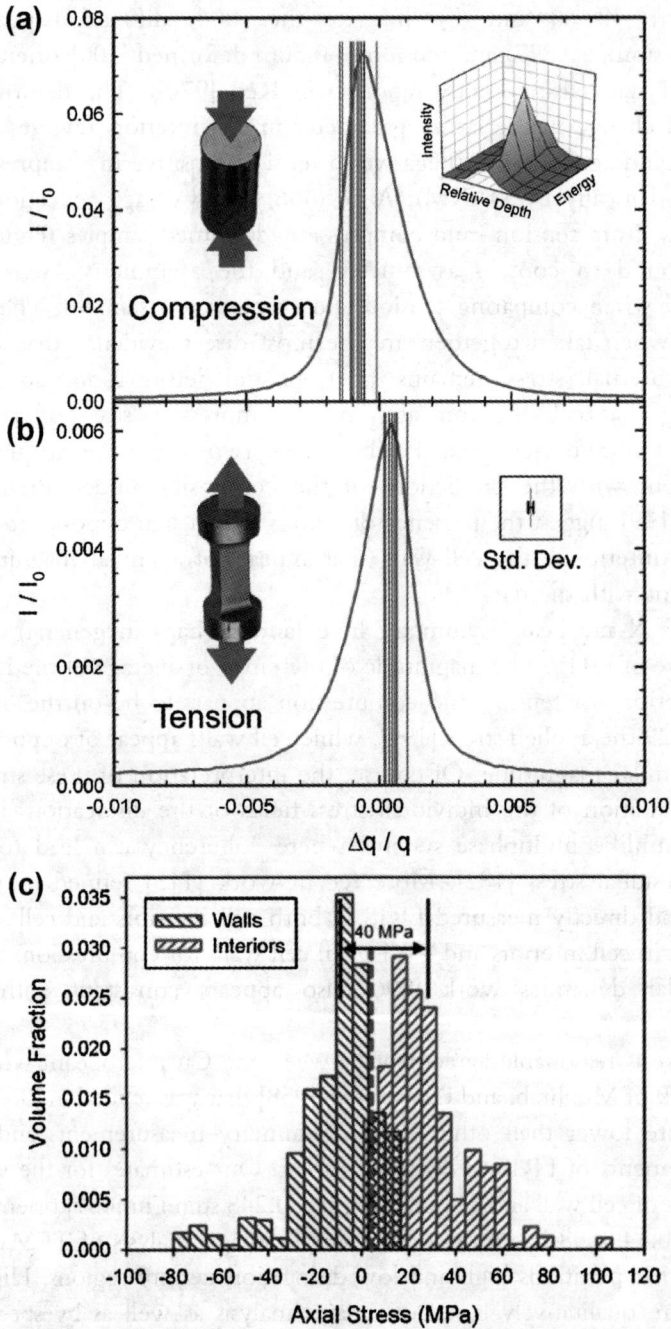

Figure 38 The (006) X-ray peak profile of (a) compression-deformed, (b) tension-deformed [001] oriented Cu single crystal, and (c) measured stress distributions for cell interiors (right) and cell walls (left) [115]. Asymmetry is observed and the vertical line represents the lattice parameters within cell interiors as determined by X-ray microbeams. *From Ref. [970].*

et al. value of 0.4 based on X-ray diffraction of unloaded specimens that include a large number of cells, if Mughrabi et al. had used more appropriate values of f_w.

Earlier work by Kassner et al. [78] failed to observe LRIS within subgrains by CBED in creep-deformed Cu and Al. The CBED results do not agree with the Al-5at%Zn tensile creep work of Morris and Martin, who found up to $+20\ \sigma_A$ LRIS at or near the subgrain walls and over $+2\ \sigma_A$ away from the walls. However, the error in the unloaded CBED pure Al measurements are about $\pm 2\ \sigma_A$. Neither did the CBED of creep-deformed Cu find LRIS which is not really consistent with the X-ray peak asymmetry work of [954] (1 σ_A in walls, 0.1 σ_A in the interiors). In this case, the $\pm 1.0\ \sigma_A$ CBED error may have left the LRIS in these specimens undetected.

The work of Mughrabi [138] and Lepinoux and Kubin [179] suggests LRIS of about a $2\ \sigma_A$ in PSB walls. (As mentioned earlier, however, Mughrabi recently revised his original estimates to lower values of about $1.3\ \sigma_A$ [962].) Similar values were found in Ni cyclically deformed with PSBs [961]. Hence, PSB laden structures may be cases of relatively high LRIS in the dislocation walls. In summary, the magnitudes of LRIS in plastically deformed materials appear smaller than generally suggested, and the magnitudes in creep are not firmly established, if present. Some materials such as those cyclically deformed to saturation (PSB formation) and equal-channel angular pressed (ECAP) may be cases of relatively high LRIS [116].

2.5.2 Other Creep Notes Regarding Long-Range Internal Stresses

High-temperature work by Hasegawa et al. [175] suggested that dissolutions of the cell/subgrains occurred with a reversal of the strain, indicating an "unraveling" of the substructure in Cu-16at%Al, perhaps consistent with the ideas of Sleeswyk and coworkers and the concept that backstress applied to high temperatures is related to dislocation configurations. This suggestion was proposed by Argon and Takeuchi [137], and subsequently adopted by Gibeling and Nix [117] and Nix and Ilschner [26]. With this model, the subgrain boundaries that form from dislocation reaction, bow under action of the shear stress, and this creates relatively high local stresses. The high stresses in the vicinity of the boundary are suggested to be roughly a factor of 3 larger than the applied stress. On unloading, a negative stress in the subgrain interior causes reverse plasticity (or anelasticity), such as illustrated in Figure 27. Whether the LRIS in creep-deformed materials really rises to levels much higher than those deformed at ambient temperature remains an important question, although it increasingly appears unlikely.

3. RATE-CONTROLLING MECHANISMS

3.1 Introduction

The mechanism for plastic flow for five-power-law creep is generally accepted to be diffusion controlled. Evidence in addition to the activation energy being essentially equal to that of lattice self-diffusion includes the Sherby and Weertman analysis showing that the activation volume for creep is also equal to that of self-diffusion [5]. More recent elegant experiments by Campbell et al. [40] showed that impurity additions that alter the self diffusivity also correspondingly affect the creep-rate. However, an established theory for five-power-law creep is not available although there have been numerous attempts to develop a fundamental mathematical description based on dislocation–climb control. This section discusses some selected attempts.

3.1.1 Weertman Model [25,187–189]

Weertman made one of the early attempts to fundamentally describe creep by dislocation climb. Here, the creep process consists of glide of dislocations across relatively large distances, $\bar{x}_g$, followed by climb at the rate-controlling velocity, v_c, over a distance, $\bar{x}_c$. The dislocations climb and annihilate at a rate predictable by a concentration gradient established between the equilibrium vacancy concentration:

$$c_v = c_0 \exp(-Q_v/kT) \tag{38}$$

and the concentration near the climbing dislocation.

The formation energy for a vacancy, Q_V, is altered in the vicinity of a dislocation in a solid under an applied stress, σ, due to work resulting from climb:

$$c_v^d = c_0 \exp(-Q_v/kT)\exp(\pm\sigma\Omega/kT) \tag{39}$$

where Ω is the atomic volume. Again, the (steady-state) flux of vacancies determines the climb velocity:

$$v_c \cong 2\pi\left(\frac{D}{b}\right)(\sigma\Omega/kT)\ln(R_0/b) \tag{40}$$

where R_0 is the diffusion distance, related to the spacing of dislocations (Weertman suggests $Pn(R_0/b) \cong 3\ Pn\ 10$).

Weertman approximates the average dislocation velocity, $\bar{v}$:

$$\bar{v} \cong v_c \bar{x}_g / \bar{x}_c \qquad (41)$$

$$\dot{\varepsilon}_{ss} = \rho_m b \bar{v} \qquad (42)$$

Weertman assumes:

$$\rho_m \cong \left(\frac{\sigma_{ss}}{Gb} \right)^2 \qquad (43)$$

Weertman appears to suggest that the density of dislocations $\rho \cong \rho_m$ and dislocation interaction suggests that ρ_m should scale with σ_{ss} by Eqn (43). This is also analogous to the phenomenological Eqn (19), leading to:

$$\dot{\varepsilon}_{ss} = K_6 \frac{D_{sd}}{b^2} (G\Omega/kT) \left(\frac{\bar{x}_g}{\bar{x}_c} \right) \left(\frac{\sigma}{G} \right)^3 \qquad (44)$$

the classic "natural" or three-power-law equation.

3.1.2 Barrett and Nix Model [190]

Several investigators considered five-power-law creep as controlled by the nonconservative (climb) motion of (edge) jogs of screw dislocations [112,190,191]. The models appear similar to the earlier description by Weertman, in that climb motion controls the average dislocation velocity and that the velocity is dictated by a vacancy flux. The flux is determined by the diffusivity and the concentration gradient established by climbing jogs.

The model by Barrett and Nix [190] is reviewed as representative of these models. Here, similar to the previous Eqn (42):

$$\dot{\gamma}_{ss} = \rho_{ms} \bar{v} b \qquad (45)$$

where $\dot{\gamma}_{ss}$ is the steady-state (shear) creep rate from screws with dragging jogs and ρ_{ms} is the density of mobile screw dislocations.

For a vacancy producing jog in a screw segment of length, j, the chemical dragging force on the jog is:

$$f_p = \frac{kT}{b} \ln \frac{c_p}{c_v} \qquad (46)$$

where c_p is the concentration of vacancies in the vicinity of the jogs. The jog is considered a moving point source for vacancies, and it is possible to express c_p as a function of D_v, and the velocity of the jog, v_p:

$$c_p^* - c_v = \frac{v_p}{4\pi D_v b^2} \tag{45}$$

where c_p^* is the steady-state vacancy concentration near the jog. Substitution of Eqn (47) into Eqn (46) leads to:

$$f_p = \frac{kT}{b} \ln\left[1 + \frac{v_p}{4\pi D_v b^2 c_v}\right] \tag{48}$$

and with $\tau b j = f_p$:

$$v_p = 4\pi D_v b^2 c_v \left[\exp\left(\frac{\tau b^2 j}{kT}\right) - 1\right] \tag{49}$$

Of course, both vacancy-producing and vacancy-absorbing jogs are present, but for convenience, the former is considered, and substituting Eqn (49) into Eqn (45) yields:

$$\dot{\gamma}_{ss} = 4\pi D \beta_2 \left(\frac{b}{a_0}\right) \rho_{ms} \left[\exp\left(\frac{\tau b^2 j}{kT}\right) - 1\right] \tag{50}$$

where a_0 is the lattice parameter.

Barrett and Nix suggest the $\rho_{ms} = A_8\sigma^3$ (rather than $\rho \propto \sigma^2$) and $\dot{\gamma}_{ss} \cong A_9\tau^4$ is obtained. One difficulty with the theory (Eqn (50)) is that all (at least screw) dislocations are considered mobile. With stress drops, the strain rate is predicted to decrease, as observed. However, ρ_{ms} is also expected to drop (with decreasing ρ_{ss}) with time and a *further* decrease in $\dot{\gamma}_{ss}$ is predicted, despite the observation that $\dot{\gamma}$ (and $\dot{\gamma}_{ss}$) increases. Jog-screw models have more recently been applied to Ti-alloys including TiAl [79] and Zr [815].

3.1.3 Ivanov and Yanushkevich [192] Model

These investigators were among the first to explicitly incorporate subgrain boundaries into a fundamental climb-control theory [192]. This model is widely referenced. However, the described model (after translation) is less than very lucid and other reviews of this theory [187] do not appear to clarify all of the details of the theory.

Basically, the investigators suggest that there are dislocation sources within the subgrains and that the emitted dislocations are obstructed by subgrains walls.

The emitted dislocations experience the stress fields of boundary and other emitted dislocations. Subsequent slip or emission of dislocations requires annihilation of the emitted dislocations at the subgrain wall, which is climb controlled. The annihilating dislocations are separated by a mean height:

$$\bar{h}_m = k_3 \frac{Gb}{\tau} \tag{51}$$

where the height is determined by equating a calculated backstress to the applied stress, τ. The creep-rate is:

$$\dot{\varepsilon}_{ss} = \frac{\lambda^2 b \bar{v} \rho'_m}{\bar{h}_m} \tag{52}$$

where $\rho'_m = 1/\bar{h}_m \lambda^2$; where ρ'_m is the number of dislocation loops per unit volume. The average dislocation velocity:

$$\bar{v} = k_4 D_v b^2 \exp(\tau b^3 / kT - 1) \tag{53}$$

is similar to Weertman's previous analysis. This yields:

$$\dot{\varepsilon}_{ss} = \frac{k_5 D_{sd} b G}{kT} \left(\frac{\tau}{G}\right)^3 \tag{54}$$

In this case, the third power is a result of the inverse dependence of the climb distance on the applied stress. Modifications to the model have been presented by Nix and Ilschner, Blum and Weertman [26,193–195].

3.1.4 Network Models (Evans and Knowles, [105] with Modifications [938])

Several investigators have developed models for five-power creep based on climb control utilizing dislocation networks, including early work by McLean and coworkers [84,196], Lagneborg and coworkers [85,197], Evans and Knowles [105], Wilshire and coworkers [104,126], Ardell and coworkers [54,86,101,102,152,153], and Mott and others [106,109,198–200,938]. There are substantial similarities between the models. A common feature is that the dislocations interior to the subgrains are in the form of a Frank network [87]. This is a three-dimensional mesh illustrated in Figure 19.

Network models generally consider coarsening of the dislocation network due to recovery of sessile links ($l < l_c$ where l is the dislocation link-length and l_c is the critical dislocation link-length required for activating dislocation sources). Refinement occurs by multiplications (e.g., Frank-Read sources) once $l_c < l$. A discussion of the network-based models by Ardell and Lee [101] in the Harper-Dorn regime (all link lengths $< l_c$) is presented elsewhere [18]. Evans and Knowles [105] presented a recovery based model for the evolution of dislocation network that may also be consistent with natural three-power law at low stresses [19]. The Evans and Knowles model is attractive but needs modification.

Here [938] the model of Evans and Knowles [105] is modified in order to develop a dislocation network theory for the creep.

It is assumed that the distribution of the dislocation link-lengths is uniform with the smallest length equal to Burgers vector, b, and the largest link length equal to the critical link length for a Frank-Read source, L_c. Shi and Northwood [154] presented a statistical model for calculating the dislocation link lengths based on the dislocation density, ρ:

$$\bar{l} = \left(\frac{2}{\pi c} \right)^{\frac{1}{2}} \frac{1}{\sqrt{\rho}} \tag{55}$$

where $\bar{l}$ is the average dislocation link length and c is the dislocation network geometry factor relating the volume of a polyhedron to its edge length, which is the dislocation link-length. Further, Shi and Northwood [154] state that for a tight, uniform, network:

$$\left(\frac{2}{\pi c} \right)^{\frac{1}{2}} = 1 \tag{56}$$

Phenomenologically, it is known approximately that [29]:

$$\sigma_{ss} = k' Gb \sqrt{\rho_{ss}} \tag{57}$$

where σ_{ss} is the steady state stress value, k' is a constant, and G is the shear modulus. Equation (57) is shown to be consistent with a verified Taylor equation for five-power law steady-state creep of aluminum [169]:

$$\sigma_{ss} = \sigma_0 + \alpha MGb \sqrt{\rho_{ss}} \tag{58}$$

where α in Eqn (58) is a value consistent with the expected value for dislocation hardening and is equal to ≈ 0.2 and M is the Taylor factor. Equations (55) and (57) give the following for $\bar{l}$:

$$\bar{l} = \frac{1}{\sqrt{\rho}} = \frac{\alpha' Gb}{S_{ave} \sigma} \tag{59}$$

where α' is a constant and S_{ave} is the average Schmid factor.

By definition, the dislocation density is equal to the product of the average dislocation link length and the total number of dislocation links per unit volume, N. Hence, the following expression for N is calculated based on Eqn (59):

$$N = \rho^{\frac{3}{2}} = \frac{1}{\bar{l}^3} \tag{60}$$

Like Evans and Knowles [105], it is assumed that the glide of dislocation links is rapid and creep is controlled by dislocation climb. For a three-dimensional dislocation network, both nodes and dislocation links may climb. The slowest of the above two will govern the creep rate. Based on the concentration of vacancies and assuming dislocation links as a perfect sink and source of vacancies, the following climb velocities are calculated [105]:

$$v_n = \frac{4\pi D_L F b}{kT} \tag{61}$$

$$v_l = \frac{2\pi D_{sd} F b}{kT \ln\left(\frac{\bar{l}}{2b}\right)} \tag{62}$$

where v_n is the climb velocity of nodes, D_{sd} is the lattice diffusion coefficient, F is the total force per unit length of the dislocations, k is Boltzmann constant, T is the temperature, and v_l is the climb velocity of dislocation links. Equation (61) assumes that the climb velocity of a single node is equivalent to that of a jog and is taken from Hirth and Lothe [347]. Equation (62) is originally from Weertman [187]:

$$v_l = \frac{2\pi D_{sd} b}{b \ln\left(\frac{R}{b}\right)} \left[\exp\left(\frac{\sigma'' \Omega}{kT}\right) - 1\right] \tag{63}$$

where R is the distance from the dislocation to the point at which the vacancy concentration is nearly equal to the equilibrium vacancy concentration in the crystal, σ'' is the stress acting on the dislocation that produces a climb force, and Ω is the atomic volume. Equation (63) reduces to the form of Eqn (62) based on the following reasonable assumptions: $R \approx \bar{l}/2$, $\Omega \approx b^3$, $F \approx \sigma b$, and the activation volume $\sigma'' \Omega / kT < 1$.

Equations (61) and (62) give the following as the ratio of v_n and v_l:

$$\frac{v_n}{v_l} = 2 \ln\left(\frac{\bar{l}}{2b}\right) \tag{64}$$

which is >1 and hence the climb of dislocation links will govern the creep.

Of course, vacancies can diffuse both through dislocations and through lattice. Since in a network, dislocations links may provide the short circuit path for the vacancies from one point to any other, it may be possible that pipe diffusion becomes dominant over lattice diffusion [105]. The governing diffusion path is assessed by the ratio:

$$P = \frac{\beta G^2 D_{sd}}{3\sigma^2 D_p \ln\left(\frac{\beta}{2\sigma}\right)} \tag{65}$$

where β is a constant of the order of unity and D_p is the pipe diffusion coefficient. Lattice diffusion will be the creep-governing process for $P \gg 1$. Based on the values of D_p and D_{sd} it can be concluded that at higher temperatures and low stresses, as in the case of five-power-law creep, lattice diffusion will be the controlling process.

It is assumed here that only the external forces acting on the dislocation links should affect the climb velocity of dislocation links. There may be two obvious external forces on the dislocations: (1) the applied stress and (2) the stresses due to other dislocations (elastic interaction). Nevertheless, there is a tendency of a dislocation link to increase its length in order to reduce the line tension (strain energy) and hence this tendency due to line tension, given by $Gb^2/\bar{l}$, also contributes to the total climb force. This force was not considered by Zbib et al. [950] in their three-dimensional dislocation network model. However, the line tension is responsible for the coarsening of the network decreasing the dislocation density during an annealing process conducted in the absence of any applied stress. Hence the following will be an approximate expression for force per unit length of the dislocations:

$$F = \sigma_n b + \frac{Gb^2}{\pi\bar{l}} + \frac{Gb^2}{\bar{l}} \tag{66}$$

where σ_n is the component of the applied stress along the direction of climb. The first term in Eqn (64) is due to the applied load whereas the second term comes from the elastic interaction of the dislocations. The third term is due to the line tension. The form of Eqn (66) is similar to the form used by Evans and Knowles [105]. Assuming $\sigma_n = C\sigma$ where C is the geometric constant correlating the applied stress and the normal stress acting upon climbing dislocations, the following expression is calculated for the climb force:

$$F = C\sigma b + \frac{Gb^2}{\bar{l}}\left(1 + \frac{1}{\pi}\right) \tag{67}$$

Using the value of $\bar{l}$ from Eqn (59) and combining all constant in one term, the following expression for F is calculated:

$$F = \gamma \sigma b \qquad (68)$$

where γ is a constant and is given by the following equation:

$$\gamma = C + \frac{S_{ave}}{\alpha'}\left(1 + \frac{1}{\pi}\right) \qquad (69)$$

Substituting the value of $\bar{l}$ from Eqn (59) and F from Eqn (68) in Eqn (62), the following expression for climb velocity is obtained:

$$v_l = \frac{2\pi D_{sd}b}{kT \ln\left(\frac{\alpha'}{S_{ave2}}\frac{G}{\sigma}\right)}\gamma \sigma b \qquad (70)$$

The swept area by a dislocation loop, ϕ, can be given by:

$$\phi = \bar{l}s \qquad (71)$$

where s is the slip distance of the dislocation loop. It is assumed here that slip is caused by the movement of the individual links.

The rate of release of the dislocation loops for the network, $\dot{N}$, can be given by:

$$\dot{N} = 2\left\{\frac{v_l}{(L_c - \bar{l})}Nn'\right\} \qquad (72)$$

where n' is the total number of dislocation loops generated by a Frank-Read source before the source length becomes subcritical by mesh refinement resulting from other dislocation sources. The form of Eqn (72) differs from the one used by Evans and Knowles [105]. Equation (72) uses $(L_c - \bar{l})$ whereas Evans and Knowles [105] use only $\bar{l}$. $(L_c - \bar{l})$ is the distance climbed by a dislocation link (of length $\bar{l}$) in order to activate a Frank-Read source and hence is more appropriate. The term in the curly bracket gives the number of dislocation loops per unit volume generated by the Frank-Read sources, which are activated purely by the climb of the dislo-cation links. It may also be possible to generate a dislocation loop if two nodes break, freeing a dislocation link. It is a viable assumption as the critical shear stress required to untangle a Lomer-Cottrell node in FCC material is equal to $\sim 0.8\ Gb/L_B$, where L_B is the dislocation link length [951] and this stress is of the same order as of the critical stress required to release a dislocation loop from a Frank–Read source. Hence, like Evans and Knowles, a multiplying factor of 2 is used in Eqn (72) to show that the probability of release of a dislocation loop from either a Frank–Read source (for which critical length is L_c) or due to the breakage/untangling of nodes (i.e., when the dislocation link length L_B is less than L_c for the resultant concentrated force on a node is more than node strength.

In order to fully quantify Eqn (72), it is required to calculate the number of dislocation loops, n', generated per Frank–Read source. It may be calculated using the basic property of the steady state:

$$\rho^+ = \rho^- \tag{73}$$

where ρ^+ is the rate of increase in the dislocation density whereas ρ^- is the rate at which dislocations are annihilated. The following gives the dislocation density generation-rate:

$$\rho^+ = \frac{N v_l}{(L_c - \bar{l})} n' c'' \bar{l} \tag{74}$$

where c'' is a geometrical constant relating average link length to the average length of the dislocation loops generated by the Frank–Read source. Dislocation annihilation can take place by climb of a dislocation link, bringing two opposite-signed dislocation links on same slip plane. In this case, two links will be annihilated and this annihilation rate can be given by:

$$\rho^- = v_l N \tag{75}$$

Substituting Eqns (74) and (75) into Eqn (73) and assuming $c'' \approx 1$ gives:

$$n' = \frac{L_c}{\bar{l}} - 1 \tag{76}$$

Taking $L_c = bG/\tau$, where τ is the resolved shear stress in the plane and in the direction of b [169,952] and for a single crystal, $\tau = S\sigma$ where S is the Schmidt factor, Eqn (76) suggests that $n' \approx 1$.

Substituting the value of v_l from Eqn (70) in Eqn (72) and assuming $n' \approx 1$ (Eqn (76)) leads to:

$$\dot{N} = \frac{4\pi\gamma}{(L_c - \bar{l})\bar{l}^3} \left[\frac{D_{sd} b^2 \sigma}{kT \ln\left(\frac{\alpha'}{S_{ave}2} \frac{G}{\sigma}\right)} \right] \tag{77}$$

Now, substituting the value for $\bar{l}$ from Eqn (59) and taking $L_c = bG/\tau$:

$$\dot{N} = \frac{4\pi\gamma}{\left(\frac{bG}{\tau} - \frac{\alpha'bG}{S_{ave}\sigma}\right)\left(\frac{\alpha'bG}{S_{ave}\sigma}\right)^3}\left[\frac{D_{sd}b^2\sigma}{kT\ln\left(\frac{\alpha'}{S_{ave2}}\frac{G}{\sigma}\right)}\right] \tag{78}$$

For a single crystal, $\tau = S\sigma$ where S is the Schmidt factor, hence the following expression:

$$\dot{N} = \frac{4\pi\gamma}{\left(\frac{1}{S_{ave}} - \frac{\alpha'}{S_{ave}}\right)\left(\frac{\alpha'}{S_{ave}}\right)^3\left(\frac{bG}{\sigma}\right)^4}\left[\frac{D_{sd}b^2\sigma}{kT\ln\left(\frac{\alpha'}{S_{ave2}}\frac{G}{\sigma}\right)}\right] \tag{79}$$

Strain rate is given by:

$$\dot{\varepsilon} = \beta\dot{N}\phi b \tag{80}$$

where β is a constant that converts the shear strain into the uniaxial strain (S for single crystals). Now substituting the values of parameters, ϕ and $\dot{N}$, in Eqn (80) from Eqns (71) and (79), respectively:

$$\dot{\varepsilon} = \beta\frac{4\pi\gamma}{\left(\frac{1}{S_{ave}} - \frac{\alpha'}{S_{ave}}\right)\left(\frac{\alpha'}{S_{ave}}\right)^3\left(\frac{bG}{\sigma}\right)^4}\left[\frac{D_{sd}b^2\sigma}{kt\ln\left(\frac{\alpha'}{S_{ave2}}\frac{G}{\sigma}\right)}\right](\bar{l}s)b \tag{81}$$

Taking $\bar{l} = 2$, i.e., hardening is independent of applied stress [15, 29]:

$$\dot{\varepsilon} = \frac{4\pi\gamma\beta}{\left(\frac{1}{S_{ave}} - \frac{\alpha'}{S_{ave}}\right)\left(\frac{\alpha'}{S_{ave}}\right)^3\left(\frac{bG}{\sigma}\right)^4}\left[\frac{D_{sd}b^2\sigma}{kt\ln\left(\frac{\alpha'}{S_{ave2}}\frac{G}{\sigma}\right)}\right]\bar{l}^2 b \tag{82}$$

Substituting the value for $\bar{l}$ from Eqn (59) and re-arranging Eqn (82) gives:

$$\dot{\varepsilon} = \frac{4\pi\gamma\beta}{\frac{(1-\alpha')\alpha'}{S_{ave}^2}}\frac{1}{\ln\left(\frac{\alpha'}{2S_{ave}}\frac{G}{\sigma}\right)}\frac{D_{sd}Gb}{kT}\left(\frac{\sigma}{G}\right)^3 \tag{83}$$

This equation is indicated in Figure 39, for the data reported earlier in Figure 15. The network model predicts the low-stress region very well, but

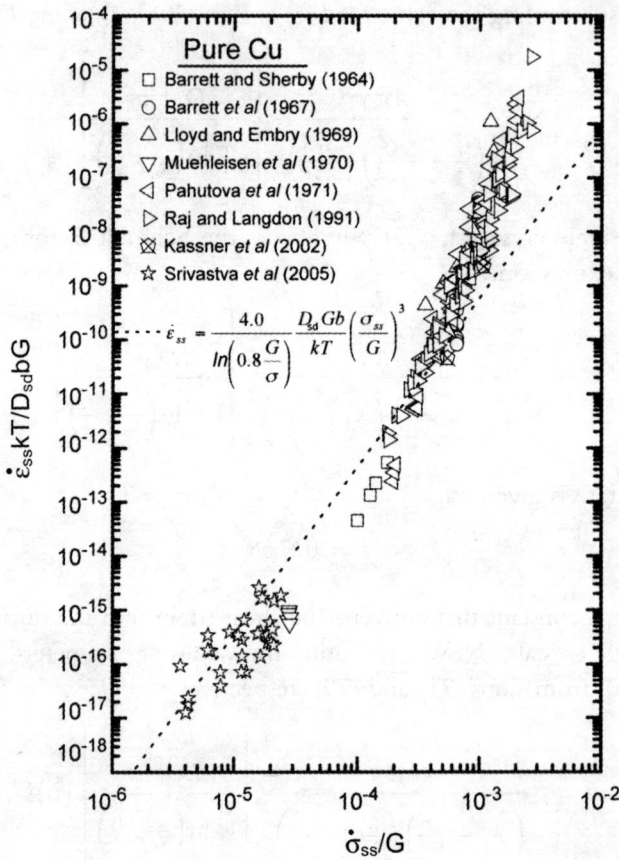

Figure 39 A comparative plot showing the prediction based on dislocation network theory for polycrystalline pure Cu. The data fit well with the present model at low and moderate stresses. *From Ref. [938].*

cannot model the five-power regime very well. It will also be shown in the Harper-Dorn section that this equation predicts the low-stress region of pure Al very well.

3.1.5 Recovery-Based Models

One shortcoming of the previously discussed models is that a recovery aspect is not included in detail. It has been argued by many (e.g., Refs [81,201]) that steady state, for example, reflects a balance between dislocation hardening processes, suggested to include strain-driven network refinements, subgrain-size refinement or subgrain-boundary mesh-size refinement, and thermally activated softening processes that result in coarsening of the latter features.

Maruyama, Karashima, and Oikawa [202] attempted to determine the microstructural feature associated with the rate-controlling (climb) process for creep by examining the hardening and recovery rates during transients in connection with the Bailey-Orowan [203,204] equation:

$$\dot{\varepsilon}_{ss} = \frac{r_r}{h_r} \tag{84}$$

where $r_r = d\sigma/dt$ is the recovery rate and $h_r = d\sigma/d\varepsilon$ is the hardening rate.

The recovery rates in several single-phase metals and alloys were estimated by stress reduction tests, while work-hardening rates were calculated based on the observed network dislocation densities within the subgrains and the average dislocation separation within the subgrain walls, d. Determinations of $\dot{\varepsilon}_{ss}$ were made as a function of σ_{ss}. The predictions of Eqn (84) were inconclusive in determining whether a subgrain wall or network hardening basis was more reasonable, although a somewhat better description was evident with the former.

More recently, Daehn et al. [205,206] attempted to formulate a more basic objective of rationalizing the most general phenomenology such as five-power-law behavior, which has not been successfully explained. Hardening rates (changes in the (network) dislocation density) are based on experimentally determined changes in ρ with strain at low temperatures:

$$\rho_{t+dt} = \rho_t + M_\rho \left(\frac{\rho}{\rho_o}\right)^c \dot{\gamma} dt \tag{85}$$

where M_ρ is the dislocation breeding constant and c and ρ_o are constants. Refinement is described by the changes in a substructural length-scale ℓ' (ρ, d, or λ) by:

$$\frac{d\ell'}{dt} = -\left(\frac{M_\rho \cdot \left(\ell_o'\right)^{2c} (\ell)^{3-2c}}{2g'^2}\right) \dot{\gamma} \tag{86}$$

where ℓ_o' is presumably a reference length scale and g' is a constant.

The flow stress is related to the substructure by:

$$\tau = \frac{\widehat{k}}{b\ell'} \tag{87}$$

Daehn et al. note that if network strengthening is relevant, the above equation should reduce to the Taylor equation.

Coarsening is assumed to be independent of concurrent plastic flow and diffusion controlled:

$$d(\ell')^{m_c} = KDdt \tag{88}$$

where m_c and K are constants and D is the diffusivity. Constants are based on microstructural coarsening observations at steady-state, refinement and coarsening are equal, and the authors suggest that:

$$\dot{\gamma} = BD\left(\frac{\tau}{G}\right)^n \tag{89}$$

results in $n = 4$–6.

This approach to understanding five-power behavior seems attractive and can potentially allow descriptions of primary and transient creep.

3.1.6 The Effect of Stacking-Fault Energy

The basic effect of stacking-fault energy on the creep rate may be explained in terms of the effect of γ on the climb rate of dislocations. Of course, the climb of edge dislocations appears to be by climb of individual jogs [87]. In the case of extended dislocations, it has often been suggested that partial dislocations must constrict before climb is possible. Basically, for an extended edge dislocation in fcc, the (extra half) {220} planes perpendicular to the Burgers vector have an ABAB stacking sequence. The climb of a partial would lead to an energetically very unfavorable AA stacking sequence. Argon and Moffatt [969] suggested a climb mechanism for extended dislocations, and suggested the climb velocity of extended dislocations is:

$$v_c = \frac{A\sigma\Omega D_{sd}}{bkT}\left(\frac{\chi}{Gb}\right)^2 \tag{90}$$

where $A = 2(24\pi(1 - v)/(2 + v))^2$. The main effect of the extension (of the jog) is an attenuation of the frequency vector. The exponent, 2, is reasonably close to the value observed by Mohammed and Langdon. Equation (90) is substituted into the earlier equations that relate creep rate to climb velocity. They proposed an atomic model where partial vacancies must coalesce and the time required for the coalescence is dependent on the stacking-fault energy.

Gottstein and Argon [894] later considered the influence of stacking-fault energy on the mobile dislocation density and derived a dependence of creep rate on $(\gamma/Gb)^3$. Li and Kong [976] considered the time needed to constrict a jog, and also derived $(\gamma/Gb)^3$ dependence on the creep rate.

3.2 Dislocation Microstructure and the Rate-Controlling Mechanism

Consistent with the earlier discussion, the details by which the dislocation climb control—which is, of course, diffusion controlled—is specifically related to the creep rate, are not clear. The existing theories (some prominent models discussed earlier) basically fall within two broad categories: (1) those that rely on the heterogeneous dislocation substructure (i.e., the subgrain boundaries) and (2) those that rely on the more uniform Frank dislocation network (not associated with dislocation heterogeneities such as cells or subgrain walls).

3.2.1 Subgrains

The way by which investigators rely upon the former approach varies, but basically theories that rely on the dislocation heterogeneities believe that one or more of the following are relevant:

1. The subgrain boundaries are obstacles for gliding dislocations, perhaps analogous to suggestions for high-angle grain boundaries in an (e.g., annealed) polycrystal described by the Hall–Petch relation. In this case, the misorientation across the subgrain boundaries, which is related to the spacing of the dislocations that constitute the boundaries, has been suggested to determine the effectiveness of the boundary as an obstacle [207]. (One complication with this line of reasoning is that it now appears well established that, although these features may be obstacles, the mechanical behavior of metals and alloys during five-power-law creep appears independent of the details of the dislocation spacing, d, or misorientation across subgrain boundaries $\theta_{\lambda,\text{ave}}$, as shown in Figure 30.)

2. It has been suggested that the boundary is a source for internal stresses, as mentioned earlier. Argon et al. [137], Gibeling and Nix [117], Morris and Martin [42,43], and Derby and Ashby [177] suggested that subgrain boundaries give rise to high local internal stresses that are relevant to the rate-controlling mechanism. Morris and Martin claim to have measured high local stresses that were 10–20 times larger than the applied stress near Al-5at%Zn subgrains walls formed within the five-power-law regime. Their stress calculations were based on dislocation loop radii measurements. Many have suggested that subgrain boundaries are

important as they may be "hard" regions, such as, according to Mughrabi [138], discussed earlier, extended to the case of creep by Blum and coworkers in a series of articles (e.g., Ref. [136]). Basically, here the subgrain wall is considered three dimensional with a high yield stress compared to the subgrain or cell interior. Mughrabi originally suggested that there is elastic compatibility between the subgrain wall and the matrix. This gives rise to a high internal stress. These investigators appear to suggest that these elevated stresses are the relevant stress for the rate-controlling process (often involving dislocation climb) for creep, usually presumed to be located in the vicinity of subgrain walls.

3. Others have suggested that the ejection of dislocations from the boundaries is the critical step [42,43,202]. The parameter that is important here is basically the spacing between the dislocations that comprise the boundary, which is generally related to the misorientation angle across the boundary. Some additionally suggest that the relevant stress is not the applied stress, but the stress at the boundary, which may be high, as just discussed above.

4. Similar to (1), it has been suggested that boundaries are important in that they are obstacles for gliding dislocations (perhaps from a source within the subgrain) and that, with accumulation at the boundary (e.g., a pile-up at the boundary), a backstress is created that "shuts off" the source, which is only reactivated once the number of dislocations within a given pile-up is diminished. It has been suggested that this can be accomplished by climb and annihilation of dislocations at the same subgrain boundary [26,192–195]. This is similar to the model discussed in Section 3.1.3. Some suggest that the local stress may be elevated as discussed in (2) above.

3.2.2 Dislocations

Others have suggested that the rate-controlling process for creep plasticity is associated with the Frank dislocation network within the subgrains, as was discussed earlier. That is, the strength associated with creep is related to the details (often the density) of dislocations in the subgrain interior [54,84–86,98,101,102,104–106,108–110,129,146,152,153,155,196–200,208].

One commonly proposed mechanism by which the dislocation network is important is that dislocation sources are the individual links of the network. As these bow, they can become unstable, leading to Frank-Read sources, and plasticity ensues. The density of links that can be activated sources depends on the link length distribution and, thus, related to the density of dislocation line length within the subgrains. The generated dislocation loops are absorbed by the network, leading to refinement or decreasing ℓ. The network also naturally coarsens at elevated temperature and plasticity is activated as links reach the critically long segment length, ℓ_c. Hence, climb (self-diffusion) control is justified. Some of the proponents of the importance of the interior dislocation density have based their judgments on experimental evidence that shows that creep strength (resistance) is associated with higher dislocation density and appears independent of the subgrain size [110,129,141,153].

3.2.3 Theoretical Strength of Obstacles

In view of the different microstructural features (e.g., $\lambda, d, \theta_{\lambda_{ave}}, \rho, \ell_c$) that have been suggested to be associated with the strength or rate-controlling process for five-power-law creep, it is probably worthwhile to assess strength associated with different obstacles. These are calculable from simple (perhaps simplistic) equations. The various models for the rate-controlling, or strength-determining process, are listed below. Numerical calculations are based on pure Al creep deforming as described in Figure 31.

1. The network stress τ_N. Assuming a Frank network, the average link length, ℓ (assumed here to be uniform $\cong \ell_c$):

$$\tau_N \cong \frac{Gb}{\ell} \qquad (91)$$

Using typical aluminum values for five-power-law creep (e.g., $l \cong 1/\sqrt{\rho_{ss}}$) $\tau_N \cong 5$ MPa (fairly close to $\tau_{ss} \cong 7$ MPa for Al at the relevant ρ_{ss}).

2. If the critical step is regarded as ejection of dislocations from the subgrain boundary:

$$\tau_B \cong \frac{Gb}{d} \qquad (92)$$

$\tau_B \cong 80$ MPa, much higher (by an order of magnitude or so) than the applied stress.

3. If subgrain boundaries are assumed to be simple tilt boundaries with a single Burgers vector, an attractive or repulsive force will be exerted on a slip dislocation approaching the boundary. The maximum stress is:

$$\tau_{bd} \cong \frac{0.44\,Gb}{2(1-\nu)d} \qquad (93)$$

from Ref. [88] based on Ref. [209]. This predicts a stress of about $\tau_{bd} = 50$ MPa, again much larger than the observed applied stress.

4. For dragging jogs resulting from passing through a subgrain boundary, assuming a spacing $\cong j \cong d$:

$$\tau_j \cong \frac{E_j}{b^2 j} \tag{94}$$

from Ref. [88]. For Al, E_j, the formation energy for a jog, $\cong 1$ eV [88] and $\tau_j \cong 45$ MPa, a factor 6–7 higher than the applied stress.

5. The stress associated with the increase in dislocation line length (jog or kinks) to pass a dislocation through a subgrain wall (assuming a wall dislocation spacing, d) from above is expected to be [88]:

$$\tau_L \cong 0.2 \frac{Gb^3}{b^2 d} \cong 0.2 \frac{Gb}{d} \cong 16 \text{ MPa} \tag{95}$$

about a factor of 2 larger than the applied stress.

Thus, it appears that stresses associated with ejecting dislocations from, or passing dislocations through, subgrain walls are typically 16–45 MPa for Al within the five-power-law regime. This is roughly two to seven times larger than the applied stress. Liu et el. [360] used dislocation dynamics simulations to show that subgrain boundaries are effective obstacles. Based on the simplified assumptions, this disparity may not be considered excessive and does not eliminate subgrain walls as important, despite the favorable agreement between the network-based (using the average link length, ℓ) strength and the applied stress (internal stresses not considered). These calculations indicate why some subgrain-based strengthening models utilize elevated internal stresses. It must be mentioned that care must be exercised in utilizing the above, athermal equations for time-dependent plasticity. These equations do not consider other hardening variables (solute, etc.) that may account for a substantial fraction of the applied stress, even in relatively pure metals. Thus, these very simple theoretical calculations do not provide obvious insight into the microstructural feature associated with the rate-controlling process, although a slight preference for network-based models might be argued as the applied stress best matches network predictions for dislocation activation. The position is particularly reasonable if there are generally sources within the subgrains, which appears to be the case based on Eqn (91).

3.3 In situ and Microstructure-Manipulation Experiments

3.3.1 In situ Experiments

In situ straining experiments, particularly those of Calliard and Martin [95], are often referenced by the proponents of subgrain (or heterogeneous dislocation arrangements) strengthening. Here thin foils (probably less than 1 μm thick) were strained at ambient temperature (about 0.32 T_m). It was concluded that the interior dislocations were not a significant obstacle for gliding dislocations; rather, the subgrain boundaries were effective obstacles. This is an important experiment, but is limited in several ways: first, it is low temperature (i.e., $\cong 0.32 \ T_m$) and may not be relevant for the five-power-law regime; also, in thin foils such as these, as McLean mentioned [84] long ago, a Frank network is disrupted as the foil thickness approaches ℓ. Finally, subgrains can be obstacles, of course, but the important event may be dislocation emission from network sources, with annihilation at the subgrain wall. Henderson-Brown and Hale [210] performed in situ high-voltage transmission electron microscope (HVEM) creep experiments on Al-1Mg (class M) at 300 °C, in thicker foils. Dislocations were obstructed by subgrain walls, although the experiments were not described in substantial detail. As mentioned earlier, Mills [157] performed in situ deformation on an Al–Mg alloy within the three-power or viscous-drag regime, and subgrain boundaries were not concluded as obstacles.

3.3.2 Prestraining Experiments

Work by Kassner et al. [98,110,111], discussed earlier, utilized ambient temperature prestraining of austenitic stainless steel to (1) show that the elevated temperature strength was independent of the subgrain size and (2) that the influence of the dislocation density on strength was reasonably predicted by the Taylor equation. Ajaja and Ardell [152,153] also performed prestraining experiments on austenitic stainless steels and showed that the creep rate was influenced only by the dislocation density. Their prestrains led to elevated ρ, without subgrains, and quasi steady-state creep rates. Presumably, this prestrain led to decreased average and critical link lengths in a Frank network. (Although, eventually, a new "genuine" steady state may be achieved [211] at the elevated temperature, this may not occur over the convenient strain/time ranges. Hence, the conclusion of a steady state being independent of the prestrain may be, in some cases, ambiguous.)

Others, including Parker and Wilshire [212], performed prestraining experiments on Cu showing that ambient temperature prestrain (cold work) reduces the elevated temperature creep rate at $410\,°C$. This was attributed by the investigators as being due to the refinement Frank network. Well-defined subgrains did not form; rather, cell walls were observed. A quantitative microstructural effect of the cold work was not clear.

3.4 Additional Comments on Network Strengthening

Previous work on stainless steel in Figure 28 showed that the density of dislocations within the subgrain interior or the network dislocations influence the flow stress at a given strain rate and temperature. The hardening in stainless stress is shown to be consistent with the Taylor relation if a linear superposition of "lattice" hardening (τ_o, or the stress necessary to cause dislocation motion in the absence of a dislocation substructure) is present and the dislocation hardening ($\alpha MGb\rho^{1/2}$) is assumed (regardless of the source of dislocation hardening, e.g., bowing stress, passing stress, etc.). The Taylor equation also applies to pure aluminum (with a steady-state structure), having both a much higher stacking fault energy than stainless steel and an absence of substantial solute additions.

If both the phenomenological description of the influence of the strength of dislocations in high-purity metals such as aluminum have the form of the Taylor equation and also have the expected values for the constants, then it would appear that the elevated temperature flow stress is actually provided by the "forest dislocations" (Frank network).

Figure 23 illustrates the well-established trend between the steady-state dislocation density and the steady-state stress. From this and from Figure 14, which plots modulus-compensated steady-state stress versus diffusion-coefficient compensated steady-state strain rate, the steady-state flow stress can be predicted at a reference strain rate (e.g., $5 \times 10^{-4}\,s^{-1}$), at a variety of temperatures, with an associated steady-state dislocation density. If Eqn (29) is valid for Al as for 304 stainless steel, then the values for α could be calculated for each temperature, by assuming that the annealed dislocation density and the σ_o values account for the annealed yield strength reported in Figure 40.

Figure 41 indicates, first, that typical values of α at 0.5 T_m are within the range of those expected for Taylor strengthening. Stated a different way, the phenomenological relationship for strengthening of (steady-state) structures suggests that the strength can be reasonably predicted based on a Taylor equation. We expect the strength we observe, based only on the (network) dislocation density, is completely independent of the

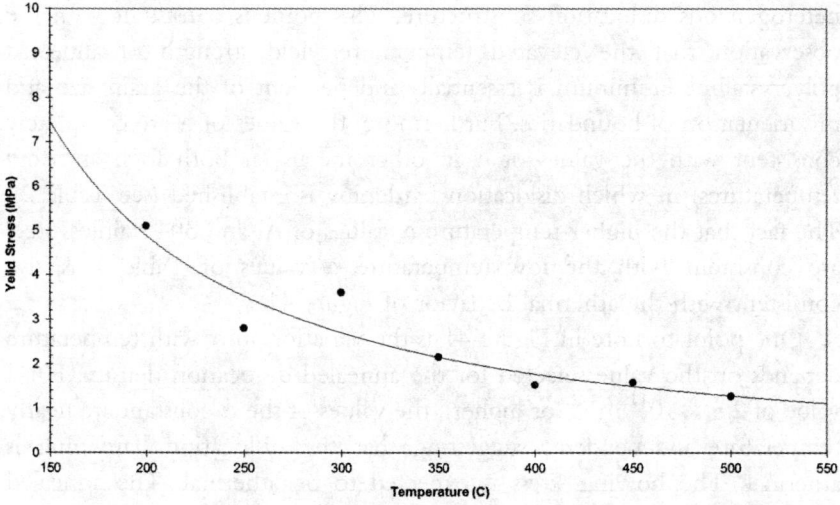

Figure 40 The yield strength of annealed 99.999% pure Al as a function of temperature. *(From Ref. [149].)* $\dot{\varepsilon} = 5 \times 10^{-4}\text{s}^{-1}$.

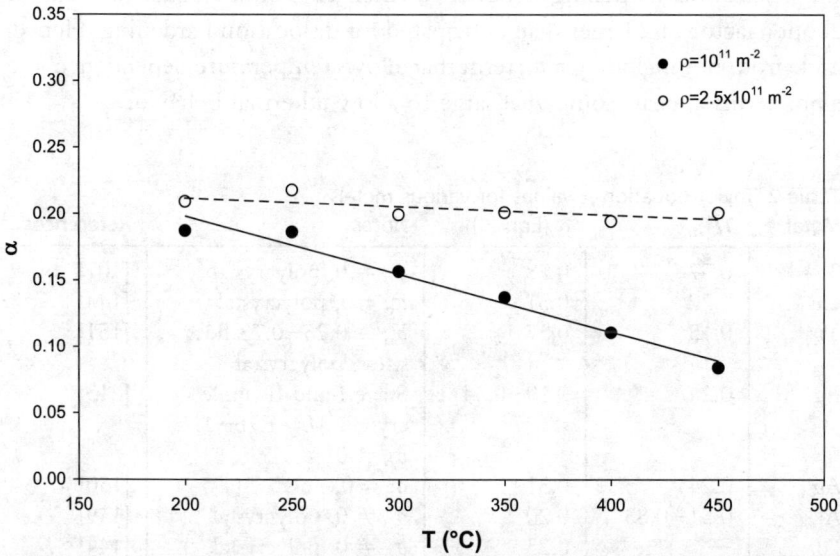

Figure 41 The values of the constant alpha in the Taylor Eqn (29) as a function of temperature. The alpha values depend somewhat on the assumed annealed dislocation density. Dark dots, $\rho = 10^{11}$ m^{-2}; hollow, $\rho = 2.5 \times 10^{11}$ m^{-2}.

heterogeneous dislocation substructure. This point is consistent with the observation that the elevated temperature yield strength of annealed polycrystalline aluminum is essentially independent of the grain size and misorientation of boundaries. Furthermore, the values of α are completely consistent with the values of α in other metals (at both high and low temperatures) in which dislocation hardening is established (see Table 2). The fact that the higher temperature α values of Al and 304 stainless steel are consistent with the low temperature α values of Table 1 is also consistent with the athermal behavior of Figure 41.

One point to note in Figure 41 is the variation in α with temperature depends on the value selected for the annealed dislocation density. For a value of 2.5×10^{11} m^{-2} (or higher), the values of the α constant are nearly temperature independent, suggesting that the dislocation hardening is athermal. The bowing stress is expected to be athermal. The annealed dislocation density for which athermal behavior is observed is very close to the observed value in Figure 30(a) and according to Blum [214]. The suggestion of athermal dislocation hardening is consistent with the model by Nes [215], where as in the present case, the temperature dependence of the flow stress is provided by the temperature-dependent σ_o term. It perhaps should be mentioned that if it is assumed both that $\sigma_o = 0$ and that the dislocation hardening is athermal, then α is about equal to 0.53, or about a factor of 2 larger than anticipated for dislocation hardening. Hence, aside from not including a σ_o term that allows temperature dependence, the alpha terms appears somewhat large to allow athermal behavior.

Table 2 Taylor equation α values for various metals

Metal	T/T_m	α (Eqn (6))	Notes	References
304	0.57	0.28	$\sigma_o \neq 0$, polycrystal	[107]
Cu	0.22	0.31	$\sigma_o = 0$, polycrystal	[144]
Ti	0.15	0.37	$\sigma_o \cong 0.25-0.75$ flow stress, polycrystal	[151]
Ag	0.24	0.19–0.34	Stage I and II single crystal, $M = 1.78-1$, $\sigma_o \neq 0$	[213]
Ag	0.24	0.31	$\sigma_o = 0$, polycrystal	[150]
Al	0.51–0.83	0.20	$\sigma_o \neq 0$, polycrystal	[149]
Fe	—	0.23	$\sigma_o \neq 0$, polycrystal	[144]

Note: α values of Al and 304 stainless stress are based on dislocation densities of intersections per unit area. The units of the others are not known, and these α values would be adjusted lower by a factor of 1.4 if line length per unit volume was utilized.

The trends in dislocation density during primary creep have been less completely investigated for the case of constant strain-rate tests. Earlier work by Kassner et al. [98,110,111] on 304 stainless steel found that at 0.57 T_m, the increase in flow stress by a factor of 3, associated with increases in dislocation density with strain, is consistent with the Taylor equation. That is, the ρ versus strain and stress versus strain give a σ versus ρ that falls on the line of Figure 28. Similarly, the aluminum primary transient in Figure 31(a) can also be shown as consistent with the Taylor equation. The dislocation density monotonically increases to the steady-state value under constant strain-rate conditions.

Challenges to the proposition of Taylor hardening for five-power-law creep in metals and Class M alloys include the microstructural observations during primary creep under constant-*stress* conditions. For example, it has nearly always been observed during primary creep of pure metals and Class M alloys that the density of dislocations not associated with subgrain boundaries increases from the annealed value to a peak value, but then gradually decreases to a steady-state value that is between the annealed and the peak density [38,92,163–165] (e.g., Figure 29). Typically, the peak value, ρ_p, measured at a strain level that is roughly one-fourth of the strain required to attain steady state ($\varepsilon_{ss}/4$), is a factor of 1.5–4 higher than the steady-state ρ_{ss} value. It was believed by many to be difficult to rationalize hardening by network dislocations if the overall density is decreasing while the strain rate is decreasing. Therefore, an important question is whether the Taylor hardening, observed under constant strain-rate conditions, is consistent with this observation [169]. This behavior could be interpreted as evidence that most of these dislocations have a dynamic role rather than a (Taylor) hardening role, since the initial strain rates in a constant stress test may require by the equation:

$$\dot{\varepsilon} = (b/M)\rho_m v \tag{96}$$

a high mobile (nonhardening) dislocation density, ρ_m, that gives rise to high initial values of total density of dislocations not associated with subgrain boundaries, ρ (v is the dislocation velocity). As steady state is achieved and the strain rate decreases, so does ρ_m and in turn, ρ. (We can suggest that $\rho_h + \rho_m = \rho$, where ρ is the total density of dislocations not associated with subgrain boundaries and ρ_h are those dislocations that at any instant are part of the Frank network and are not mobile.)

More specifically, Taylor hardening during primary (especially during constant stress) creep may be valid based on the following argument. From Eqn (96) $\dot{\varepsilon} = \rho_m vb/M$, we assume [216]:

$$v = k_7 \sigma^1 \qquad (97)$$

and, therefore, for constant strain-rate tests:

$$\dot{\varepsilon}_{ss} = [k_7 b/M]\rho_m \sigma \qquad (98)$$

In a constant strain-rate test at yielding ($\dot{\varepsilon} = \dot{\varepsilon}_{ss}$), ε_p (plastic strain) is small, there is only minor hardening, and the mobile dislocation density is a fraction f_m^o of the total density:

$$f_m^o \rho_{(\varepsilon_p=0)} = \rho_{m(\varepsilon_p=0)}$$

Therefore, for aluminum (see Figure 31(a)):

$$\rho_{m(\varepsilon_p=0)} = f_m^o 0.64\rho_{ss} \quad \left(\text{based on } \rho \text{ at } \varepsilon_p = 0.03\right) \qquad (99)$$

where f_m^o is basically the fraction of dislocations in the annealed metal that are mobile at yielding (half the steady-state flow stress) in a constant strain-rate test. Also from Figure 4, $\sigma_y/\sigma_{ss} = 0.53$. Therefore, at small strains:

$$\dot{\varepsilon}_{ss} = f_m^o 0.34[k_7 b/M]\rho_{ss}\sigma_{ss} \qquad (100)$$

(constant strain rate at $\varepsilon_p = 0.03$).

At steady state, $\sigma = \sigma_{ss}$ and $\rho_m = f_m^s \rho_{ss}$, where f_m^s is the fraction of the total dislocation density that is mobile and:

$$\dot{\varepsilon}_{ss} = f_m^s[k_7 b/M]\rho_{ss}\sigma_{ss} \qquad (101)$$

(constant strain rate at $\varepsilon_p > 0.20$.)

By combining Eqns (100) and (101) we find that f_m at steady state is about one-third the fraction of mobile dislocations in the annealed poly-crystals ($0.34 f_m^o = f_m^s$). This suggests that during steady state only one-third or less of the total dislocations (not associated with subgrain boundaries) are mobile and the remaining two-thirds or more participate in hardening. The finding that a large fraction are immobile is consistent with the observation that increased dislocation density is associated with increased strength for steady-state and constant strain-rate testing deformation. Of course, there is the assumption that the stress acting on the dislocations as a function of strain (microstructure) is proportional to the applied flow stress. Furthermore, we have presumed a 55% increase in ρ over primary creep with some uncertainty in the density measurements.

For the constant stress case we again assume:

$$\dot{\varepsilon}_{\varepsilon_p \cong 0} = f_m^p [k_7 b/M] \rho_p \sigma_{ss} \quad \text{(constant stress)} \tag{102}$$

where f_m^p is the fraction of dislocations that are mobile at the peak (total) dislocation density of ρ_p, the peak dislocation density, which will be assumed equal to the maximum dislocation density observed experimentally in a ρ–ε plot of a constant stress test. Since at steady-state:

$$\dot{\varepsilon}_{ss} \cong 0.34 f_m^o [k_7 b/M] \rho_{ss} \sigma_{ss} \tag{103}$$

by combining with Eqn (102):

$$\dot{\varepsilon}_{\varepsilon_p \cong 0} / \dot{\varepsilon}_{ss} = \left(\frac{f_m^p}{f_m^o} \right) 3 \rho_p / \rho_{ss} \quad \text{(constant stress)} \tag{104}$$

(f_m^p / f_m^o) is not known but if we assume that at macroscopic yielding, in a constant strain-rate test, for annealed metal, $f_m^o \cong 1$, then we might also expect at small strain levels and relatively high dislocation densities in a constant-stress test, $f_m^p \cong 1$. This would suggest that fractional decreases in $\dot{\varepsilon}$ in a constant stress test are not equal to those of ρ. This apparent contradiction to purely dynamic theories, i.e., based strictly on Eqn (96), is reflected in experiments [92,162–165] where the kind of trend predicted in this last equation is, in fact, observed. Equation (104) and the observations of $\dot{\varepsilon}$ against ε in a constant stress test at the identical temperature can be used to predict roughly the expected constant-stress ρ–ε curve in aluminum at 371 °C and about 7.8 MPa, the same conditions as the constant strain-rate test. If we use small plastic strain levels, $\varepsilon \cong \varepsilon_{ss}/4$ (where ρ values have been measured in constant stain-rate tests), we can determine the ratio (e.g., $\dot{\varepsilon}_{\varepsilon=(\varepsilon_{ss}/4)} / \dot{\varepsilon}_{\varepsilon=\varepsilon_{ss}}$) in constant stress tests. This value seems to be roughly 6 at stresses and temperatures comparable to the present study [92,165,212]. This ratio was applied to Eqn (102) assuming $(f_m^p / f_m^c) \cong 1$; the estimated ρ–ε tends are shown in Figure 42. This estimate, which predicts a peak dislocation density of 2.0 ρ_{ss}, is consistent with the general observations discussed earlier for pure metals

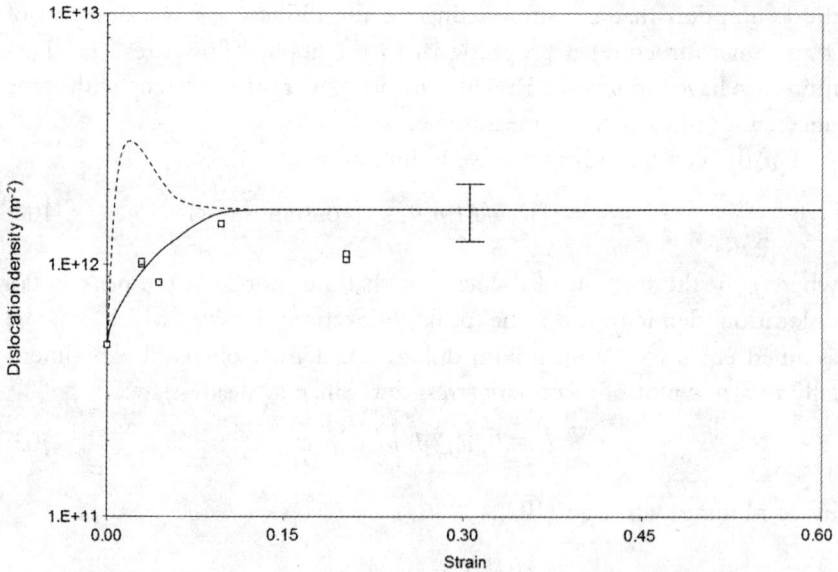

Figure 42 The predicted dislocation density (- - -) in the subgrain interior against strain for aluminum deforming under constant stress conditions is compared with that for constant strain-rate conditions (——). The predicted dislocation density is based on Eqn (74), which assumes Taylor hardening.

and Class M alloys that ρ_p is between 1.5 and 4 ρ_{ss}(1.5–2.0 for aluminum [92]). Thus, the peak behavior observed in the dislocation density versus strain-rate trends, which at first glance appears to impugn dislocation network hardening, is actually consistent, in terms of the observed ρ values, to Taylor hardening.

One imprecision in the argument above is that it was assumed (based on some experimental work in the literature) that the stress exponent for the elevated temperature (low stress) dislocation velocity, v, is 1. This exponent may not be well known and may be greater than 1. The ratio ρ_p/ρ_{ss} increases from a value of 3 in Eqn (19) to higher values of 3 (2^{n-1}), where n is defined by $v = \sigma^n$. This means that the observed strain rate peaks would predict smaller dislocation peaks or even an absence of peaks for the observed initial strain rates in constant-stress tests. In a somewhat circular argument, the consistency between the predictions of Eqn (104) and the experimental observations may suggest that the exponents of 1–2 may be reasonable. Also, the values of the peak dislocation densities and strain rates are not unambiguous, and this creates additional uncertainty in the argument.

4. OTHER EFFECTS ON FIVE-POWER-LAW CREEP

4.1 Large Strain Creep Deformation and Texture Effects

Traditionally, creep has been associated with tensile tests, and accordingly, with relatively small strains. Of course, elevated temperature creep plasticity can be observed in torsion or compression, and the phenomenological expressions presented earlier are still valid, only with modification due to different texture evolution (or changes in the average Taylor factors) with the different deformation modes. These differences in texture evolution have been discussed in detail by several investigators [38,217] for lower temperature deformation. Some lower temperature deformation texture trends may be relevant to five-power-law creep trends. A fairly thorough review of elevated temperature torsion tests and texture measurements on aluminum is presented by Kassner et al. and McQueen et al. [218,219]. Some of the mechanical results are illustrated in Figure 43. Basically, the figure shows that with torsion deformation, the material hardens to a genuine steady state that is a balance between dislocation hardening and dynamic recovery. However, with the relatively large strain deformation that is permitted by torsion, the flow stress decreases, in this case, about 17% to a new stress that is invariant even with very large strains to 100 or so. (Perhaps there is an increase in torque of 4% in some cases with $\varepsilon > 10$ of uncertain origin.) These tests were performed on particularly precise testing equipment. The essentially invariant stress over the extraordinarily large strains suggests a genuine mechanical steady state. The cause of this softening has been carefully studied, and dynamic recrystallization and grain-boundary sliding (GBS) were considered. Creep measurements as a function of strain through the "softened" regime [220], and microstructural analysis using both polarized light optical (POM) and transmission electron microscopy (TEM) [218] reveal that five-power-law creep is occurring throughout the softened regime and that the modest decrease in flow stress is due to a decrease in the average Taylor factor, $\overline{M}$.

This has been confirmed by X-ray texture analysis [37] and is also consistent in magnitude with theoretical texture modeling of deformation in torsion [38,217]. If compression specimens are extracted from the torsion specimen deformed into the softened regime, the flow stress of the compression specimen is actually higher than the torsion flow stress, again confirming the texture conclusion [218].

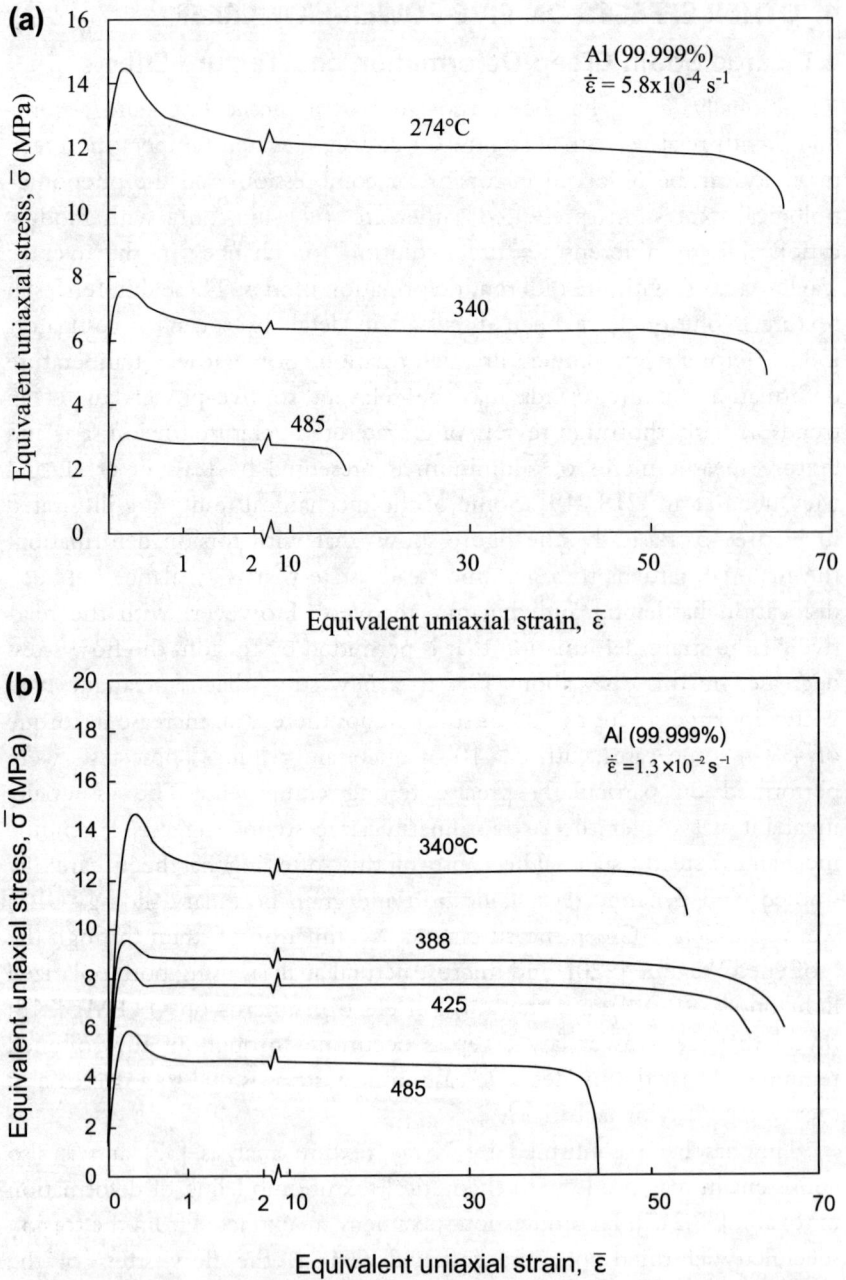

Figure 43 The stress versus strain behavior of Al deformed in torsion to very large strains at two strain rates, (a) and (b). *Based on Ref. [39].*

The microstructural evolution of specimens deformed to large strains, not achievable in tension or compression, is quite interesting and has also been extensively researched in a few metals and alloys. The initial high-angle grain boundaries of the aluminum polycrystalline aggregate spiral about the torsion axis with deformation within the five-power-law regime. At least initially, the total number of grains in the poly-crystalline aggregate remains constant and the grains quickly fill with subgrains with low misorientation boundaries. The grains thin until they reach about twice the average subgrain diameter with increasing strain in torsion. Depending on the initial grain size and the steady-state subgrain size, this may require substantial strain, typically about 10. The high-angle grain boundaries of the polycrystalline aggregate are serrated (triple points) as a result of subgrain boundary formation. As the serrated spiraling grains of the polycrystalline aggregate decrease in width to about twice the subgrain size, there appears to be a pinching off of impinging serrated grains. At this point, the area of high-angle boundaries, which was gradually increasing with torsion, reaches a constant value with increasing strain. Despite the dramatic increase in high-angle boundaries, no change in flow properties has been observed so far (e.g., to diffusional creep or enhanced strain rate due to the increased contribution of grain boundary sliding). Figure 44 is a series of POM micrographs illustrating this progression. Interestingly, the subgrain size is about constant from the peak stress at about 0.2 strain to the very large torsion strains. This, again, suggests that subgrain boundaries are mobile and annihilate to maintain the equiaxed structure and modest misorientation. Examination of those boundaries that form from dislocation reaction (excluding the high-angle boundaries of the starting polycrystal) reveals that the average misorien-tation at the onset of steady state was, as stated earlier, only 0.5°. However, by a strain of between 1 and 1.5 it had tripled to 1.5°—see also Figure 31(b)— but appears to be fixed beyond this strain. This is, again, consistent with earlier work referenced that indicates that $\theta_{\lambda_{ave}}$ may increase (d decreases) during at least early steady state. Furthermore, at the onset of steady state, nearly all of the subgrain boundaries formed are low-θ_λ dislocation boundaries. However, with very large strain deformation there is an increase in high-angle boundary area (geometric dynamic recrystallization or GDX). Nearly one-third of the subgrain boundaries are high-angle boundaries, but these appear to have ancestry back to the initial, or starting, polycrystal. Notwithstanding, the flow stress is un-changed. That is, at a strain of 0.2, at about 0.7 T_m and a modest strain

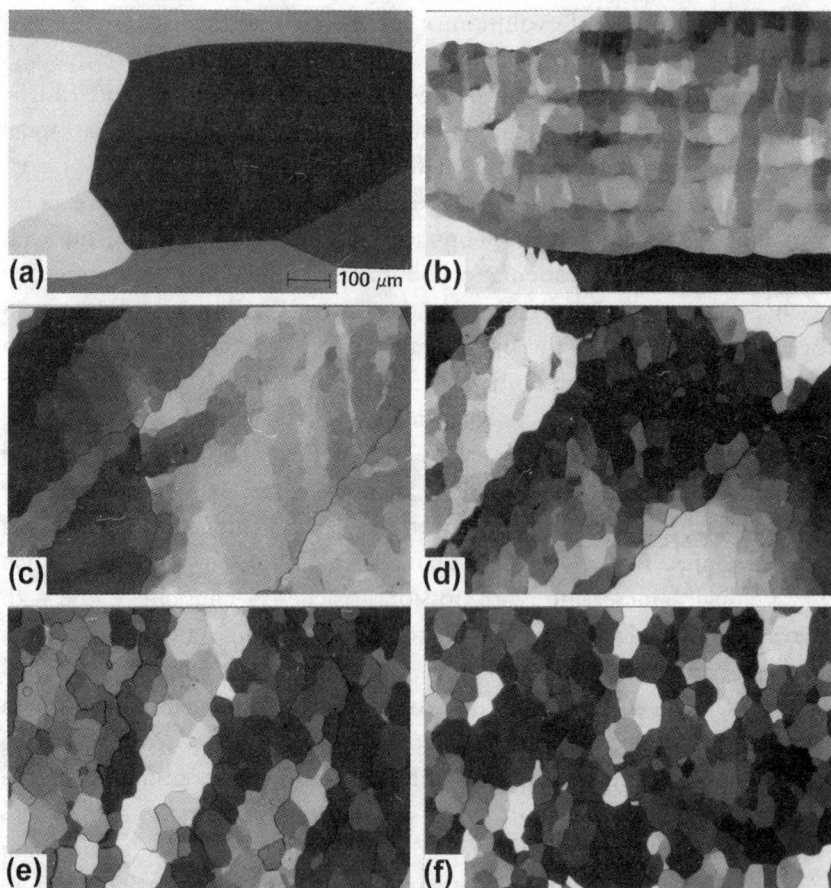

Figure 44 Polarized light optical micrographs of aluminum deformed at 371 °C at $5.04 \times 10^{-4}\,s^{-1}$ (Figure 31(b)) to equivalent uniaxial strains of (a) 0, (b) 0.2, (c) 0.60, (d) 1.26, (e) 4.05, and (f) 16.33. Geometric dynamic recrystallization (GDX) is observed. *From Ref. [18].*

rate, the average subgrain size is about 13 μm and the average misorientation angle of subgrain boundaries is about 0.5°. If we increase the plastic strain by nearly two orders of magnitude to about 16, the subgrain size and interior or network dislocation density is unchanged, but we have "replaced" nearly one-third of the subgrain facets with high-angle boundaries (through GDX) and tripled the misorientation of the remaining two-thirds. However, the flow stress is unchanged. This, again, suggests that the details of the subgrain boundaries are not an important consideration in the rate-controlling process for five-power-law creep.

Other elevated temperature torsion tests on other high-stacking fault energy alloys in the five-power-law regime have shown a similar softening as theoretically predicted [221]. The cause of softening was not ascribed to texture softening by those investigators but (probably incorrectly) rather to continuous reactions (continuous dynamic recrystallization) [222].

Recent work by Hughes et al. [141] showed that polycrystals deformed at elevated temperature may form geometrically necessary boundaries (GNBs) from dislocation reactions to accommodate differences in slip within a single grain. Whether these form in association with GDX is unclear, although it appears that the grain boundary area with large strain deformation is at least approximately consistent with grain thinning in the case of Al just discussed. HABs, however, have been observed to form in single crystals at elevated temperature from dislocation reaction [142], and the possibility that these form from dislocation reaction in polycrystals should also be considered.

It should be also mentioned that it has been suggested that in at least Al and some Al-alloys [130], slip on {110} planes (or non-octahedral slip) can occur, leading to nontraditional textures such as the cube {001} type. Finally, it should be mentioned that the decrease in the Taylor factor is expected to decrease the flow stress at low temperatures, but it is only a "climb stress" decrease that would be expected to decrease the five-power-law stress. An analysis should be done to assess whether the decrease in M is accompanied with a corresponding decrease in the dislocation climb stress.

4.2 Effect of Grain Size

First, of course, it has been suggested that with fine-grain size refinement, the mechanism of plastic flow may change from five-power behavior to Coble creep [52], which, as discussed earlier, is a diffusion creep mechanism relying on short-circuit diffusion along grain boundaries. Grain boundary sliding comprising a more significant role is also possible. However, the influence of high-angle grain boundaries on five-power-law creep is less clear. Some work has been performed on the Hall–Petch relationship in copper [224] and aluminum [147] at elevated temperatures. Some results from these studies are illustrated in Figure 45. Basically, both confirm that decreasing grain size results in increased elevated temperature strength in predeformed copper and annealed aluminum (a constant dislocation density for each grain size was not confirmed in Cu). The temperature and applied strain rates correspond to five-power-law creep in these pure metals.

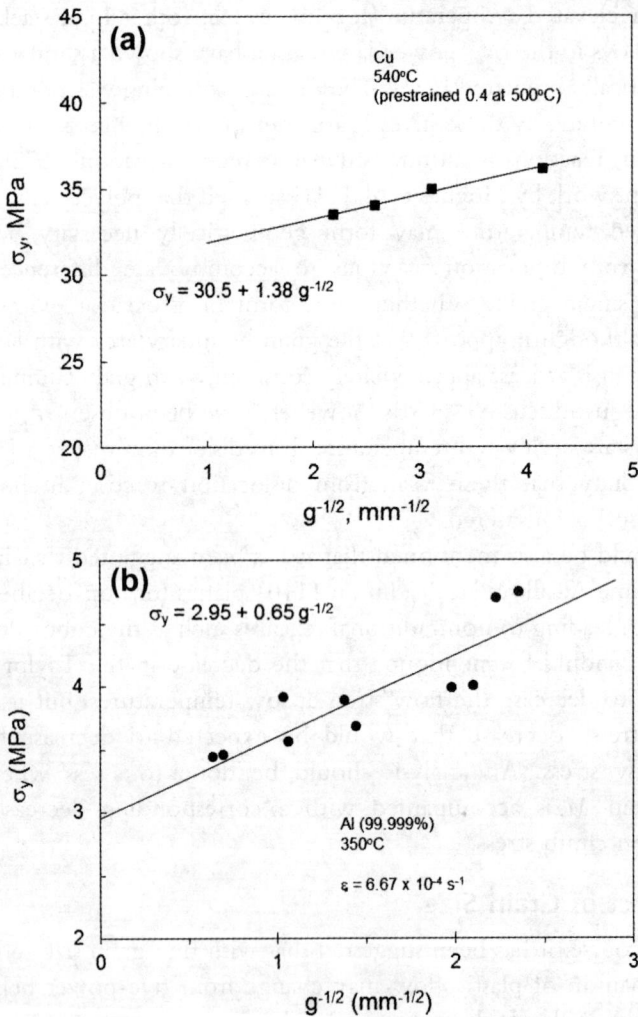

Figure 45 The effect on grain size on the elevated temperature strength of (a) pre-strained Cu and (b) annealed Al. *From Refs [147,224].*

Interestingly, though, the effect of diminishing grain size may decrease with increasing temperature. Figure 46 shows that the Hall–Petch constant, k_y, for high-purity aluminum significantly decreases with increasing temperature. The explanation for this in unclear. First, if the effect of decreasing grain size at elevated temperature is purely the effect of a Hall–Petch strengthening (e.g., not GBS), the explanation for decreasing Hall–Petch constant would require knowledge of the precise strengthening mechanism. It is possible, for instance, that increased strengthening

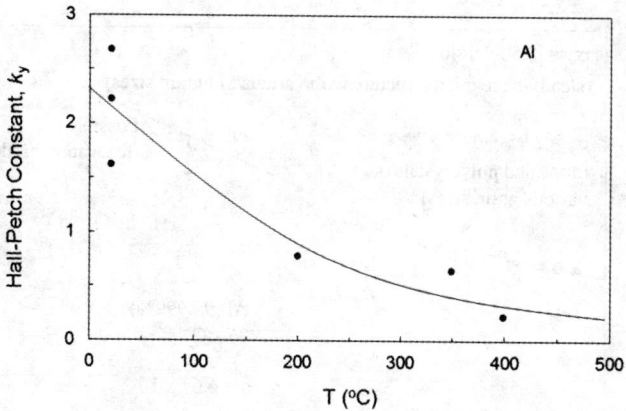

Figure 46 The variation of the Hall–Petch constant in Al with temperature. *Based on Ref. [147].*

with smaller grain sizes is associated with the increased dislocation density in the grain interiors due to the activation of dislocation sources [225]. Therefore, thermal recovery may explain a decreased density and less pronounced strengthening. This is, of course, speculative and one must be careful that other effects such as grain boundary sliding are not becoming important. For example, it has been suggested that in aluminum, grain boundary sliding becomes pronounced above about $0.5\ T_m$ [226,227]. Thus, it is possible that the decreased effectiveness of high-angle boundaries in providing elevated temperature strength may be the result of GBS, which would tend to decrease the flow stress. However, the initial Al grain size decreased from about 250 μm to only about 30 μm through GDX in Figure 44, but the flow properties at $0.7\ T_m$ appear unchanged since the stress exponent, n, and activation energy, Q, appear to be unchanged [218,220].

The small effect of grain size changes on the elevated-temperature flow properties is consistent with some earlier work reported by Barrett et al. on Cu and Garafalo et al. on 304 stainless steel, where the steady-state creep rate appeared at least approximately independent of the starting grain size in the former case and not substantially dependent in the latter case [228,229]. Thus, it appears that decreasing grain size has a relatively small effect on increasing the flow stress at high temperatures over the range of typical grain sizes in single-phase metals and alloys.

Figure 47 plots the effect of grain size on the yield stress of annealed polycrystalline aluminum with the effect of (steady-state structure)

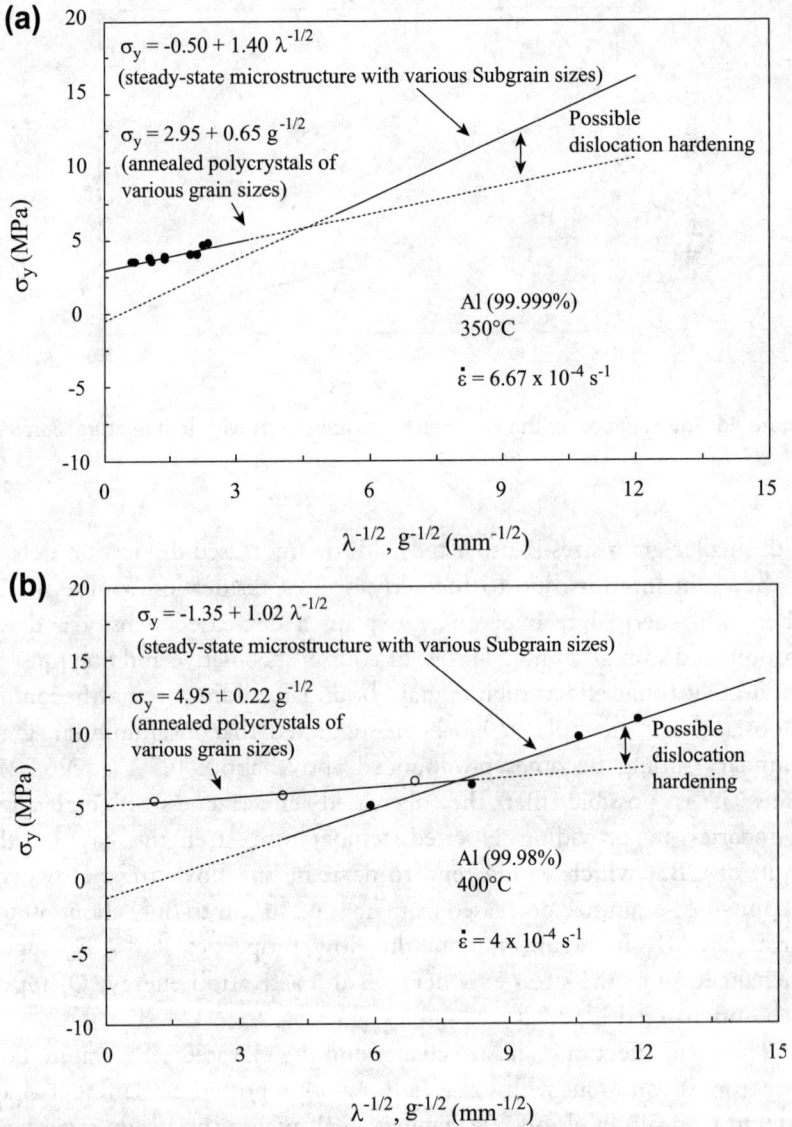

Figure 47 (a) The variation of the yield strength of annealed aluminum with various grain sizes, g, and creep-deformed aluminum with various subgrain sizes, λ, at 350 °C. Both λ and g data are described by the Hall–Petch equation. The annealed aluminum data is from Figure 45(b) and the subgrain containing Al strength data at a fixed T, $\dot{\varepsilon}$ is based on interpolation of data from [4,230] and also summarized in [147,148,231]. (b) As in (a) but at 400 °C and less pure Al, based on [147,148,232]. The subgrain containing metal here and in (a), above, is stronger than expected based on Hall–Petch strengthening by the subgrains alone.

subgrain size on the elevated flow stress all at the same temperature and strain rate, as well. Of course, while polycrystalline samples had an annealed dislocation density, the steady-state substructures with various subgrain sizes had various elevated dislocation densities that increased with decreasing subgrain size. Nonetheless, the figure reveals that for identical, small sizes, the subgrain substructure (typical $\theta_{\lambda_{ave}} \cong 0.5 - 1°$) had higher strength than polycrystalline annealed aluminum (typical $\theta = 30°-35°$). There might be an initial inclination to suggest that subgrain boundaries, despite the very low misorientation, are more effective in hardening than high-angle grains of identical size. However, as discussed earlier, the extra strength may be provided by the network dislocations that are of significantly higher density in steady-state structures as compared to the annealed metal. This increase in strength appeared at least approximately predictable based solely on Eqn (29) for dislocation strengthening assuming appropriate values for the constants, such as α [107,148,231]. Wilshire [82] also recently argued that subgrains are unlikely sources for strength in metals and alloys due to the low strength provided by high-angle boundaries at elevated temperature. More recently, there has been work on the creep behavior of materials subjected to severe plastic deformation (SPD), generally through equal channel angular pressing [604, 1100]. SPD can reduce the grain size of materials [721]. However, there are complications in terms of isolating the hardening effects of grain boundaries as HABs may be "non-equilibrium" with long-range internal stresses [782] and elevated dislocation densities.

4.3 Impurities and Small Quantities of Strengthening Solutes

It appears that the same solute additions that strengthen at ambient temperature often provide strength at five-power-law temperatures. Figure 48 shows the relationship between stress and strain rate of high-purity (99.99%) and lower-purity (99.5%) aluminum. The principal impurities were not specified, but probably included a significant fraction of Fe and Si, and some second phases may be present. The strength increases with decreasing purity for a fixed strain rate. Interestingly, Figure 49 shows that the subgrain size is approximately predictable mostly on the basis of the stress, independent of composition for Al.

Straub and Blum [90] also showed that the subgrain size depends only on the modulus-compensated stress in Al and several dilute Al alloys, although the stress/strain rate may change substantially with the purity at a

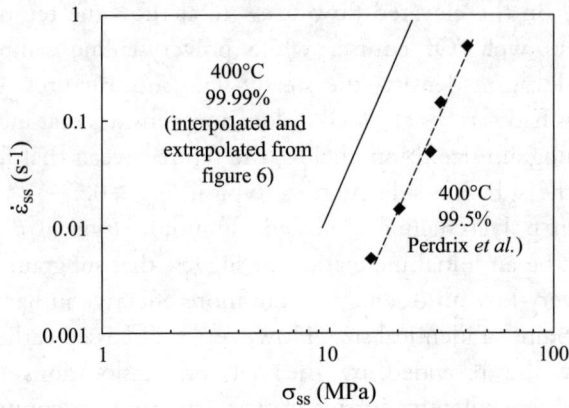

Figure 48 The steady-state strain rate versus steady-state stress for Al of different purities. *Data from Figure 23 and Perdrix et al. [233].*

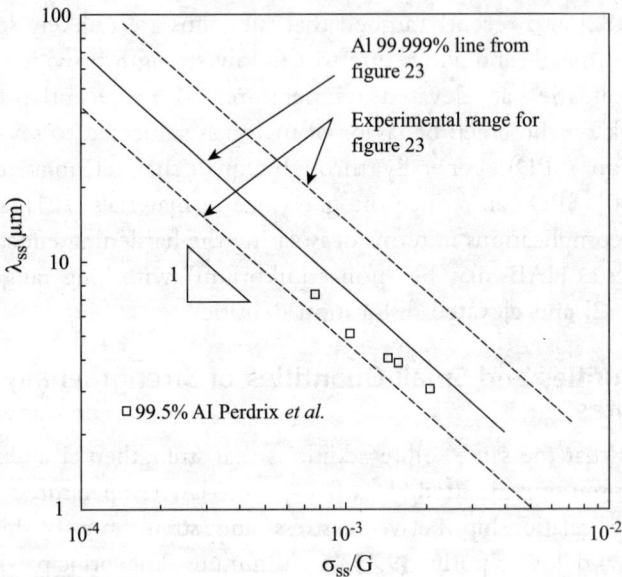

Figure 49 A plot of the variation of the steady-state subgrain size versus modulus-compensated steady-state flow stress for Al of different purities. The relationship between σ_{ss} and λ_{ss} for less pure Al is at least approximately described by the high-purity relationship.

specific strain rate/stress and temperature. If the λ_{SS} versus σ_{SS}/G in Figures 29 and 30 were placed in the same graph, it would be evident that, for identical subgrain sizes, the aluminum, curiously, would have higher strength. (The opposite is true for fixed ρ_{SS}.) Furthermore, it appears that λ

is not predictable only on the basis of σ_{SS}/G for dispersion-strengthened Al [90]. Steels do not appear to have λ_{SS} values predictable on the basis of αFe λ_{SS} versus σ_{SS}/G trends, although this failure could be a result of carbides present in some of the alloys. Thus, the aluminum trends may not be as evident in other metals. It should, however, be mentioned that for a wide range of single-phase metals, there is a rough relationship between the subgrain size and stress [77]:

$$\lambda = 23Gb/\sigma \tag{105}$$

Of course, sometimes ambient-temperature strengthening interstitials (e.g., C in γ-Fe) can weaken at elevated temperatures. In the case of C in γ-Fe, D_{sd} increases with C concentration and $\dot{\varepsilon}_{SS}$ correspondingly increases [16].

4.4 Sigmoidal Creep

Sigmoidal creep behavior occurs when in, e.g., a single-phase alloy, the creep rate decreases with strain (time) but with further strain curiously increases. This increase is followed by, again, a decrease in creep rate. An example of this behavior is illustrated in Figure 50 taken from Evans and

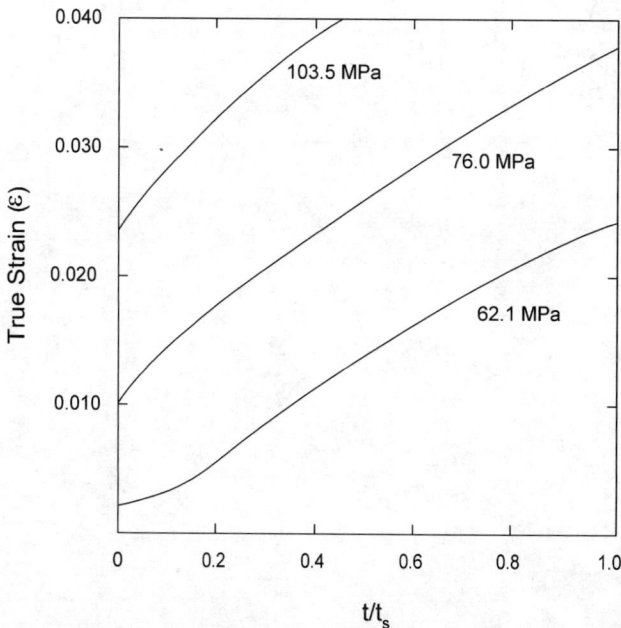

Figure 50 Transient creep curves obtained at 324 °C for 70–30 α-brass, where t_s is the time to the start of steady-state creep. *From Ref. [235].*

Wilshire for 70–30 α-brass [234,235]. This behavior was also observed in the Cu–Al alloy described elsewhere [175] and also in Zr of limited purity [236]. The sigmoidal behavior in all of these alloys appears to reside within certain (temperature)/(stress/strain-rate) regimes. The explanations for sigmoidal creep are varied. Evans and Wilshire suggest that the inflection is due to a destruction in short-range order in α-brass leading to higher creep rates. Hasegawa et al. [175], on the other hand, suggest that changes in the dislocation substructure in Cu–Al may be responsible. More specifically, the increase in strain rate prior to the inflection is associated with an increase in the total dislocation density, with the formation of cells or subgrains. The subsequent decrease is associated with cellular tangles (not subgrains) of dislocations. Evans and Wilshire suggest identical dislocation substructures with and without sigmoidal behavior, again, without subgrain formation. Warda et al. [236] attributed the behavior in Zr to dynamic strain aging. In this case oxygen impurities give rise to solute atmospheres. Eventually, the slip bands become depleted and normal five-power behavior resumes. Dramatic increases in the activation energy are suggested to be associated with the sigmoidal behavior. Thus, the explanation for sigmoidal behavior is unclear. One common theme may be very planar slip at the high temperatures.

CHAPTER 3

Diffusional Creep

M.-T. Perez-Prado, M.E. Kassner

Non-dislocation-based diffusional creep at high temperatures ($T \approx T_{\mathrm{m}}$) and very low stresses in fine-grained materials was qualitatively suggested 50 years ago by Nabarro [237]. This was rigorously (quantitatively) proposed and described by Herring [51]. Mass transport of vacancies through the grains from one grain boundary to another was described. Excess vacancies are created at grain boundaries perpendicular to the tensile axis with a uniaxial tensile stress. The concentration may be calculated using [23]

$$c = c_{\mathrm{v}} \left[\exp\left(\frac{\sigma b^3}{kT} \right) - 1 \right] \tag{106}$$

where c_{v} is the equilibrium concentration of vacancies. Usually $(\sigma b^3 / kT) \gg 1$, and therefore Eqn (106) can be approximated by

$$c = \left[c_{\mathrm{v}} \left(\frac{\sigma b^3}{kT} \right) \right] \tag{107}$$

These excess vacancies diffuse from the grain boundaries lying normal to the tensile direction towards those parallel to it, as illustrated in Figure 51.

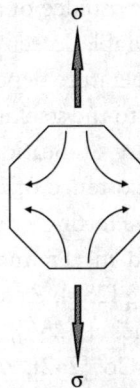

Figure 51 Nabarro-Herring model of diffusional flow. Arrows indicate the flow of vacancies through the grains from boundaries lying normal to the tensile direction to parallel boundaries. Thicker arrows indicate the tensile axis.

Fundamentals of Creep in Metals and Alloys
ISBN 978-0-08-099427-7
http://dx.doi.org/10.1016/B978-0-08-099427-7.00003-7

Grain boundaries act as perfect sources and sinks for vacancies. Thus, grains would elongate without dislocation slip or climb. The excess concentration of vacancies per unit volume is, then, $c_v\sigma/kT$. If the linear dimension of a grain is g, the concentration gradient is $c_v\sigma/kTg$. The steady-state flux of excess vacancies can be expressed as $D_v c_v\sigma/kTg$, where g is the grain size. The resulting strain rate is given by,

$$\dot{\varepsilon}_{ss} = \frac{D_{sd}\sigma b^3}{kTg^2} \tag{108}$$

In 1963, Coble [52] proposed a mechanism by which creep was instead controlled by grain boundary diffusion. He suggested that, at lower temperatures ($T < 0.7\ T_m$), the contribution of grain boundary diffusion is larger than that of self-diffusion through the grains. Thus, diffusion of vacancies along grain boundaries controls creep. The strain rate suggested by Coble is

$$\dot{\varepsilon}_{ss} = \frac{\alpha_3 D_{gb}\sigma b^4}{kTg^3} \tag{109}$$

where D_{gb} is the diffusion coefficient along grain boundaries and α_3 is a constant of the order of unity. The strain rate is proportional to g^{-2} in the Herring model, whereas it is proportional to g^{-3} in the Coble model. Greenwood [238] more recently formulated expressions that allow an approximation of the strain rate in materials with nonequiaxed grains under multiaxial stresses for both lattice and grain-boundary diffusional creep.

Several studies reported the existence of a threshold stress for diffusional creep below which no measurable creep is observed [239–242]. This threshold stress has a strong temperature dependence that Mishra et al. [243] suggest is inversely proportional to the stacking fault energy. They proposed a model based on grain boundary dislocation climb by jog nucleation and movement to account for the existence of the threshold stress.

The occurrence of Nabarro-Herring creep has been reported in poly-crystalline metals [244–247] and in ceramics [248–252]. Coble creep has also been claimed to occur in Mg [251], Zr and Zircaloy-2 [253], Cu [254], Cd [255], Ni [255], copper–nickel [256], copper–tin [256], iron [257], magnesium oxide [258,259], βCo [242], αFe [240], and other ceramics [260]. The existence of diffusional creep must be inferred from indirect experimental evidence, which includes agreement with the rate equations developed by Herring and Coble, examination of marker lines visible at the specimen surface that lie approximately parallel to the tensile axis [261], or by the observation of some microstructural effects such as precipitate-denuded zones (Figure 52). These zones are predicted to develop adjacent to the grain boundaries normal to the tensile axis in dispersion-hardened alloys. Denuded zones were first reported by Squires et al. [263] in a Mg-0.5 wt% Zr alloy. They suggested that magnesium atoms would diffuse into the grain boundaries perpendicular to the tensile axis. The inert zirconium hydride precipitates act as grain boundary markers. The investigators proposed a possible relation between the appearance of these zones and diffusional creep. Since then, denuded zones have been observed on numerous occasions in the same alloy and suggested as proof of diffusional creep.

The existence of diffusional creep has been questioned [264] during the past two decades by some investigators [59,61,265–270] and defended by others [56–58,60,261,271,272]. One major point of disagreement is the relationship between denuded zones and diffusional creep. Wolfenstine et al. [59] suggested that previous studies on the Mg-0.5 wt% Zr alloy [273] are sometimes inconsistent and incomplete since they do not give information regarding the stress exponent or the grain-size exponent. By

25 μm

Figure 52 Denuded zones formed perpendicular to the tensile direction in a hydrated Mg-0.5 wt% Zr alloy at 400 °C and 2.1 MPa [262].

analyzing data from those studies, Wolfenstine et al. [59,265] suggested that the stress exponents corresponded to a higher-exponent power-law creep regime. Wolfenstine et al. also suggested that the discrepancy in creep rates calculated from the width of denuded zones and the average creep rates (the former being sometimes as much as six times lower than the latter) as evidence of the absence of correlation between denuded zones and diffusional creep. Finally, the same investigators [59,265,266] claim that denuded zones can also be formed by other mechanisms including the redissolution of precipitates due to grain boundary sliding accompanied by grain-boundary migration and the drag of solute atoms by grain-boundary migration.

Several responses to the critical report of Wolfenstine et al. [59] were published defending the correlation between denuded zones and diffusional creep [57,58,271]. Greenwood [57] suggests that the discrepancies between theory and experiments can readily be interpreted on the basis of the inability of grain boundaries to act as perfect sinks and sources for vacancies. Bilde-Sørensen and Smith [58] agree that denuded zones may be formed by other mechanisms than diffusional creep but they claim that, if the structure of the grain boundary is taken into consideration, the asymmetrical occurrence of denuded zones is fully compatible with the theory of diffusional creep. Similar arguments were presented by Kloc [271].

Recently, McNee et al. [274] claim to have found additional evidence of the relationship between diffusional creep and denuded zones. They studied the formation of precipitate-free zones in a fully hydrided magnesium plate around a hole drilled in the grip section. The stress state around the hole is not uniaxial, as shown in Figure 53. They have observed a clear dependence of the orientation of denuded zones on the direction of the stress in the region around the hole. Precipitate-free zones were mainly

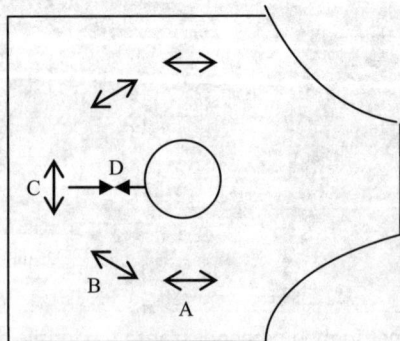

Figure 53 Orientation of stresses around a hole.

observed in boundaries perpendicular to the loading direction at each location. They claim that this relationship between the orientation of the denuded zones and the loading direction is consistent with the mechanism of formation of these zones being diffusional creep.

Ruano et al. [266–268], Barrett et al. [269], and Wang [270] suggest that the dependence of the creep rate on stress and grain size is not always in agreement with that of the diffusional creep theory. A reinterpretation of several data reported in previous studies led Ruano et al. to propose that the creep mechanism is that of Harper–Dorn creep in some cases and grain boundary sliding in others, reporting a better agreement between experiments and theory using these models.

This suggestion has been contradicted by Burton and Reynolds [60], Owen and Langdon [56], and Fiala and Langdon [272].

McNee et al. [275] recently reported that what they suggest is direct microstructural evidence of diffusional creep in an oxygen-free high conductivity copper tensile tested at temperatures between 673 and 773 K and stresses between 1.6 and 8 MPa. The temperature and stress dependencies were found to be consistent with diffusional creep. Scanning electron microscope surface examination revealed, first, displacement of scratches at grain boundaries and, second, widened grain boundary grooves on grain boundaries transverse to the applied stress in areas associated with scratch displacements. In principle, both diffusional creep and some alternative mechanism involving grain boundary sliding could be responsible for the observed scratch displacements. The use of atomic force microscopy to profile lines traversing boundaries both parallel and perpendicular to the tensile axis led to the conclusion that the scratch displacements originated from the deposition of material at grain boundaries transverse to the tensile axis and the depletion of material at grain boundaries parallel to the tensile axis. The investigators claimed that these features can only be attributed to the operation of a diffusional flow mechanism. However, a strain rate an order of magnitude higher than that predicted by Coble creep was found. Thus, the investigators questioned the direct applicability of the diffusional creep theory.

Nabarro himself, perhaps the principal champion of diffusional creep, recently suggested that diffusional creep may or must be accompanied by Harper-Dorn creep [276,277]. This may be a case of Nabarro "hopping from the frying pan and into the fire," because Harper-Dorn creep, as will be discussed subsequently, may be tenuous. Lifshitz [278] in 1963 pointed out the necessity of grain boundary sliding for maintaining grain coherency during diffusional creep in a polycrystalline material. More recent theoretical studies have also emphasized the essential role of grain boundary sliding for continuing steady-state diffusional creep [279–282]. The observations reported by McNee et al. [275] may, in fact, reflect the cooperative operation of both mechanisms. Many studies have been devoted to assess the separate contributions from diffusional creep and grain boundary sliding to the total strain [283–292]. Some claim that both diffusional creep and grain boundary sliding contribute to the overall strain and that they can be distinctly separated [284–288]; others claim that one of them is an accommodation process [289–292]. Many of these studies are based on several simplifying assumptions, such as the equal size of all grains and that the total strain is achieved in a single step. Sahay and Murty [282] claimed that when the dynamic nature of diffusional creep is taken into account (changes in grain size, etc. that take place during deformation), separation of the strain contributions from diffusion and sliding becomes impossible.

CHAPTER 4

Harper-Dorn Creep

M.E. Kassner

Contents

1. INTRODUCTION

The steady-state, time-dependent plasticity, or creep, at high and intermediate temperatures of pure metals, type M alloys, and many ceramics and minerals over a fairly wide range of stress, that usually comprises conventional creep regimes, follows a classic 5-power-law behavior and power-law breakdown that is illustrated in Figure 14. At low stress (often at high temperatures) the steady-state creep rate is often suggested to evince Newtonian, or 1-power, behavior. The figure has some, but certainly not

Fundamentals of Creep in Metals and Alloys
ISBN 978-0-08-099427-7
http://dx.doi.org/10.1016/B978-0-08-099427-7.00004-9

all, of the steady-state creep data of Al but describes part of the general trends of Al and many other metals and ceramics to start the discussion of this review.

A new low-stress mechanism for creep at high temperatures and low stresses in Figure 14 was originally proposed in a 1957 study by Harper and Dorn [50]. This mechanism has since been termed "Harper-Dorn creep." By performing creep tests on aluminum of high purity and large grain sizes, these investigators found that the steady-state creep rate increased linearly with the applied stress and the activation energy was that of self-diffusion. The observed creep process could not be ascribed diffusional creep discussed in the previous chapter. They reported creep rates as high as a factor of 1400 greater than the theoretical rates calculated by the Herring and Coble models. The same observations were reported some years later by Barrett et al. [269], Mohamed et al. [294,334], and Ardell and Lee [101], as summarized in Figure 54. The right-hand portion shows a few

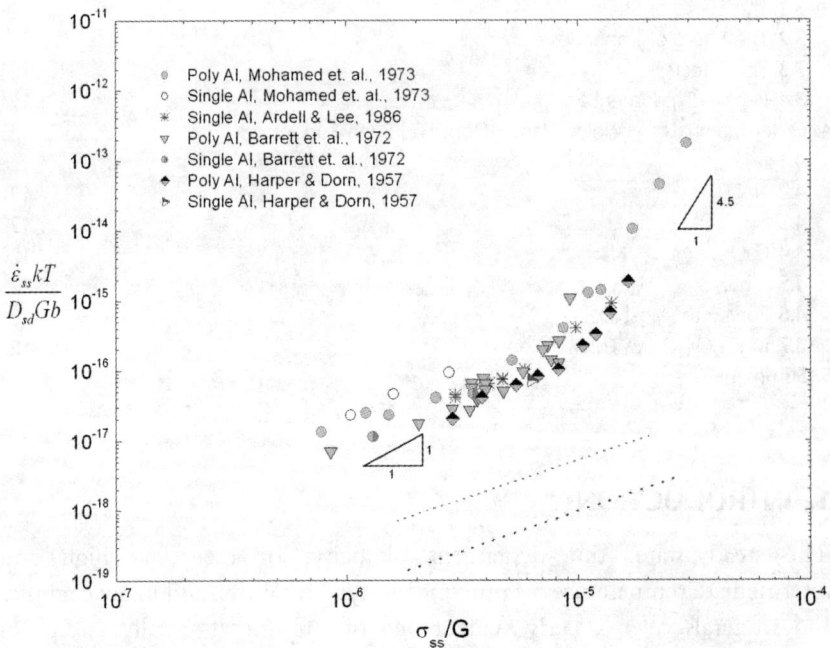

Figure 54 A comparison between the diffusion-coefficient compensated strain-rate versus modulus-compensated stress for pure aluminum based on early data [50,101,294,324], with theoretical predictions for Nabarro-Herring creep [295] (dashed lines) for 3.3- and 9 mm grain sizes. Harper-Dorn creep was presumed in the low-stress exponent regime, a different mechanism than 5-power-law creep at higher stresses.

5-power-law data points, but mostly 1-power, at the left hand portion of Figure 14. A primary stage of creep was observed, which would not be expected according to the Herring diffusional model because the concentration of vacancies immediately upon stressing cannot exceed the steady state value. Furthermore, grain boundary shearing was reported to occur during creep and similar steady-state creep rates were observed in aluminum single crystals and in polycrystalline specimens with a 3.3 mm grain size. Diffusional creep is not expected, of course, in single crystals. This evidence led these investigators to conclude that low-stress creep at high temperatures in materials of large grain sizes occurred via a dislocation-climb mechanism. (It should be mentioned that a "threshold" stress is subtracted from the applied stress by Harper and Dorn, which may be problematic. This fact appears to often be overlooked by subsequent investigators.)

The relationship between the applied stress and the steady-state creep rate for Harper-Dorn creep is phenomenologically described by [295]

$$\dot{\varepsilon}_{ss} = A_{HD} \left(\frac{D_{sd} G b}{kT} \right) \left(\frac{\sigma}{G} \right)^1 \tag{110}$$

where A_{HD} is a constant. Since these early observations [50], Harper-Dorn creep has been reported to occur in a large number of metals and alloys as well as a variety of ceramics and ice (see Table 3). Interests in ceramics and minerals lie in predicting low-stress creep plasticity in geological systems such as the lower crust, lower mantle, and inner core of the Earth [910]. Figures 55–59 illustrate Harper-Dorn in CaO, MgO, Mg_2SiO_4, olivine, and NaCl. Metallic systems are sometimes chosen on the basis of expected satisfactory service in long-term structural applications. The importance of Harper-Dorn may have been enhanced by the suggestions that diffusional creep does not occur and that rather (rare) cases of (Newtonian) diffusional creep are actually Harper-Dorn (dislocation Newtonian creep) [322].

Several studies have been published during the past 30 years that distinguished Harper-Dorn from classic 5-power-law creep (and diffusional creep). Yavari et al. [295] provided more evidence that the Harper-Dorn creep rate is independent of the specimen grain size. Similar rates were observed both in polycrystalline materials and in single crystals. They determined, by etch-pits, that the dislocation density was relatively low and *independent* of the applied stress, unlike 5-power-law creep. Owen and Langdon [889] suggested that the values of the dislocation density of

Table 3 Materials for which Harper-Dorn creep has been suggested to operate with relevant references

Metals		Ceramics (and minerals)	
Material	References	Material	References
Al	[50,55,86,101, 199,269,294–296, 900,905]	CaO	[300,301,898]
Pb	[294]	UO_2	[302]
α-Ti	[297]	MgO	[303,304, 911–915,925]
α-Fe	[240]	TiO_2	[887]
α-Zr	[298]	$Mn_{0.5}Zn_{0.5}Fe_2O_4$	[305]
βCo	[299]	BeO	[305]
Sn	[294]	Al_2O_3	[305]
Cu	[78,228,244, 890–893, 895,897,899[a]]	$Co_{0.5}Mg_{0.5}O$	[84[a],308]
		NaCl	[309,310,916[a], 917[a],918[a]]
		$MgCl_2 \cdot 6H_2O$	[307]
		$KZnF_3$	[31]
		$KTaO_3$	[312]
		$CaTiO_3$	[307]
		Ice	[306]
		$CaCO_3$	[305]
		SiO_2	[31]
		$(Mg,Fe)_2SiO_4$	[315,888,903[a], 919[a],920[a],921[a], 922[a]]
		$NaAlSi_3O_8$–$CaAl_2Si_2O_8$	[315,888]

[a]For basic creep articles making no reference to Harper-Dorn creep.

Al–5at% Mg are near $10^9 \, m^{-2}$. (Dislocations were found to be predominantly close to edge orientation.) Ardell [34] suggested, using etch-pit analysis, that the dislocation density would not reach values less than a "frustration" level, effectively rendering the density independent of stress in aluminum with a density of about $10^8 \, m^{-2}$. Barrett et al. also found ρ stress-independent in Al [269] at $7 \times 10^7 \, m^{-2}$ using etch pits. Nes [215], however, suggested that the dislocation density using X-ray topography apparently showed that ρ was dependent of stress by $\sigma^{1.3}$ although the creep

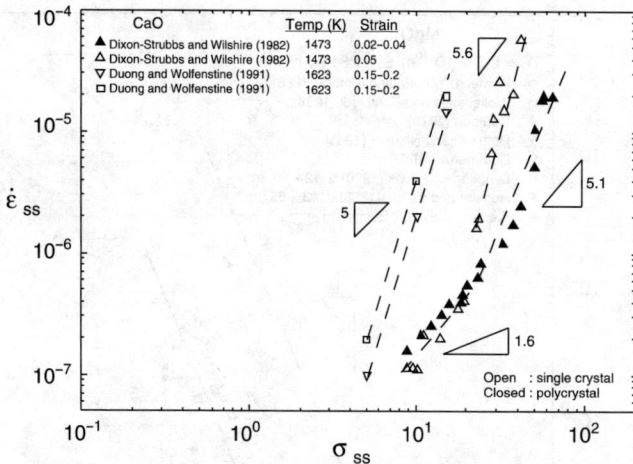

Figure 55 Creep behavior at a single temperature for CaO.

behavior was not well defined. His technique may be the ideal method to assess the low dislocation densities.

The fact that the activation energy for Harper-Dorn creep is about equal to that of self-diffusion suggests that Harper-Dorn creep occurs by climb of edge dislocations. Weertman and Blacic [320] suggested that creep is not observed at constant temperatures, but only with low amplitude temperature fluctuations, where the vacancy concentration would not be in thermal equilibrium, thus leading to climb stresses on edge dislocations of the order of 3–6 MPa. This explanation does not appear widely accepted, partly due to the observation that Harper-Dorn creep is consistently observed by a wide assortment of investigators, presumably with different temperature control abilities [321].

In summary, the early low-stress experiments, primarily in metals, indicated that Harper-Dorn includes:

1. activation energy about equal to lattice self-diffusion,
2. grain-size independence, with grain boundary shearing,
3. steady-state stress exponent of about one,
4. dislocation density that appears independent of stress, and
5. primary creep stage.

These combined aspects distinguish the phenomenon from 5-power-law creep, and low-stress exponent grain-boundary sliding (superplasticity), and the Herring diffusional creep model.

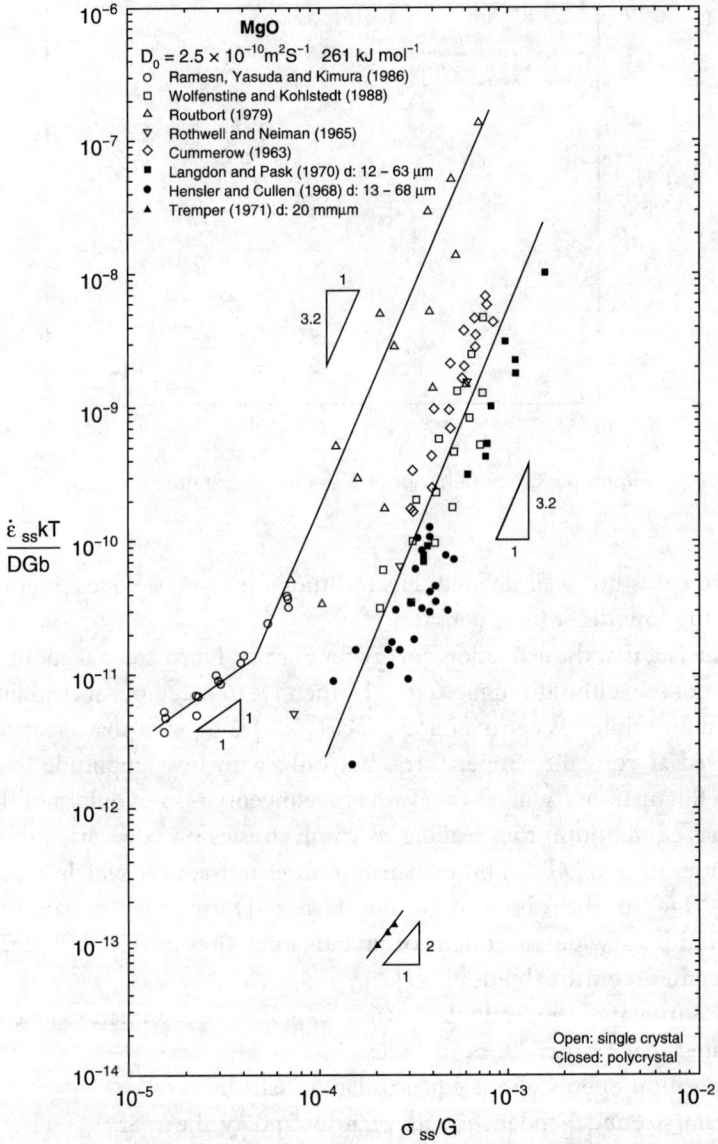

Figure 56 Steady-state creep behavior of MgO.

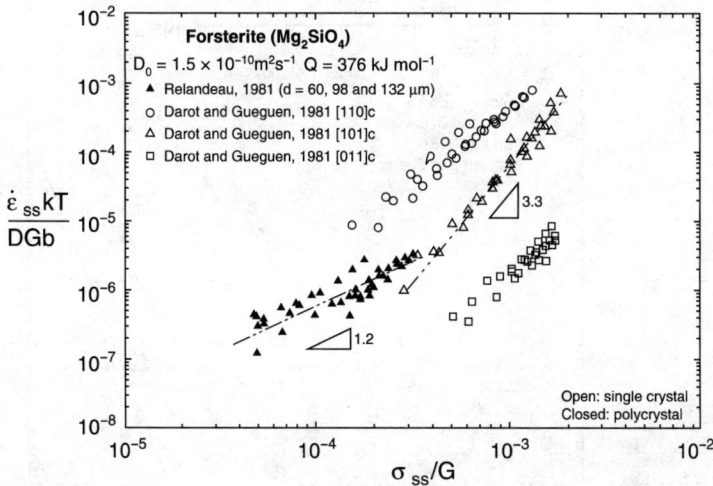

Figure 57 The steady-state creep behavior of Mg_2SiO_4.

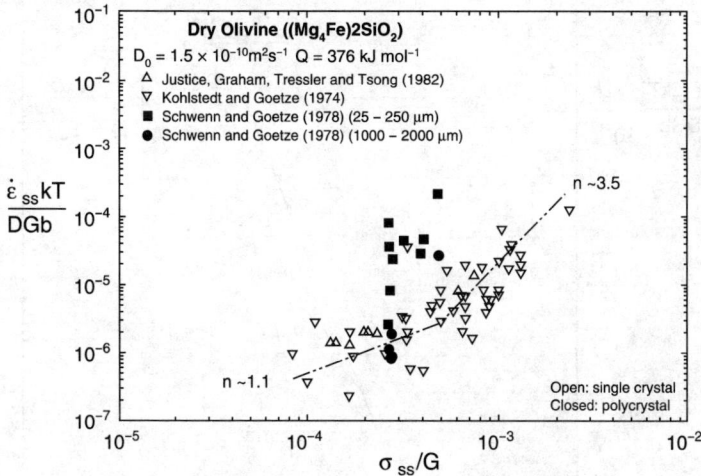

Figure 58 The steady-state creep behavior of olivine.

2. THEORIES FOR HARPER-DORN

The theories of Harper–Dorn creep will be, briefly, reviewed.

2.1 Harper and Dorn [50,199,900]

As discussed by Langdon [318], Harper–Dorn was initially described by the motion of jogged screw dislocations, analogous to that described by Mott

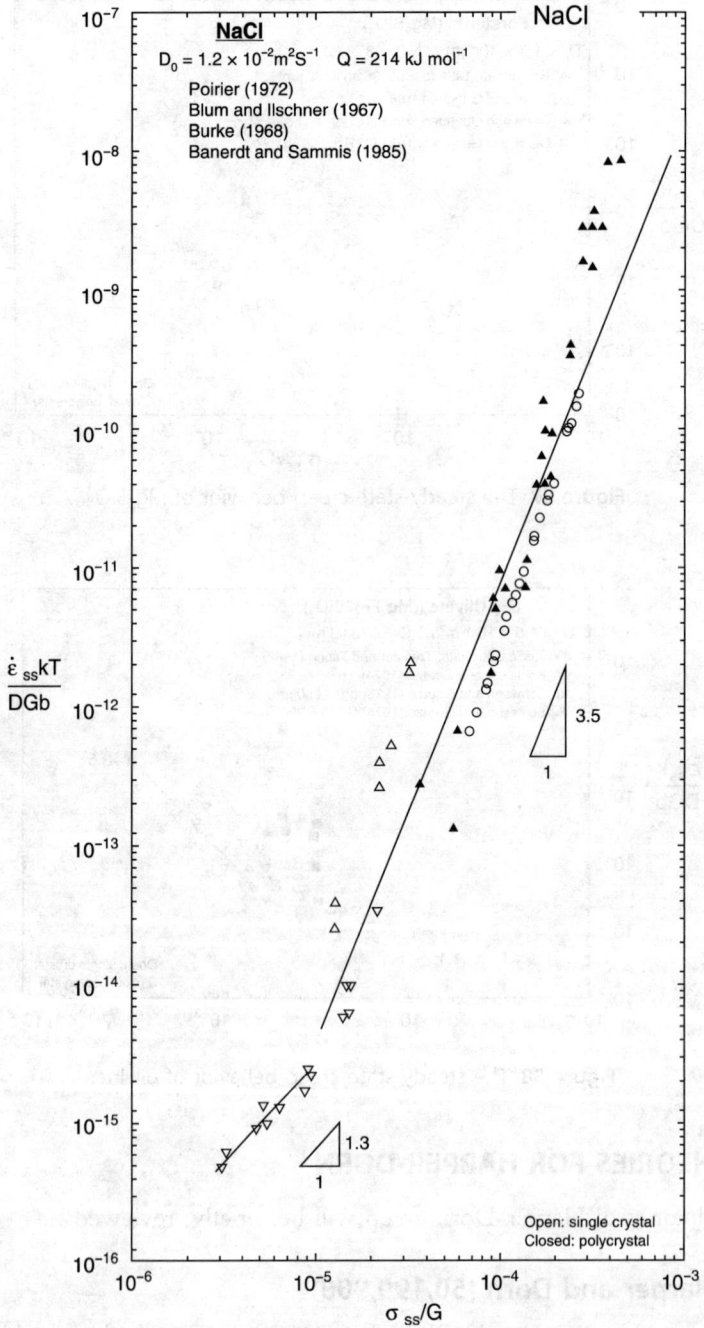

Figure 59 Steady-state creep behavior of the NaCl single and polycrystals.

[198] and later by Barrett and Nix [190]. This leads to a steady-state strain rate.

$$\dot{\gamma} = 12\pi\rho\ell_{js}b\left(\frac{D_{sd}Gb}{kT}\right)\left(\frac{\tau}{G}\right)^{1.0} \tag{111}$$

where ρ is the dislocation spacing and ℓ_{js} is the jog spacing in screw dislocations. Langdon criticized this model on the basis that it requires unrealistically small jog spacings.

2.2 Friedel [87]

Langdon also states that Friedel suggested that the "Harper-Dorn" is actually diffusion creep where vacancies diffuse between the relatively small subgrain boundaries.

$$\dot{\gamma} = A_{NH}\left(\frac{D_{sd}Gb}{kT}\right)\left(\frac{b}{\lambda}\right)^2\left(\frac{\tau}{G}\right)^{1.0} \tag{112}$$

As Langdon points out, subgrains are not always observed in the Harper-Dorn region. When subgrains do form, the size, according to Barrett et al. [269], tends to be stress dependent and this would increase the stress-dependence in Eqns (110) and (112) beyond that of the observed value of 1.0.

2.3 Barrett, Muehleisen, and Nix [269]

Barrett et al. suggested that, as with many other diffusion-controlled creep processes, dislocation generation occurs with dislocation climb (at an assumed fixed number of sources). The rate of dislocation generation $\dot{\rho}_+$ is

$$\dot{\rho}_+ = \rho_0\frac{v_C}{x} \tag{113}$$

Where ρ_0 is the fixed dislocation length per unit volume, v_C is the climb velocity, and x is the distance over which climb must occur to create glide dislocations. The climb velocity under a climb stress σ is [347]

$$v_C = \frac{Db^2\sigma}{kT} \tag{114}$$

They assume $x = \frac{Gb}{\sigma}$ (which appears to assume Taylor hardening) leading to

$$\dot{\rho}_+ = \frac{\rho_0 Db\sigma^2}{kTG}. \tag{115}$$

Dislocation annihilation is assumed to occur only at subgrain boundaries. (This is a complication as subgrain boundaries may not always exist.) The annihilation rate is

$$\dot{\rho}_- = \frac{\rho v_g}{\lambda} \tag{116}$$

Where v_g is the glide velocity and λ is the subgrain size (phenomenologically $\lambda = \lambda_0/\sigma$). It was assumed.

$$v_g = v_0 \sigma \tag{117}$$

(and a stress exponent of one for dislocation velocity, here, is a critical assumption) leading to

$$\dot{\rho}_- = \frac{\rho v_0 \sigma^2}{\lambda_0} \tag{118}$$

at steady state

$$\dot{\rho} = \dot{\rho}_+ - \dot{\rho}_- = 0 \quad \text{and} \tag{119}$$

$$\frac{\rho_0 D b \sigma^2}{kTG} = \frac{\rho v_0 \sigma^2}{\lambda_0} \quad \text{or} \tag{120}$$

$$\rho = \frac{\rho_0 D b \lambda_0}{kTG v_0} \tag{121}$$

and ρ is independent of σ, another critical assumption. Combining with the Orowan equation,

$$\dot{\varepsilon}_{ss} = \rho b v_g \tag{122}$$

$$\dot{\varepsilon}_{ss} = \frac{\rho_0 D b^2 \sigma \lambda_0}{kTG} \tag{123}$$

Again, critical to this derivation is a stress exponent of one for v_g and the assumption $x = (Gb)/\sigma$, which appear tenuous.

2.4 Langdon et al. [318]

Langdon et al. suggested that the rate-controlling processes for Harper-Dorn is based on the climb of jogged edge-dislocations under vacancy saturation conditions, and

$$\dot{\gamma} = \frac{6\pi\rho b^2}{\ell n(1/\rho^{1/2}b)} \left(\frac{D_{sd} Gb}{kT} \right) \left(\frac{\tau}{G} \right)^{1.2} \tag{124}$$

Here, vacancy saturation implies that the steady-state vacancy concentration near and away from the climbing jog is fixed by a steady state between vacancy emission/absorption from the jog and longer-range diffusion to and from the jog.

2.5 Wu and Sherby [53]

The fact that within both the Harper-Dorn and the 5-power-law regimes, the underlying mechanism of plastic flow appeared to be diffusion controlled, led Wu and Sherby to propose a unified relation that describes the creep behavior over both ranges. This model incorporates an internal stress that arises from the presence of random stationary dislocations present within subgrains. At any time during steady-state flow, they assume half of the dislocations moving under an applied stress are aided by the internal stress field (the internal stress adds to the applied stress), whereas the motion of the other half is inhibited by the internal stress. The internal stress is calculated from the dislocation density by the dislocation hardening equation ($\tau = \alpha Gb\sqrt{\rho}$, where $\alpha \cong 0.5$). The unified equation is [322]

$$\dot{\varepsilon}_{ss} = \frac{1}{2} A_{WS} \frac{D_{eff}}{b^2} \left\{ \left(\frac{\sigma + \sigma_i}{E} \right)^n + \frac{|\sigma - \sigma_i|}{(\sigma - \sigma_i)} \left| \frac{\sigma - \sigma_i}{E} \right|^n \right\} \tag{125}$$

where A_{WS} is a constant and σ_i is the internal stress. At high stresses, where $\sigma \gg \sigma_i$, σ_i is negligible compared with σ and Eqn (125) reduces to the (5-power-law) relation

$$\dot{\varepsilon}_{ss} = A_{10} \frac{D_{eff}}{b^2} \left(\frac{\sigma}{E} \right)^n \left(\text{with } n = 4 - 7 \right) \tag{126}$$

At low stresses, where $\sigma \ll \sigma_i$ (Harper-Dorn regime), Eqn (125) reduces to Eqn (110). A reasonable agreement has been suggested between the predictions from this model and experimental data [53,322] for pure aluminum, γ-Fe, and β-Co. The internal stress model was criticized by Nabarro [321], who claimed a unified approach to both 5-power-law and Harper-Dorn creep is not possible because none of these processes are, in themselves, well understood and unexplained dimensionless constants were introduced in order to match theoretical predictions with experimental data.

Also, the dislocation density in Harper-Dorn creep is constant, whereas it increases with the square of the stress in the power-law regime. Thus, the physical processes occurring in both regimes must be different (although Ardell [54] attempts to rationalize this using network-creep models).

2.6 Wang [323,325,326,328]

The internal stress model of Sherby and Wu was also criticized by Wang [323], who proposed that the transition between power-law creep and Harper-Dorn creep takes place instead at a stress (σ) equal to the "Peierls stress (σ_p)" [325,328]. Wang [326] suggests that the steady-state dislocation density is related to the Peierls stress. In equilibrium, the stress due to the mutual interaction of moving dislocations is in balance not only with the applied stress but also with lattice friction, which fluctuates with an amplitude of the Peierls stress. As a result, the steady-state dislocation density ρ in dislocation creep can be written as

$$b\rho^{1/2} = 1.3\left[\left(\frac{\tau}{G}\right)^2 + \left(\frac{\tau_p}{G}\right)^2\right]^{1/2} \tag{127}$$

where τ is the applied shear stress. When $\tau \gg \tau_p$, the dislocation density is proportional to the square of the applied stress, and 5- (or 3-) power-law creep is observed. This is consistent with the recent work of Kassner [901]. Conversely, when $\tau \ll \tau_p$, the dislocation density is independent of the applied stress and Harper-Dorn occurs. As will be discussed later, it appears that the dislocation density does vary with stress within the so-called Harper-Dorn regime.

2.7 Ardell [54,86,101,401]

A different and fairly extensive approach to Harper-Dorn is based on the dislocation network theory by Ardell et al. The dislocation link length distribution contains no segments that are long enough to glide freely. That is, the longest links of length L_m are smaller than the critical link length to activate a (e.g., Frank-Read) dislocation source. Harper-Dorn is, therefore, a phenomenon in which all the plastic strain in the crystal is a consequence of dislocation network coarsening. The recovery of the dislocation density during Harper-Dorn creep is comparable to static recovery in the absence of an applied stress; climb of nodes is facilitated by line tension of dislocation links. The stress dependence of the $\dot{\varepsilon}_{ss}$ arises because the applied stress biases the collisions, since the lengths of all the links must increase as σ_{ss} increases,

thereby increasing the collision possibilities. The climb velocity of the
nodes is mostly affected by the resolved force arising from the line tensions
of the dislocations at the nodes. Accidental collisions between these links
can refine the network and stimulate further coarsening.

$$\dot{\varepsilon} = \frac{\pi C b^3 D}{2kT} \rho \sigma \qquad (128)$$

where

$$C = \frac{\alpha}{\langle u \rangle^2} \int_0^{u_c} u^{2-m}(u-1)\Phi(u)\mathrm{d}u \qquad (129)$$

and Φ_u is the scaled-link-length (u) distribution function, $\alpha \cong 0.5$ and
m is a phenomenological exponent. The independence of ρ with σ is a
consequence of the frustration of the dislocation network coarsening that
arises because of the exhaustion of Burgers vectors that can satisfy Frank's
rule at the nodes.

3. MORE RECENT DEVELOPMENTS

3.1 The Effect of Strain

Aluminum is clearly the most extensively studied material in the
Harper-Dorn regime. Blum et al. [55] recently questioned the existence of
Harper-Dorn creep, not having been able to observe the decrease of the
stress exponent to a value of one when performing compression tests with
changes in stress in pure aluminum (99.99% purity), in a low-stress regime
where Harper-Dorn had been observed by others. These results are illustrated, later, in Figure 61. Nabarro [330] responded to these reservations
claiming that the lowest stress used by Blum et al. (0.093 MPa) was still too
high to observe Harper-Dorn. Blum et al. [331] subsequently performed
compression tests using even lower stresses (as low as 0.06 MPa), failing again
to observe $n = 1$ stress exponents. Instead, exponents close to five were
measured, indicating normal, 5-power-law creep extending into the
so-called Harper-Dorn regime. Basically, Blum suggested that earlier
Harper-Dorn studies did not accumulate sufficient strain (hardening) to
achieve steady state. Strain-rates in the Harper-Dorn regime are so low that
unusually long testing periods are required to achieve modest strains. In some
Al cases, Blum is correct that creep-rates reported by other investigators are

too high based on a failure to achieve steady state. However, in other cases where Harper-Dorn is suggested, steady state appears to have been achieved.

3.2 The Effect of Impurities

Recently, Mohamed et al. [334–336] suggested that impurities may play an essential role in Harper-Dorn creep. They performed relatively large strain (up to 10%) creep tests at stresses lower than 0.06 MPa in Al polycrystals of 99.99 and 99.9995 purity. They observed Harper-Dorn creep only in the latter (most pure) metal. Accelerations in the creep curve corresponding to the high purity Al are apparent in Figure 60. These accelerations are absent in the less-pure 99.99 Al creep curve, at an identical temperature and stress.

Mohamed et al. also reported that the microstructure of the 99.9995 Al includes wavy grain boundaries, an inhomogeneous dislocation density distribution as determined by etch-pits, small new grains forming at the specimen surface, and large dislocation density gradients across grain boundaries. Well-defined subgrains were not observed. However, the microstructure of the deformed 99.99 Al is formed by a well-defined array of subgrains. These observations led Mohamed et al. to conclude that the restoration mechanism taking place during so-called Harper-Dorn creep includes discontinuous dynamic recrystallization (DRX) rather than the contended dynamic recovery. Nucleation of recrystallized grains would

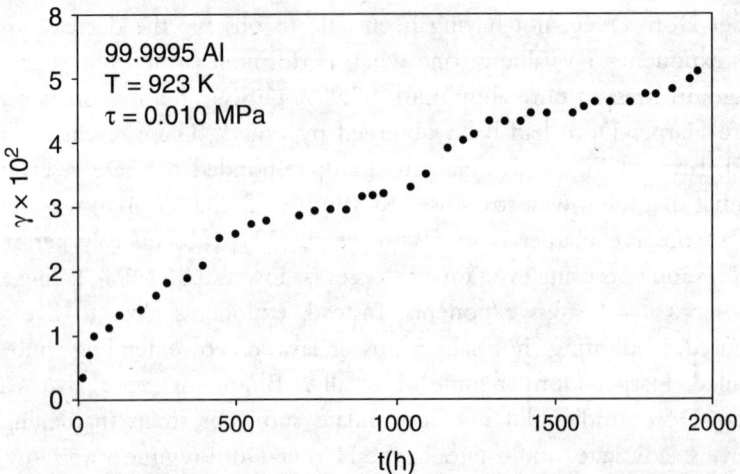

Figure 60 Creep curve corresponding to very pure 99.9995% Al, *from Mohamed et al. [334]*. The undulations are suggested to result from new restoration mechanisms.

take place at the specimen surfaces and, due to the low amount of impurities, highly mobile boundaries would migrate toward the specimen interior. This restoration mechanism would give rise to the periodic accelerations observed in the creep curve, by which much of the strain is produced. Therefore, Mohamed et al. believed that high purity leads to DRX. It is difficult to accurately determine the stress exponent due to the appearance of periodic accelerations in the creep curves. However, Mohamed et al. [335] claimed that $n = 1$ exponent are only obtained if creep curves up to small strains (1–2%) are analyzed, as was done in the past. Mohamed et al. estimated stress exponents of about 2.5 at larger strains for high-purity DRX specimens of Al.

The work by Mohamed et al. has received some criticism. Langdon [337] argues that the jumps in the creep curves are not very clearly defined. Also, Mohamed claims that DRX occurs during creep of very high purity metals at regular strain increments, whereas the incremental strains corresponding to the accelerations tend to be relatively non-uniform. Grain growth might be a more appropriate restoration mechanism. It is certainly true that 99.999% pure Al has a greater propensity for (static) recrystallization (and presumably DRX and grain growth) than 99.99% pure Al, but Figure 54 shows the so-called Harper-Dorn present in just 99.99% pure Al in other studies, in contradiction to the suggestion by Mohamed.

Additional recent experiments were performed by McNee et al. [905] on, generally, 99.999% pure Al. These investigators generally, consistent with Blum, observed the extension of 5-power-law creep into Harper-Dorn regime. A few tests, however, show relatively high strain-rates (creep-rates) and might be more consistent with the early experiments supportive of a low-stress exponent. Despite these new experiments, Langdon [904] appears to have suggested that Harper-Dorn is, nonetheless, an independent mechanism.

3.3 Size Effects

Raj et al. [923,924] suggeested the surfaces are dislocation sources which leads to a size effect. Nes et al. [201] suggested that, under conditions typical of Harper-Dorn creep, the statistical slip-length may become comparable to or even exceed the specimen diameter (a size effect). That is, it was suggested that under Harper-Dorn conditions, the size influences the rates of generation and loss of dislocations. The role of dislocation generation is reduced and the loss of dislocations is no longer controlled by dynamic recovery, but by static recovery. The result is that creep rate scales linearly with the applied stress. Size effects with specimen dimensions approaching the obstacles spacing was demonstrated by Uchic et al. [909].

3.4 Recent Experiments

Figure 61 summarizes the data of the more recent experiments just described. The 99.999% pure Al data of Mohamed and coworkers show lower stress exponents of 1–2.5 due to the suggested additional restoration mechanisms (e.g., grain growth (GG) or DRX). The data of Blum and coworkers and Mohammed and coworkers on lower purity, 99.99% where the impurities presumably suppress DRX and GG, show normal, five-power-law behavior into the so-called Harper-Dorn regime. The polycrystalline data of McNee et al., which include 99.999% and 99.99% pure Al, do not appear to evince low stress exponent behavior, although the high purity of the McNee tests may have been sometimes compromised. Rather, some threshold behavior is observed, with an exponent above 4.5.

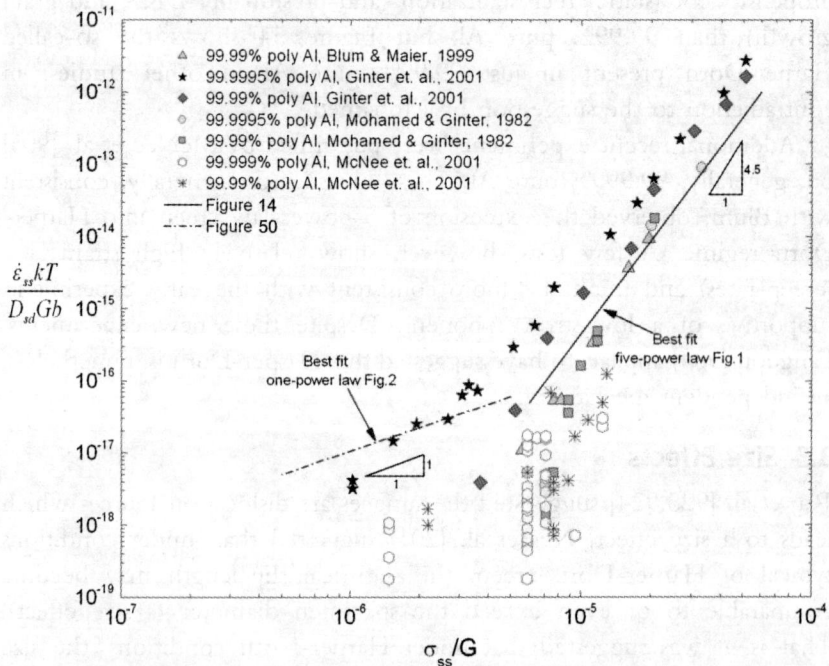

Figure 61 The trends of Figure 14 and Figure 54 are plotted with straight lines with the additional, in particular, recent data (mostly 1999–2001) relevant to Harper-Dorn.

The curious aspect of the more recent data of Figure 61 is that the low stress Al data are more consistent with 5-power-law behavior than "Harper-Dorn." In summary, the aluminum data are ambiguous as to whether Harper-Dorn is, in fact, a separate creep mechanism. One complication is that polycrystalline specimens were used in these more recent studies and average Taylor factors may be variable in coarse grain (e.g., 5 mm) in small-dimension (e.g., <0.5 mm) specimens.

Creep experiments were recently conducted using relatively large single crystals of high-purity aluminum at temperatures and stresses within the range where it is reasonable to anticipate the occurrence of Harper-Dorn creep [139, 953]. A stress exponent of greater than three was observed. These are illustrated in Figure 62. Additionally, the results from these experiments suggest that, contrary to several earlier reports, the dislocation substructure is *not* independent of the applied stress and instead the network dislocation density varies with stress as a direct extension of the behavior anticipated within the conventional 5-power creep regime. This is illustrated in Figure 63. It should also be emphasized that Harper and Dorn report their data *after* a threshold stress correction (based on surface and interface energies (tension)). If these corrections are not made, then the data appear very close to those of Figure 62. Another interesting observation of this work is that subgrains are not observed.

The network-based creep model proposed by Evans and Knowles [105] in Chapter 2 was modified [938] and was compared with the recent experimental results in the Harper-Dorn regime. Despite the simplicity of the present model, the experimental results on pure Al single crystals and polycrystalline copper show an excellent match with the theory. Thus, there is a possibility that Harper-Dorn as a separate mechanism does not exist, although new experiments (particularly on ceramics and minerals) are clearly warranted.

4. OTHER MATERIALS FOR WHICH HARPER-DORN HAS BEEN SUGGESTED

Table 3 lists materials for which Harper-Dorn has been suggested to occur, either by the (original) experiments or by subsequent reinterpretation of original data by subsequent investigators. However, the Harper-Dorn conclusions are ambiguous in several instances. As was discussed previously, the large volume of Al work is particularly ambiguous. It should also be mentioned that ceramics and minerals deforming within the

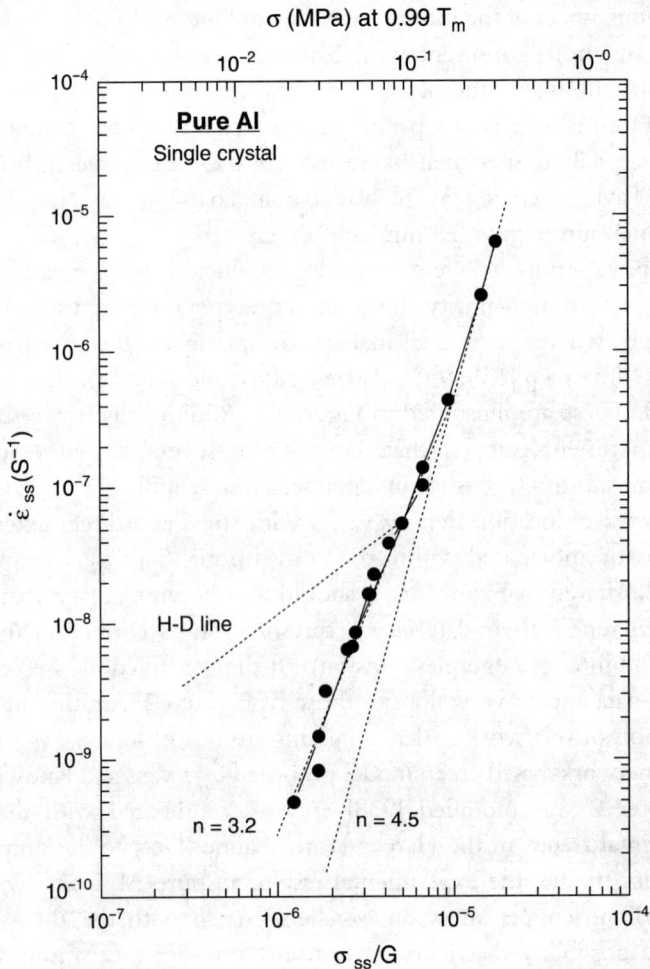

Figure 62 The stress-strain rate behavior of pure Al single crystals [139,953]. Lines with slopes 4.5 and 3 are shown in the plot.

5-power-law creep regime tend to have lower stress exponents than metals; values are closer to 3 rather than 5.

4.1 α-Zr

Harper-Dorn was suggested for helical specimens of α-Ti, α-Zr, and β-Co [297–299]. The α-Zr data are particularly ambiguous as there is a grain-size dependence in the purported Harper Dorn regime. For example, in an earlier review by Hayes et al. [80], it was shown that the creep rate of

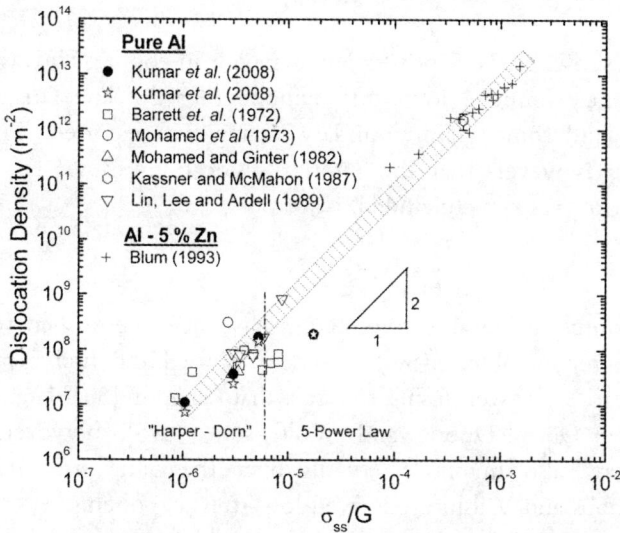

Figure 63 The dislocation density (subgrain boundaries not present) in pure (99.999%) Al single crystal at 923 K at low stresses.

zirconium at low values of σ_{ss}/G varies approximately proportional to the applied stress. The rate-controlling mechanism (s) for creep within this regime is unclear. A grain-size dependency may exist, particularly at small (<90 μm) sizes, suggesting a diffusional or perhaps a grain-boundary sliding mechanism. A grain-size independence at larger grain sizes supports, by itself, Harper-Dorn, but the low observed activation energy ($\cong 90$ kJ mol^{-1}) is not consistent with those observed at similar temperatures at higher stresses in the 5-power-law regime (270 kJ mol^{-1}) where creep is also believed to be lattice self-diffusion controlled. The stress dependence in this regime is not consistent with traditional grain-boundary sliding mechanisms.

4.2 NaCl

Banerdt and Sammis [309] suggested Harper-Dorn in NaCl, although required temperature variation corrections were substantial at the lower stresses. NaCl shows 5 (3.5)-power-law behavior and one study appears to have shown low stress exponent behavior. Bandert and Sanmis, however, appears to have unusually low total strains (<1%) and steady states are not clearly established.

4.3 Cu

Figure 15 shows classic 5-power-law behavior in copper. Shrivastava et al. [891] recently found a low stress exponent (~ 2). The Cu was polycrystalline and some grain boundary sliding was reported. The author concluded, however, that the interior dislocations cause slip, although Harper-Dorn was not concluded.

4.4 CaO

CaO is confusing, because two studies have been performed within the low strain regime; Dixon-Stubbs and Wilshire [300] find 5-power-law transitions to 1.6 power in single crystals that Langdon [301] later suggested was Harper-Dorn. Duong and Wolfenstine [898], however, observe 5-power-law also in single crystals over the same stress range that Dixon-Stubbs and Wilshire observed low-stress exponents.

4.5 MgO

MgO appears to have contradictory data. Five (actually 3.2)-power data of Routbort [925] is of higher strain- rates than other studies, while others appear unusually low [915]. Ramesh et al. [304] observed a low-stress exponent at lower stresses in single crystals.

4.6 Forsterite (Mg$_2$SiO$_4$)

Forsterite shows three or four stress exponent behavior in single crystals of different orientation. The low stress tests by Relendeau [919] on relatively fine-grained material show low stress-exponent behavior.

4.7 MgCl$_2$6H$_2$O(CO$_{0.5}$Mg$_{0.5}$)O and CaTiO$_3$

The Harper-Dorn creep behaviors of these were recently questioned by Berbon and Langdon [316]. Discontinuous DRX was suggested in MgCl$_2$6H$_2$O(CO$_{0.5}$Mg$_{0.5}$)O, while it was suggested that a transition in mechanism is not evident for CaTiO$_3$ on a double logarithmic plot.

5. SUMMARY

In summary, there is real question as to whether Harper-Dorn creep, as classically defined, exists. Whether there is a stress exponent decrease from 5-power-law creep with decreasing stress without a dramatic change in creep mechanisms, is an open question that requires further study.

CHAPTER 5

The 3-Power-Law Viscous Glide Creep

M.-T. Perez-Prado, M.E. Kassner

Creep of solid solution alloys (designated Class I [16] or Class A alloys [338]) at intermediate stresses and under certain combinations of materials parameters, which will be discussed later, can often be described by three regions [36,339,340]. This is illustrated in Figure 64. With increasing stress, the stress exponent, n, changes in value from 5 to 3 and again to 5 in regions I, II, and III, respectively. This section will focus on region II, the so-called

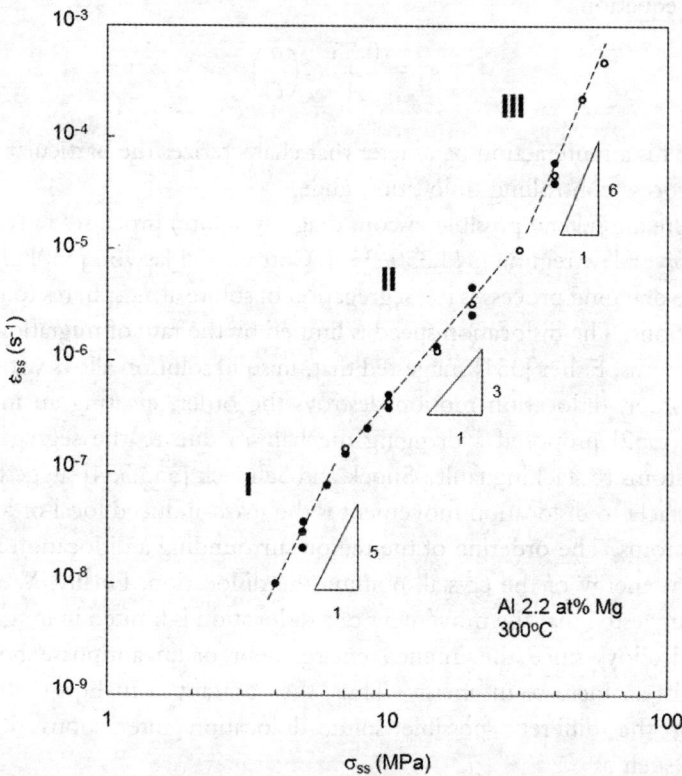

Figure 64 Steady-state creep rate versus applied stress for an Al–2.2 at% Mg alloy at 300 °C. Three different creep regimes, I, II, and III, are evident. *Based on Refs [341,342].*

Fundamentals of Creep in Metals and Alloys
ISBN 978-0-08-099427-7
http://dx.doi.org/10.1016/B978-0-08-099427-7.00005-0

129

3-power-law regime. The mechanism of deformation in region II is often described as viscous glide of dislocations [36]. This is due to the fact that the dislocations interact in several possible ways with the solute atoms, and their movement is impeded [343]. There are two competing mechanisms over this stress range: dislocation climb and glide, and glide is slower and thus rate controlling. A 3-power-law may follow naturally then from Eqn (16) [24,344,345],

$$\dot{\varepsilon} = 1/2\bar{v}b\rho_{\mathrm{m}}$$

It has been theoretically suggested that $\bar{v}$ is proportional to σ [346,347] for solute-drag viscous glide. It has been determined empirically that ρ_{m} is proportional to σ^2 for Al–Mg alloys [76,93,118,318,341,348]. Weertman [344,345] and Horiuchi et al. [349] have suggested a possible theoretical explanation for this relationship. Thus, $\dot{\varepsilon} \propto \sigma^3$. More precisely, following the original model of Weertman [344,345], viscous glide creep is described by the equation

$$\dot{\varepsilon}_{\mathrm{ss}} \cong \frac{0.35}{A} G\left(\frac{\sigma}{G}\right)^3 \qquad (130)$$

where A is an interaction parameter that characterizes the particular viscous drag process controlling dislocation glide.

There are several possible viscous drag (by solute) processes in region II, or 3-power-law regime [344,350–353]. Cottrell and Jaswon [350] proposed that the dragging process is the segregation of solute atmospheres to moving dislocations. The dislocation speed is limited by the rate of migration of the solute atoms. Fisher [351] suggested that, in solid solution alloys with short-range order, dislocation motion destroys the order, creating an interface. Suzuki [352] proposed a dragging mechanism due to the segregation of solute atoms to stacking faults. Snoek and Schoeck [353,354] suggested that the obstacle to dislocation movement is the stress-induced local ordering of solute atoms. The ordering of the region surrounding a dislocation reduces the total energy of the crystal, pinning the dislocation. Finally, Weertman [344] suggested that the movement of a dislocation is limited in long-range-ordered alloys since the implied enlargement of an antiphase boundary results in an increase in energy. Thus, the constant A in Eqn (130) is the sum of the different possible solute-dislocation interactions described earlier, such as

$$A = A_{\mathrm{C-J}} + A_{\mathrm{F}} + A_{\mathrm{S}} + A_{\mathrm{Sn}} + A_{\mathrm{APB}} \qquad (131)$$

Several investigators proposed different 3-power models for viscous glide where the principal force retarding the glide of dislocations was due to Cottrell–Jaswon interaction $(A_{C-J} + A_F + A_S + A_{Sn} + A_{APB})$ [87,118,345,355]. In one of the first theories, Weertman [345] suggested that dislocation loops are emitted by sources and sweep until they are stopped by the interaction with the stress field of loops on different planes, and dislocation pile-ups form. The leading dislocations can, however, climb and annihilate dislocations on other slip planes. Mills et al. [118] modeled the dislocation substructure as an array of elliptical loops, assuming that no drag force exists on the pure screw segments of the loops. Their model intended to explain transient 3-power creep behavior. Takeuchi and Argon [355] proposed a dislocation glide model based on the assumption that once dislocations are emitted from the source, they can readily disperse by climb and cross-slip, leading to a homogeneous dislocation distribution. They suggested that both glide and climb are controlled by solute drag. The final relationship is similar to that by Weertman. Mohamed and Langdon [73] derived the following relationship that is frequently referenced for 3-power-law viscous creep when only a Cottrell–Jaswon dragging mechanism is considered

$$\dot{\varepsilon}_{ss} \cong \frac{\pi(1-\nu)kT\tilde{D}}{6e^2 Cb^5 G}\left(\frac{\sigma}{G}\right)^3 \tag{132}$$

where e is the solute–solvent size difference, C is the concentration of solute atoms, and $\tilde{D}$ is the diffusion coefficient for the solute atoms, calculated using Darken's [357] analysis. Later, Mohamed [358] and Soliman et al. [359] suggested that Suzuki and Fischer interactions are necessary to accurately predict the 3-power-law creep behavior of several Al–Zn, Al–Ag, and Ni–Fe alloys.

Region II has been reported to occur preferentially in materials with a relatively large atom size mismatch [361,362]. Higher solute concentrations also favor the occurrence of 3-power-law creep [73,338,340,344]. As illustrated in Figure 65, for sufficiently high concentrations, region III can even be suppressed. The difference between the creep behaviors corresponding to Class I (A) and Class II (M) is evident in Figure 66 [349] where strain rate increases with time with the former and decreases with the latter. Others have observed even more pronounced primary creep features in Class I (A) Al–Mg [156,363]. Alloys with 0.6% and 1.1 at% Mg

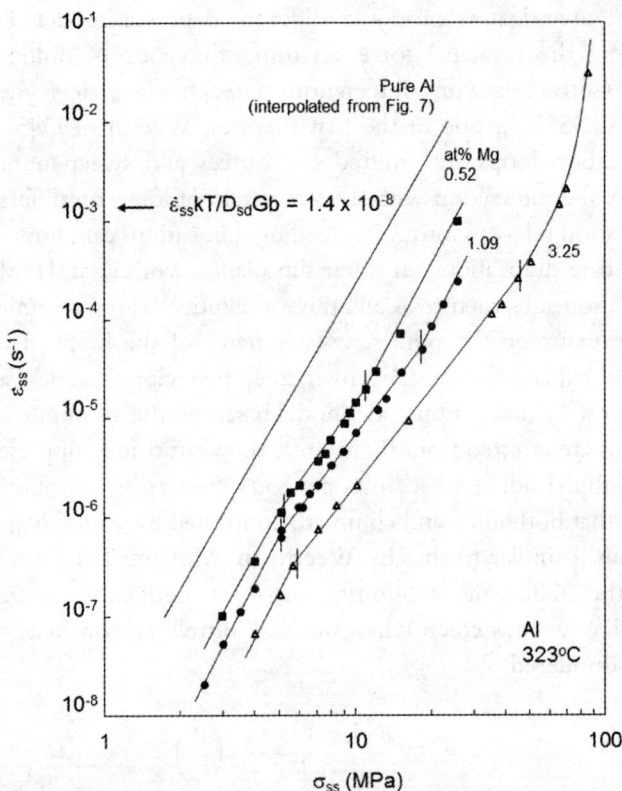

Figure 65 Steady-state creep rate versus applied stress for three Al–Mg alloys (Al–0.52 at% Mg, ▪; Al–1.09 at% Mg, •; Al–3.25 at% Mg, ▲) at 323 °C [356].

are Class II (M) alloys and those with 3.0, 5.1, and 6.9% are Class I (A) alloys. Additionally, inverse creep transient behavior is observed in Class I (A) alloys [118,349,364,365] and is illustrated in Figure 67 [349]. A drop in stress is followed by a decrease in the strain rate in pure aluminum, which then increases with a recovering dislocation substructure until steady state at the new, lower, stress. However, with a stress decrease in Class I (A) alloys (Figures 67(b) and (c)), the strain rate continually decreases until the new steady state. Analogous disparities are observed with stress increases (i.e., decreasing strain rate to steady state in Class II (M) while increasing rates with Class I (A)). Horiuchi et al. [349] argued that this is explained by the strain rate being proportional to the dislocation density and the dislocation velocity. The latter is proportional to the applied stress, while the square root of the former is proportional to the stress. With a stress drop, the dislocation velocity decreases to the value

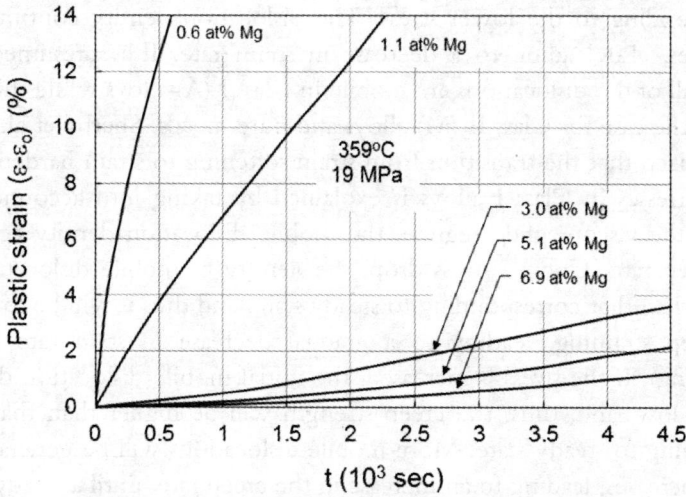

Figure 66 Creep behavior of several aluminum alloys with different magnesium concentrations 0.6 and 1.1 at% (Class II (M)) and 3.0, 5.1, and 6.9 at% (Class I (A)). The tests were performed at 359 °C and at a constant stress of 19 MPa [349].

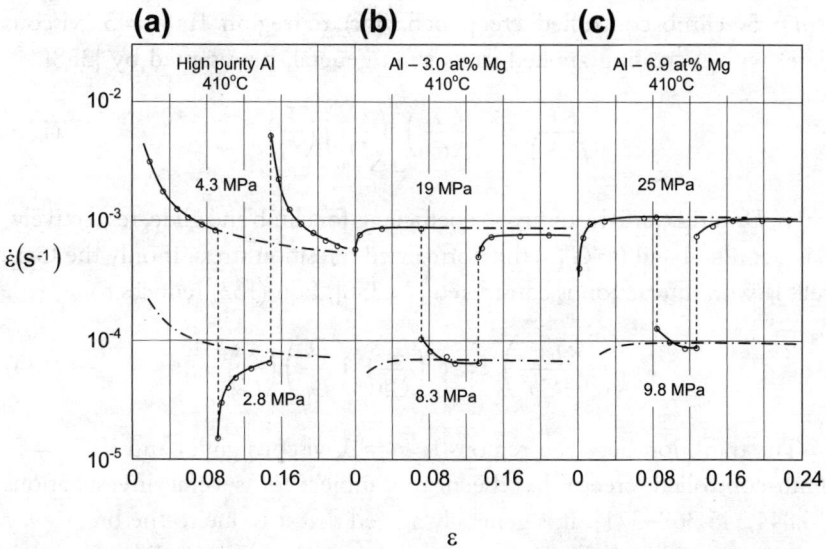

Figure 67 Effect of changes in the applied stress to the creep rate in (a) high-purity aluminum, (b) Al–3.0 at% Mg (Class I (A) alloy), and (c) Al–6.9 at% Mg (Class I (A) alloy), at 410 °C [349].

corresponding to the lower stress. The dislocation density continuously decreases, also leading to a decrease in strain rate. It is presumed that nearly all of the dislocations are mobile in Class I (A) alloys while this may not be the case for Class II (M) alloys and pure metals. Sherby et al. [366] emphasized that the transition from strain softening to strain hardening at lower stresses in Class I alloys is explained by taking into account that, within the viscous glide regime, the mobile dislocation density controls the creep rate. Upon a stress drop, the density of mobile dislocations is higher than that corresponding to steady state, and thus it will be lowered by creep straining, leading to a gradual decrease in stain rate (strain hardening). If the stress is increased, the initial mobile dislocation density will be low, and, thus, the creep strength will be higher than that corresponding to steady state. More mobile dislocations will be generated as strain increases, leading to an increase in the creep rate, until a steady-state structure is achieved (strain softening). The existence of internal stresses during 3-power-law creep is also not clearly established. Some investigators have reported internal stresses as high as 50% the applied stress [367]. Others, however, suggest that internal stresses are negligible compared to the applied stress [121,349].

The transitions between regions I and II and between regions II and III are now well established [361]. The condition for the transition from region I ($n = 5$, climb-controlled creep behavior) to region II ($n = 3$, viscous glide) with increasing applied stress is, in general, represented by [358]

$$\frac{kT}{D_g bA} = C\left(\frac{\chi}{Gb}\right)^3 \left(\frac{D_c}{D_g}\right)\left(\frac{\tau}{G}\right)_t^2 \qquad (133)$$

where D_c and D_g are the diffusion coefficients for climb and glide, respectively, C is a constant and $(\tau/G)_t$ is the normalized transition stress. If only the Cottrell–Jaswon interaction is considered [73,358], Eqn (133) reduces to

$$\left(\frac{kT}{ec^{1/2}Gb^3}\right)^2 = C\left(\frac{c}{Gb}\right)^3 \left(\frac{D_c}{D_g}\right)\left(\frac{\tau}{G}\right)_t^2 \qquad (134)$$

The transition between regions II ($n = 3$, viscous glide) and III ($n = 5$, climb-controlled creep) has been the subject of several investigations [36,344,345,368–371]. It is generally agreed that it is due to the breakaway of dislocations from solute atmospheres and are thus able to glide at a much faster velocity. The large difference in dislocation speed between dislocations with and without clouds has been measured experimentally [372].

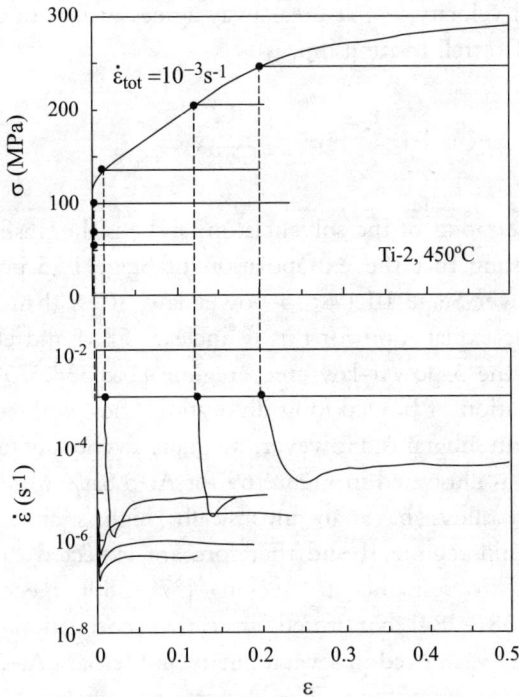

Figure 68 The σ versus ε and $\dot{\varepsilon}$ versus ε curves corresponding to Ti of commercial purity, creep tested at 450 °C at different stresses after initial loading at $10^{-3}\,s^{-1}$. The circles mark the end of loading/beginning of creep testing. *From Ref. [372].*

Figure 68 (from Ref. [372]) shows the σ versus ε and $\dot{\varepsilon}$ versus ε curves corresponding to Ti of commercial purity, creep tested at 450 °C at different stresses after initial loading at $10^{-3}\,s^{-1}$. Instead of a smooth transition to steady-state creep, a significant drop in strain rate occurs at the beginning of the creep test, which is associated with the significant decrease in dislocation speed due to the formation of solute clouds around dislocations. Thus, after a critical break-away stress, glide becomes faster than climb and the latter is, then, rate controlling in region III. Friedel [87] predicted the break-away stress for unsaturated dislocations as

$$\tau_b = A_{11}\left(\frac{W_m^2}{kTb^3}\right)C \tag{135}$$

where A_{11} is a constant, W_m is the maximum interaction energy between a solute atom and an edge dislocation, and C is the solute concentration. Endo et al. [342] showed, using modeling of mechanical experiments, that the critical velocity, v_{cr}, at breakaway agrees well with the value predicted by the Cottrell relationship, is

$$v_{cr} = \frac{DkT}{eGbR_S^3} \tag{136}$$

where R_S is the radius of the solvent atom and e is the misfit parameter.

It is interesting that the extrapolation of Stage I (5-power) predicts significantly lower Stage III (also 5-power law) stress than observed (see Figure 64). The explanation for this is unclear. TEM and etch pit observations within the 3-power-law creep regime [118,338,373–376] show a random distribution of bowed long dislocation lines, with only a sluggish tendency to form subgrains. However, subgrains eventually form in Al–Mg [121,156,377], as illustrated in Figure 69 for Al–5.8at% Mg.

Class I (A) alloys have an intrinsically high strain-rate sensitivity ($m = 0.33$) within regime II and therefore are expected to exhibit high elongations due to resistance to necking [378–380]. Recent studies by Taleff et al. [381–384] confirmed an earlier correlation between the extended ductility achieved in several binary and ternary Al–Mg alloys and their high strain-rate sensitivity. The elongations to failure for single phase Al–Mg can range from 100% to 400%, which is sufficient for many ("superplastic") manufacturing operations, such as the warm stamping of automotive body panels [384]. These elongations can be achieved at lower temperatures than those necessary for conventional superplasticity in the same alloys and still at reasonable strain rates. For example, enhanced ductility has been reported to occur at $10^{-2}\,s^{-1}$ in a coarse-grained Al–Mg alloy [384] at a temperature of 390 °C, whereas a temperature of 500 °C is necessary for superplasticity in the same alloy with a grain size ranging from 5 to 10 μm. The expensive grain refinement processing routes necessary to fabricate superplastic microstructures are unnecessary. The solute concentration in a binary Al–Mg alloy does not affect significantly mechanical properties such as tensile ductility, strain-rate sensitivity, or flow stress [385]. For example, under conditions of viscous-glide creep, variations in Mg concentration ranging from 2.8 to 5.5 wt% only change the strain-rate sensitivity from 0.29 to 0.32, which does not have a substantial effect on the elongation to failure [381]. McNelley et al. [377] attributed this observation to the saturation effect of Mg atoms in the core of the moving dislocation. However, ternary additions of Mn, Fe, and Zr seem to significantly affect the mechanical behavior of Al–Mg alloys [381]. The

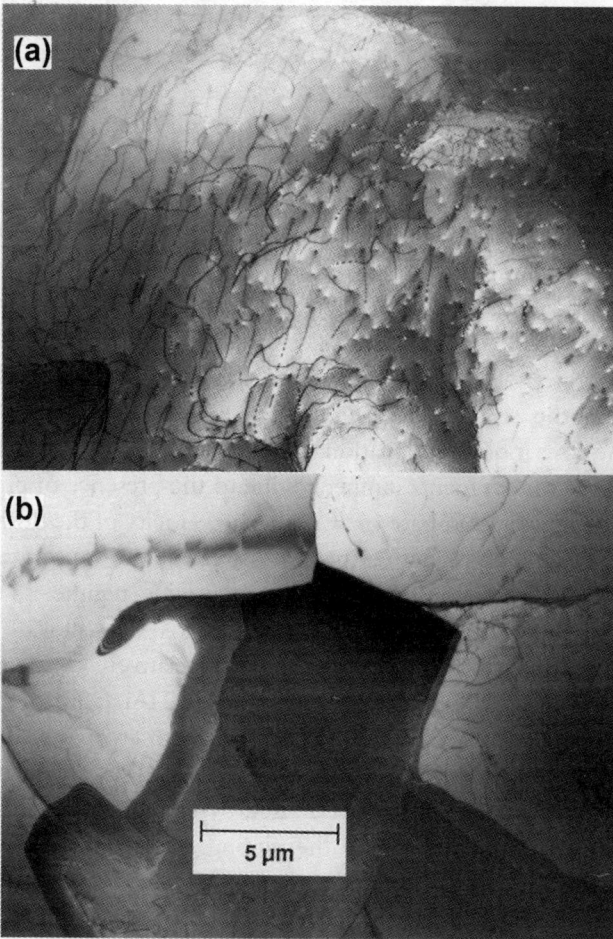

Figure 69 Al–5.8 at% Mg deformed in torsion at 425 °C to (a) 0.18 and (b) 1.1 strain in the 3-power regime.

stress exponent increases and ductility decreases significantly, especially for Mn concentrations higher than 0.46 wt%. Ternary additions above the solubility limit favor the formation of second-phase particles around which cavities tend to preferentially nucleate. Thus, a change in the failure mode from necking-controlled to cavity-controlled may occur, accompanied by a decrease in ductility [381]. Also, Mn atoms may interfere with the solute drag. The hydrostatic stress by which an atom interacts with a dislocation is determined by the volumetric size factor (Ω). The Ω values corresponding to Mg and Mn in Al are [362] $\Omega^{Al-Mg} = +40.82$ and $\Omega^{Al-Mn} = -46.81$. Both factors are nearly equal in magnitude and of opposite sign. Therefore, each added Mn atom acts as a sink for one atom of Mg, thus reducing the effective Mg concentration [381]. If Mn is added in sufficient quantities that the effective Mg concentration is lower than that required for viscous-drag creep, the stress exponent would increase and the ductility would, consequently, significantly decrease. This effect seems to be less important than the change in failure mode described earlier [381].

Class I behavior has been reported to occur in a large number of metallic alloys. These include Al–Mg- [16,73,378,386], Al–Zn- [361], Al–Cu- [387], Cu–Al- [388], Au–Ni- [371], Mg- [324, 385,389–391], Pb- [371], In- [371], and Nb- [392,393] based alloys. Viscous glide creep has also been observed in dispersion strengthened alloys. Sherby et al. [366] attributed the differences in the creep behavior between pure Al–Mg and DS Al–Mg alloys to the low mobile dislocation density of the latter. Dislocations are pinned and their movement is impeded due to the presence of precipitates. Therefore, at a given strain rate ($\dot{\gamma} = b\rho v$), the velocity of the dislocations is very high. Thus, much higher temperatures are required for solute atoms to form clouds around dislocations. The viscous glide regime is, therefore, observed at higher temperatures than in the pure Al–Mg alloys. A value of $n = 3$ has been observed in intermetallics with relatively coarse grain sizes ($g > 50$ μm), such as Ni_3Al [394], Ni_3Si [395], TiAl [396], Ti_3Al [397], Fe_3Al [398,399], and FeAl [400]. The mechanism of creep, here, is still not clear. Yang [402] argued that, since the glide of dislocations introduces disorder, the steady-state velocity is limited by the rate at which chemical diffusion can restore order behind the gliding dislocations. Finally, dislocation drag has also been attributed to lattice friction effects [401]. Viscous glide has also been reported to occur in metal matrix composites [403], although this is still controversial [404].

CHAPTER 6

Superplasticity

M.-T. Perez-Prado, M.E. Kassner

Contents

1. INTRODUCTION

Superplasticity is the ability of a polycrystalline material to exhibit, in a generally isotropic manner, very high tensile elongations prior to failure ($T > 0.5 \, T_m$) [405]. The first observations of this phenomenon were made as early as 1912 [406]. Since then, superplasticity has been extensively studied in metals. It is believed that both the arsenic bronzes used in Turkey during the Bronze Age (2500 BC) and the Damascus steels utilized from 300 BC to the end of the nineteenth century were already superplastic materials [407]. One of the most spectacular observations of superplasticity is perhaps that reported by Pearson in 1934 of a Bi–Sn alloy that underwent nearly 2000% elongation [408]. He also claimed, for the first time, that grain boundary sliding was the main deformation mechanism responsible for superplastic deformation. The interest in superplasticity has increased due to the recent observations of this phenomenon in a wide range of materials, including some materials (such as nanocrystalline materials [409],

Fundamentals of Creep in Metals and Alloys
ISBN 978-0-08-099427-7
http://dx.doi.org/10.1016/B978-0-08-099427-7.00006-2

ceramics [410,411,589], metal-matrix composites [412], and intermetallics [413]) that are difficult to form by conventional forming processes. Extensive reviews on superplasticity are available [414–417].

There are two types of superplastic behavior. The best known and studied, fine-structure superplasticity (FSS), will be briefly discussed in the following sections. The second type, internal stress superplasticity, refers to the development of internal stresses in certain materials, which then deform to large tensile-strains under relatively low externally applied stresses [417].

2. CHARACTERISTICS OF FSS

FSS materials generally exhibit a high strain-rate sensitivity exponent (m) during tensile deformation. Typically, m is larger than 0.33. Thus, n in Eqn (3), is usually smaller than three. In particular, the highest elongations have been reported to occur when $m \sim 0.5$ ($n \sim 2$) [417]. Superplasticity in conventional materials usually occurs at low strain rates ranging from 10^{-5} s^{-1} to 10^{-3} s^{-1}. However, it has been reported in recent work that large elongations to failure may also occur in selected materials at strain rates substantially higher than 10^{-2} s^{-1} [418]. This phenomenon, termed high-strain-rate superplasticity (HSRS), has been observed in some conventional metallic alloys, in metal-matrix composites, and in mechanically alloyed (MA) materials [419], among others. This will be discussed in Section 5. Very recently, HSRS has been observed in cast alloys prepared by ECAP (equal channel angular pressing) [420–423]. In this case, very high temperatures are not required and the grain size is very small (<1 μm). The activation energies for FSS tend to be low, close to the value for grain boundary diffusion, at intermediate temperatures. At high temperatures, however, the activation energy for superplastic flow is about equal to that for lattice self-diffusion.

The microscopic mechanism responsible for superplastic deformation is still not thoroughly understood. However, since Pearson's first observations [408], the most widely accepted mechanisms involve grain boundary sliding (GBS) [424–431]. GBS is generally modeled assuming sliding takes place by the movement of extrinsic dislocations along the grain boundary. This would account for the observation that the amount of sliding is variable from point to point along the grain boundary [432]. Dislocation pile-ups at grain boundary ledges or triple points may lead to stress concentrations. In order to avoid extensive cavity nucleation and growth, GBS must be aided by an accommodation mechanism [290]. The latter must ensure rearrangement of grains during deformation in order to achieve strain compatibility and relieve any stress concentrations resulting from GBS. The accommodation mechanism may include grain boundary migration, recrystallization, diffusional flow or slip. The accommodation process is generally believed to be the rate-controlling mechanism.

A large number of models have emerged in which the accommodation process is either diffusional flow or dislocation motion [433]. The best-known model for GBS accommodated by diffusional flow is depicted schematically in Figure 70, and was proposed by Ashby and Verral [434]. This model explains the experimentally observed switching of equiaxed grains throughout deformation. However, it fails to predict the stress dependence of the strain rate. According to this model:

$$\dot{\varepsilon}_{ss} = K_1 (b/g)^2 D_{eff}(\sigma - \sigma_{TH_s}/E) \tag{137}$$

where $D_{eff} = D_{sd}9[1 + (3.3 \ w/g)(D_{gb}/D_{sd})]$, K_1 is a constant, σ_{TH_s} is the threshold stress, and w is the grain boundary width. The threshold stress arises since there is an increase in boundary area during grain switching when clusters of grains move from the initial position (Figure 70(a)) to the intermediate one (Figure 70(b)).

Several criticisms of this model have been reported [435–440]. According to Spingarn and Nix [435], the grain rearrangement proposed by Ashby–Verral cannot occur purely by diffusional flow. The diffusion paths

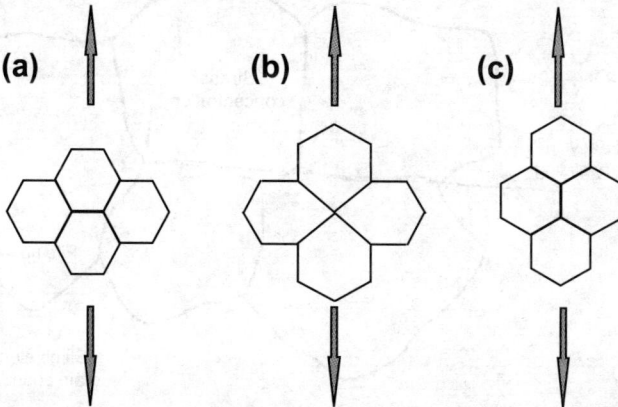

Figure 70 Ashby–Verral model of grain boundary sliding accommodated by diffusional flow [434].

are physically incorrect. The first models of GBS accommodated by diffusional creep were proposed by Ball and Hutchison [441], Langdon [442], and Mukherjee [443]. Among the most cited are those proposed by Mukherjee and Arieli [444] and Langdon [432]. According to these investigators, GBS involves the movement of dislocations along the grain boundaries, and the stress concentration at triple points is relieved by the generation and movement of dislocations within the grains (Figure 71). Figure 72 illustrates the model proposed by Gifkins [445], in which the accommodation process, which also consists of dislocation movement, only occurs in the "mantle" region of the grains, i.e., in the region close to the grain boundary. According to all of these GBS accommodated-by-slip models, $n = 2$ in a relationship such as:

$$\dot{\varepsilon}_{ss} = K_2(b/g)^{p'} D(\sigma/E)^2 \tag{138}$$

where $p' = 2$ or 3 depending on whether the dislocations move within the lattice or along the grain boundaries, respectively. K_2 is a constant, which varies with each of the models, and the diffusion coefficient, D, can be D_{sd} or D_{gb}, depending on whether the dislocations move within the lattice or along the grain boundaries to accommodate stress concentrations from GBS. In order to rationalize the increase in activation energy at high temperatures, Fukuyo et al. [446] proposed a model based on the GBS mechanism in which the dislocation accommodation process takes place by

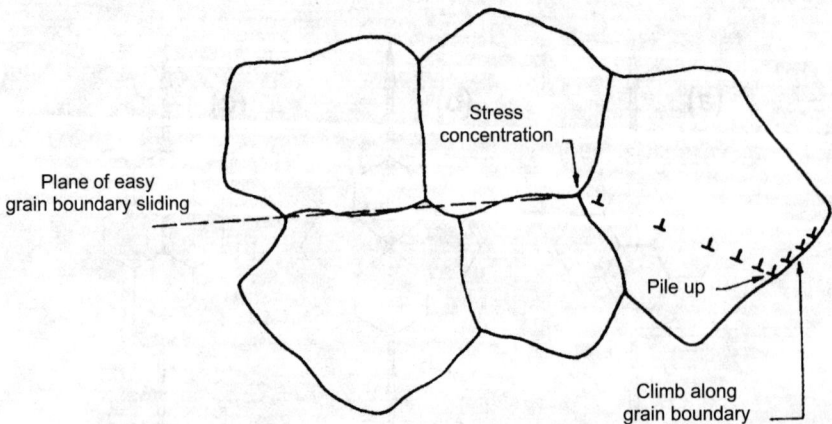

Figure 71 Ball–Hutchinson model of grain boundary sliding (GBS) accommodated by dislocation movement [441].

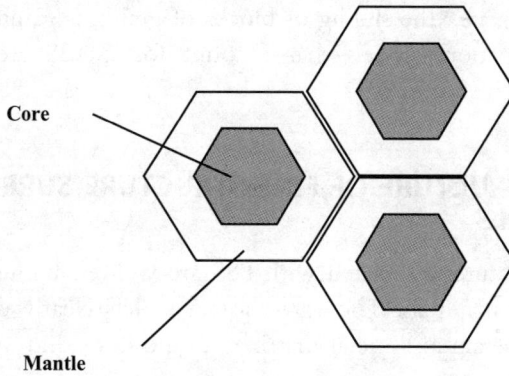

Core

Mantle

Figure 72 Gifkins "core and mantle" model [445].

sequential steps of climb and glide. At intermediate temperatures, climb along the grain boundaries is the rate-controlling mechanism due to the pile-up stresses. Pile-up stresses are absent and the glide of dislocations within the grain is the rate-controlling mechanism at high temperatures. It is believed that slip in superplasticity is accommodating and does not contribute to the total strain [447]. Thus, GBS is traditionally believed to account for all of the strain in superplasticity [448]. However, recent studies, based on texture analysis, indicate that slip may contribute to the total elongation [449–464].

The proposed mechanisms predict some behavior but have not suc-ceeded in fully predicting the dependence of the strain rate on σ, T, and g during superplastic deformation. Ruano and Sherby [465,466] formulated the following phenomenological equations, which appear to describe the experimental data from metallic materials:

$$\dot{\varepsilon}_{ss} = K_3 (b/g)^2 D_{sd} (\sigma/E)^2 \tag{139}$$

$$\dot{\varepsilon}_{ss} = K_4 (b/g)^3 D_{gb} (\sigma/E)^2 \tag{140}$$

where K_3 and K_4 are constants. These equations, with $n = 2$, correspond to a mechanism of GBS accommodated by dislocation movement. Equation (139) corresponds to an accommodation mechanism in which the disloca-tions would move within the grains (g^2) and Eqn (140) corresponds to an accommodation mechanism in which the dislocations would move along the grain boundaries (g^3). Only the sliding of individual grains has been considered. However, currently the concept of cooperative grain boundary sliding (CGBS), i.e., the sliding of blocks of grains, is gaining acceptance. Several deformation models that account for CGBS are described in Ref. [420].

3. MICROSTRUCTURE OF FINE-STRUCTURE SUPERPLASTIC MATERIALS

The microstructures associated with FSS are well established for conventional metallic materials. They are, however, less clearly defined for intermetallics, ceramics, metal-matrix composites, and nanocrystalline materials.

3.1 Grain Size and Shape

GBS in metals is favored by the presence of equiaxed small grains that should generally be smaller than 10 μm. Consistent with Eqns (137)–(140), the strain rate is usually inversely proportional to grain size, according to:

$$\dot{\varepsilon}_{ss} = K_5 g^{-p'} \tag{141}$$

where $p' = 2$ or 3 depending, perhaps, on the accommodation mechanism, and K_5 is a constant. Also, for a given strain rate, the stress decreases as grain size decreases. Grain size refinement is achieved during the thermomechanical processing by successive stages of warm and cold rolling [465–470]. However, the present understanding of microstructural control in engineering alloys during industrial processing by deformation and recrystallization is still largely empirical.

3.2 Presence of a Second Phase

The presence of small second-phase particles uniformly distributed in the matrix prevents rapid grain growth that can occur in single-phase materials within the temperature range over which superplasticity is observed.

3.3 Nature and Properties of Grain Boundaries

GBS is favored along disordered high-angle (not CSL, or coincident lattice site) boundaries. Additionally, sliding is influenced by the grain boundary composition. For example, a heterophase boundary (i.e., a boundary that separates grains with different chemical composition) slides more readily than a homophase boundary. Stress concentrations develop at triple points and at other obstacles along the grain boundaries as a consequence of GBS.

Mobile grain boundaries may assist in relieving these stresses. Grain boundaries in the matrix phase should not be prone to tensile separation.

4. TEXTURE STUDIES IN SUPERPLASTICITY

Texture analysis has been utilized to further study the mechanisms of superplasticity [432,440–464,471], using both X-ray texture analysis and computer-aided EBSP techniques [472]. Commonly, GBS, involving grain rotation, is associated with a decrease in texture [416], whereas crystallographic slip leads to the stabilization of certain preferred orientations, depending on the number of slip systems that are operating [473,474].

A large number of investigations based on texture analysis have led to the conclusion that crystallographic slip (CS) is important in superplastic deformation. According to these studies, CS is not merely an accommodation mechanism for GBS but also operates in direct response to the applied stress. Some investigators [450–455,471] affirm that both GBS and CS coexist at all stages of deformation; other investigators [456–458] conclude that CS only operates during the early stages of deformation, leading to a microstructure favoring GBS. Others [459–464] even suggest that CS is the principal deformation mechanism responsible for superplastic deformation.

5. HIGH-STRAIN-RATE SUPERPLASTICITY

HSRS has been defined by the Japanese Standards Association as superplasticity at strain rates equal to or greater than 10^{-2} s^{-1} [417,475,476]. This field has generated considerable interest in the last 25 years since these high strain rates are close to the ones used for commercial applications (10^{-2}–10^{-1} s^{-1}). Higher strain rates can be achieved by reducing the grain size (see Eqn (138)) or by engineering the nature of the interfaces in order to make them more suitable for sliding [475,476]. HSRS was first observed in a 20% SiC whisker-reinforced 2124 Al composite [418]. Since then, it has been observed in several metal-matrix composites, MA materials, conventional alloys that undergo continuous reactions (or continuous dynamic recrystallization), and alloys processed by power consolidation, by physical vapor deposition, and by intense plastic straining [423] (for example, ECAP, high-pressure torsion, or by friction stir processing [477]). The details of the microscopic mechanism responsible for HSRS are not yet well understood, but some recent theories are reviewed next.

5.1 HSRS in Metal-Matrix Composites

HSRS has been achieved in a large number of metal–matrix composites. Some of them are listed in Table 4, and more complete lists can be found elsewhere [475,478]. The microscopic mechanism responsible for HSRS in metal–matrix composites is still a matter of controversy. Any theory must account for several common features of the mechanical behavior of metal–matrix composites that undergo HSRS, such as [487]:

1. Maximum elongations are achieved at very high temperatures, sometimes even slightly higher than the incipient melting point.
2. The strain-rate sensitivity exponent changes at such high temperatures from ~ 0.1 ($n \sim 10$) (low strain rates) to ~ 0.3 ($n \sim 3$).
3. High apparent activation energy values are observed. Values of 920 kJ mol^{-1} and 218 kJ mol^{-1} have been calculated for SiC$_w$/2124 Al at low and high strain rates, respectively. These values are significantly higher than the activation energy for self-diffusion in Al (140 kJ mol^{-1}).

Both grain boundary sliding and interfacial sliding have been proposed as the mechanisms responsible for HSRS. The significant contribution of interfacial sliding is evidenced by extensive fiber pull-out that is apparent on the fracture surfaces [487]. However, an accommodation mechanism has to operate simultaneously in order to avoid cavitation at such high strain rates. The nature of this accommodation mechanism, which enables the boundary and interface mobility, is still uncertain.

Table 4 Superplastic characteristics of some metal-matrix-composites exhibiting high-strain-rate superplasticity

Material	Temperature (°C)	Strain rate (s^{-1})	Elongation (%)	References
SiC$_w$/2124 Al	525	0.3	~ 300	[418]
SiC$_w$/2024 Al	450	1	150	[479]
SiC$_w$/6061 Al	550	0.2	300	[480]
SiC$_p$/7075 Al	520	5	300	[481]
SiC$_p$/6061 Al	580	0.1	350	[482]
Si$_3$N$_{4w}$/6061 Al	545	0.5	450	[483]
Si$_3$N$_{4w}$/2124 Al	525	0.2	250	[484]
Si$_3$N$_{4w}$/5052 Al	545	1	700	[485]
AlN/6061 Al	600	0.5	350	[486]

w = whisker; p = particle.

A fine matrix grain size is necessary but not sufficient to explain HSRS. In fact, HSRS may or may not appear in two composites having the same fine-grained matrix and different reinforcements. For example, it has been found that a 6061 Al matrix with β-Si_3N_4 whiskers experiences HSRS, whereas the same matrix with β-SiC does not [487]. The nature, size, and distribution of the reinforcement are critical to the onset of HSRS.

5.1.1 Accommodation by a Liquid Phase: Rheological Model

Nieh and Wadsworth [487] have proposed that the presence of a liquid phase at the matrix–reinforcement interface and at grain boundaries within the matrix is responsible for accommodation of interface sliding during HSRS and thus for strain-rate enhancement. The presence of this liquid phase would be responsible for the observed high activation energies. A small grain size would favor HSRS, since the liquid phase would then be distributed along a larger surface area and thus can have a higher capillarity effect, preventing decohesion. The occurrence of partial melting even during tests at temperatures slightly below solvus has been explained in two different ways. First, as a consequence of solute segregation, a low melting point region could be created at the matrix–reinforcement interfaces. Alternatively, local adiabatic heating at the high strain rates used could contribute to a temperature rise that may lead to local melting.

It has been suggested [488] that HSRS with the aid of a liquid phase can be modeled in rheological terms in a similar way to semi-solid metal forming. A fluid containing a suspension of particles behaves like a non-Newtonian fluid, for which the strain-rate sensitivity and the shear strain rate are related by:

$$\tau = K_7 \cdot \dot{\gamma}^m \tag{142}$$

where τ is the shear stress, and K_7 and m are both material constants, m being the strain-rate sensitivity of the material. The shear stress and strain rate of a semi-solid that behave like a non-Newtonian fluid are related to the shear viscosity by the following equations:

$$\eta = K_7 \cdot \dot{\gamma}^{-u} \tag{143}$$

$$\eta = \tau/\dot{\gamma} \tag{144}$$

where η (is the shear viscosity and u is a material constant, related to the strain-rate sensitivity by the expression $m = 1 - u$. The viscosity of several

Al-6.5% Si metal–matrix composites was measured experimentally [417] at 700 °C as a function of shear rate. High-strain-rate sensitivity values, similar to those reported for MMCs ($\sim$0.3–0.5) in the HSRS regime, were obtained at very high shear strain rates ($200-1000$ s^{-1}). These data support the rheological model. The temperature used, however, is higher than the temperatures at which HSRS was observed.

The role of a liquid phase as an accommodation mechanism for interfacial and grain boundary sliding has been supported by other investigators [478,489–493] as well. It is suggested that the liquid phase acts as an accommodation mechanism, relieving stresses originated by sliding and thus preventing cavity formation. However, in order to avoid decohesion, it is emphasized that the liquid phase must either be distributed discontinuously or be present in the form of a thin layer. The optimum amount of liquid phase may depend on the nature of the grain boundary or interface. Direct evidence of local melting at the reinforcement–matrix interface was obtained using in situ transmission electron microscopy by Koike et al. [491] in a Si_3N_{4p}/6061 Al. The rheological model was criticized by Mabuchi et al. [489], arguing that testing the material at a temperature within the solid–liquid region is not sufficient to achieve HSRS. For example, an unreinforced 2124 alloy fails to exhibit high tensile ductility when tested at a temperature above solvus.

5.1.2 Accommodation by Interfacial Diffusion

Mishra et al. [494–496] rationalized the mechanical behavior of HSRS metal-matrix composites by taking into account the presence of a threshold stress. This analysis led them to conclude that the mechanism responsible for HSRS in metal-matrix composites is grain boundary sliding accommodated by interfacial diffusion along matrix–reinforcement interfaces. It is important to note that the particle size is often comparable to the grain size, and therefore interfacial sliding is geometrically necessary, as illustrated in Figure 73. Partial melting, especially if it is confined to triple points, may be beneficial for superplastic deformation, but it is not necessary to account for the superplastic elongations observed.

Threshold stresses are often used to explain the variation of the strain-rate sensitivity exponent with strain rate in creep studies. The presence of a threshold stress would explain the transition to a lower strain-rate sensitivity value (and thus to a higher n) at low strain rates that occur during HSRS in metal-matrix composites. Calculating threshold stresses and a (true) stress exponent, n_{hsrs}, that describes the predominant deformation

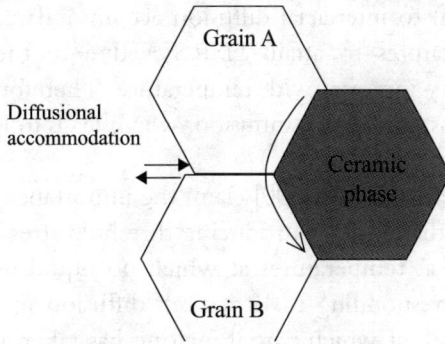

Figure 73 Interfacial diffusion controlled grain boundary sliding. The ceramic phase would not allow slip accommodation.

mechanism, is a nontrivial process, as explained in [495]. Mishra et al. [495,496] concluded that a true-stress exponent of 2 would give the best fit for their data, suggesting the predominance of grain boundary sliding as a deformation mechanism responsible for HSRS in metal-matrix composites. Additionally, activation energies (Q_{hsrs}) of the order of 300 kJ mol^{-1} were obtained from this analysis. Both dependencies ($n_{hsrs} = 2$ and $Q_{hsrs} \cong 300$ kJ mol^{-1}) are best predicted by Arzt's model for "interfacial diffusion-controlled diffusional creep" [497]. Mishra et al. [498] suggested that both diffusional creep and grain boundary sliding are induced by the movement of grain boundary dislocations, and the atomic processes involved are similar for both processes. Therefore it is reasonable to conclude that n and Q would be similar in both interfacial diffusion-controlled diffusional creep and interfacial diffusion-controlled super-plasticity. Thus, the latter is invoked to be responsible for HSRS. Figure 73 illustrates this deformation mechanism.

The phenomenological constitutive equation proposed by Mishra et al. for HSRS in metal-matrix-composites is:

$$\dot{\varepsilon}_{ss} = A_{12} \frac{D_i G b}{kT} \left(\frac{b^2}{g_m g_p} \right) \left(\frac{\sigma - \sigma_{TH_{hsrs}}}{E} \right)^2 \tag{145}$$

where D_i is the coefficient for interfacial diffusion, g_m is the matrix grain size, g_p is the particle/reinforcement size, $\sigma_{TH_{hsrs}}$ is the threshold stress for HSRS, and A_{12} is a material constant. An inverse grain size and reinforcement size dependence is suggested.

According to this model [496], as temperature rises, the accommodation mechanism would change from slip accommodation (at temperatures lower than the optimum) to interfacial diffusion accommodation. The need for very high temperatures to attain HSRS is due to the fact that grain boundary diffusivity increases with temperature. Therefore, the higher the temperature, the faster interface diffusion, which leads to less cavitation and thus higher ductility.

Mabuchi and Higashi [499,500] claim the importance of a liquid phase in HSRS arguing that, when introducing threshold stresses, the activation energy for HSRS at temperatures at which no liquid phase is present is similar to that corresponding to lattice self-diffusion in Al. However, at higher temperatures, at which partial melting has taken place, the activation energy increases dramatically. It is at these temperatures that the highest elongations are observed. The origin of the threshold stress for superplasticity is not well known. Its magnitude depends on the shape and size of the reinforcement and it generally decreases with increasing temperature.

5.1.3 Accommodation by Grain Boundary Diffusion in the Matrix: The Role of Load Transfer

The two theories described above were critically examined by Li and Langdon [501–503]. First, the rheological model was questioned, since HSRS had been recently found in Mg–Zn metal-matrix composites at temperatures below the incipient melting point, where no liquid phase is present [504]. Second, Li and Langdon [501] claim that it is hard to estimate interfacial diffusion coefficients at ceramic–matrix interfaces, and therefore validation of the interfacial diffusion-controlled grain boundary sliding mechanism is difficult. These investigators used an alternative method for computing threshold stresses, described in detail in Ref. [501], which does not require an initial assumption of the value of n. This methodology also rendered a true-stress exponent of 2 and true activation energy values that were higher than those for matrix lattice self-diffusion and grain boundary self-diffusion. These results were explained by the occurrence of a transfer of load from the matrix to the reinforcement. Following this approach, which was used before to rationalize creep behavior in metal-matrix composites [505], a temperature-dependent load-transfer coefficient α' was incorporated in the constitutive equation:

$$\dot{\varepsilon}_{ss} = \frac{A''''DGb}{kT}\left(\frac{b}{g}\right)^{p'}\left[\frac{(1-\alpha')(\sigma - \sigma_{TH_{hsrs}})}{G}\right]^{n} \tag{146}$$

where A'''' is a dimensionless constant. In their calculations, Li and Langdon assumed that D is equal to D_{gb} and the remaining constants and variables have the usual meaning. Load-transfer coefficients are expected to vary between zero (no load transfer) and one (all of the load is transferred to the reinforcement). It was found that the load-transfer coefficients obtained decreased with increasing temperature, becoming zero at temperatures very close to the incipient melting point. This indicates that load transfer would be inefficient in the presence of a liquid phase. The effective activation energies $Q^\star$ calculated by introducing the load–transfer coefficient into the rate equation for flow are similar to those corresponding to grain boundary diffusion within the matrix alloys (until up to a few degrees from the incipient melting point). Therefore, Li and Langdon proposed that the mechanism responsible for HSRS is grain boundary sliding controlled by grain boundary diffusion in the matrix. This mechanism, which is characteristic of conventional superplasticity at high temperatures, would be valid up to temperatures close to the incipient melting point.

The origin of the threshold stress is still uncertain. It has been shown that it decreases with increasing temperature, and that it depends on the shape and size of the reinforcement [498]. The temperature-dependence of the threshold stress may be expressed by an Arrhenius-type equation of the form:

$$\frac{\sigma_{TH_{hsrs}}}{G} = B \exp\left(\frac{Q_{TH_{hsrs}}}{RT}\right) \tag{147}$$

where $\sigma_{TH_{hsrs}}$ is the threshold stress for HSRS, B is a constant, and $Q_{TH_{hsrs}}$ is an energy term that seems to be associated with the process by which the mobile dislocations surpass the obstacles in the glide planes. (The threshold stress concept will be discussed again in Chapter 8.)

Li and Langdon [503] claim that the threshold stress values obtained in metal-matrix composites tested under HSRS and under creep conditions may have the same origin. They showed that similar values of $Q_{TH_{hsrs}}$ are obtained under these two conditions when, in addition to load transfer, substructure strengthening is introduced into the rate equation for flow. Substructure strengthening may arise, for example, from an increase in the dislocation density due to the thermal mismatch between the matrix and the reinforcement or to the resistance of the reinforcement to plastic flow. The "effective stress" acting on the composite in the presence of load-transfer and substructure strengthening is given by:

$$\sigma_e = (1 - \Phi)\sigma - \sigma_{TH_{hsrs}} \qquad (148)$$

where Φ is a temperature-dependent coefficient. At the low temperatures at which creep tests are performed, the value of Φ may be negligible, but since HSRS takes place at very high temperatures, often close to the melting point, the temperature dependence of Φ must be taken into account to obtain accurate values of $Q_{TH_{hsrs}}$. In fact, when the temperature dependence of Φ is considered, $Q_{TH_{hsrs}}$ values close to 20–30 kJ mol^{-1}, typical of creep deformation of MMCs, are obtained under HSRS conditions.

5.2 HSRS in MA Materials

HSRS has also been observed in some MA materials that are listed in Table 5.

As can be observed in Table 5, MA materials attain superplastic elongations at higher strain rates than metal-matrix composites. Such high strain rates are often attributed to the presence of a very fine microstructure (with average grain size of about 0.5 μm) and oxide and carbide dispersions approximately 30 nm in diameter that have an interparticle spacing of about 60 nm [417]. These particle dispersions impart stability to the microstructure. The strain-rate sensitivity exponent (m) increases with temperature, reaching values usually higher than 0.3 at the temperatures where the highest elongations are observed. Optimum superplastic elongations are often obtained at temperatures above solvus.

Table 5 Superplastic properties of some mechanically alloyed materials

Material	Temperature (°C)	Strain rate (s^{-1})	Elongation (%)	References
IN9021	450	0.7	300	[506]
IN90211	475	2.5	505	[507,508]
IN9052	590	10	330	[509]
IN905XL	575	20	190	[510]
SiC/ IN9021	550	50	1250	[510]
MA754	1100	0.1	200	[511]
MA6000	1000	0.5	308	[511]

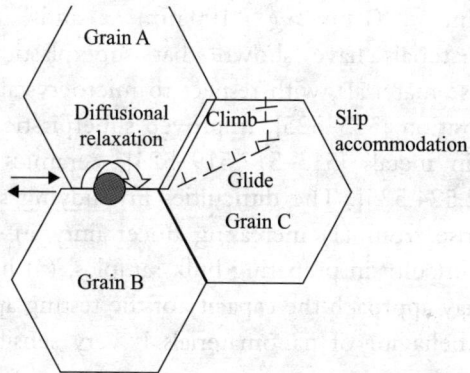

Figure 74 Grain boundary sliding accommodated by boundary-diffusion controlled dislocation slip.

After introducing a threshold stress, $n = 2$ and the activation energy is equal to that corresponding to grain boundary diffusion. These values are similar to those obtained for conventional superplasticity and would indicate that the main deformation mechanism is grain boundary sliding accommodated by dislocation slip. The rate-controlling mechanism would be grain boundary diffusion [498,501,512]. Mishra et al. [498] claim that the small size of the precipitates allows for diffusion relaxation of the stresses at the particles by grain boundary sliding, as illustrated in Figure 74. Higashi et al. [512] emphasized the importance of the presence of a small amount of liquid phase at the interfaces that contributes to stress relaxation and thus enhanced superplastic properties at temperatures above solvus. Li and Langdon [501] state that, given the small size of the particles, no load transfer takes place, and thus the values obtained for the activation energy after introducing a threshold stress are the true activation energies. According to Li and Langdon, the same mechanism (GBS rate controlled by grain boundary diffusion) predominates during HSRS in both metal-matrix composites and MA materials.

6. SUPERPLASTICITY IN NANOCRYSTALLINE AND SUBMICROCRYSTALLINE MATERIALS

The development of grain-size reduction techniques in order to produce microstructures capable of achieving superplasticity at high strain rates and low temperatures has been the focus of significant research in recent years [423,513–517]. Some investigations on the mechanical behavior of

submicrocrystalline   (1 µm > g > 100 nm)   and   nanocrystalline
(g < 100 nm) materials have shown that superplastic properties are
enhanced in these materials, with respect to microcrystalline materials of
the same composition [513–525]. Improved superplastic properties have
been reported in metals [513–517,519–524], ceramics [518], and intermetallics [522,524,525]. The difficulties in studying superplasticity in
nanomaterials arise from (1) increasing uncertainty in grain size measurements, (2) difficulty in preparing bulk samples, (3) high flow stresses
may arise that may approach the capacity of the testing apparatus, and (4)
the mechanical behavior of nanomaterials is very sensitive to the processing details.

The microscopic mechanisms responsible for superplasticity in nanocrystalline and submicrocrystalline materials are still not well understood.
Together with superior superplastic properties, significant work hardening
and flow stresses larger than those corresponding to coarser microstructures
have often been observed [521–523]. Figure 75 shows the stress–strain
curves corresponding to Ni$_3$Al deformed at 650 °C and 725 °C at a strain
rate of 1×10^{-3} s^{-1} (Figure 75(a)) and to Al-1420 deformed at 300 °C at
1×10^{-2} s^{-1}, 1×10^{-1} s^{-1}, and 5×10^{-1} s^{-1} (Figure 75(b)). It is observed
in Figure 75(a) that nanocrystalline Ni$_3$Al deforms superplastically at tem-
peratures that are more than 400 °C lower than those corresponding to the
microcrystalline material [527]. The peak flow stress, which reaches 1.5 GPa
at 650 °C, is the highest flow stress ever reported for Ni$_3$Al. Significant strain
hardening can be observed. In the same way, Figure 75(b) shows that the
alloy Al-1420 undergoes superplastic deformation at temperatures about
150 °C lower and at strain rates several orders of magnitude higher
(1×10^{-1} s^{-1} vs 4×10^{-4} s^{-1}) than the microcrystalline material [528].
High flow stresses and considerable strain hardening are also apparent.

The origin of these anomalies is still unknown. Mishra et al. [523,526]
attributed the presence of high flow stresses to the difficulty in slip
accommodation in nanocrystalline grains. Islamgaliev et al. [524] support
this argument. The difficulty of dislocation motion in nanomaterials
has also been previously reported in [529]. The stress necessary to generate
the dislocations responsible for dislocation accommodation is given
by [523]:

$$\tau = \frac{Gb}{4\pi\lambda(1 - \nu)} \left(\ln\left(\frac{\lambda_{\text{p}}}{b}\right) - 1.67 \right)$$ (149)

(a)

(b)

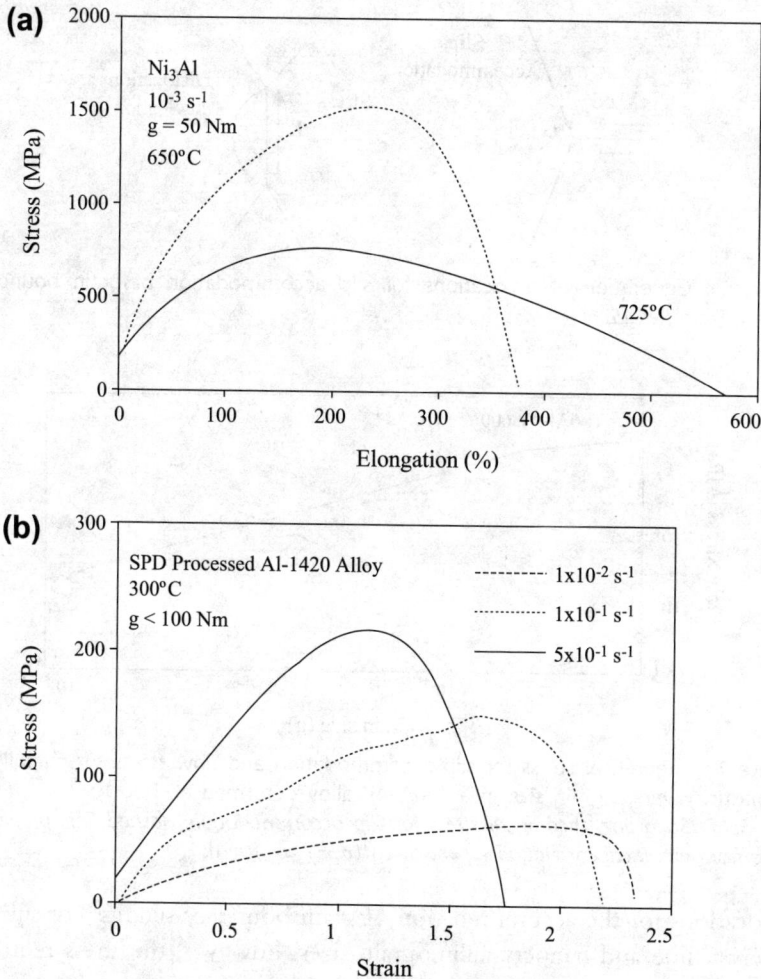

Figure 75 Stress–strain curves corresponding to (a) Ni$_3$Al deformed at 650 °C (dotted line) and 725 °C (full line) at a strain rate of 1×10^{-3} s^{-1} and (b) to Al-1420 deformed at 300 °C at 1×10^{-2} s^{-1}, 1×10^{-1} s^{-1}, and 5×10^{-1} s^{-1}. *From Refs [524] and [526].*

where λ_p is the distance between the pinning points and τ is the shear stress required to generate the dislocations (see Figure 76). Figure 77 is a plot showing the variation with grain size of the stress calculated from Eqn (149) and the flow stress required for overall superplastic deformation obtained from Eqn (138) assuming the main deformation mechanism is GBS accommodated by lattice–diffusion controlled slip. It can be observed that, for coarser grain sizes, the flow stress is high enough to generate

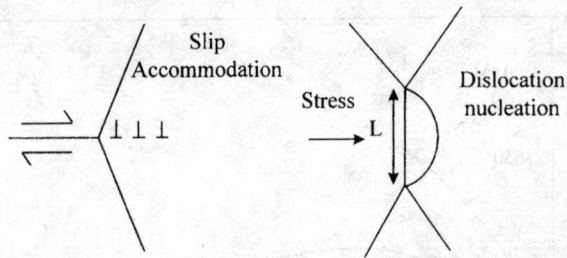

Figure 76 Generation of dislocations for slip accommodation of grain boundary sliding. *From Ref. [526].*

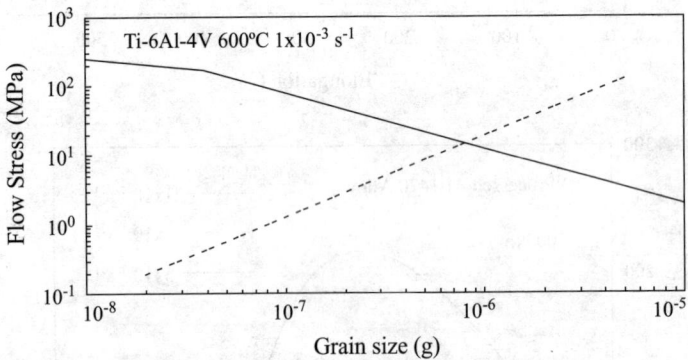

Figure 77 Theoretical stress for slip accommodation and flow stress for overall superplasticity versus grain size in a Ti-6Al-4V alloy deformed at 1×10^{-3} s^{-1}. *From Ref. [526]. (Solid line, theoretical stress for slip accommodation; dashed line, predicted stress from empirical correlation $\dot{\varepsilon}_{ss} = 5 \times 10^9 (\sigma/E)^2 (D_{sd}/g^2)$).*

dislocations for the accommodation of grain boundary sliding. For submicrocrystalline and nanocrystalline grain sizes, however, the stress required for slip accommodation is higher than the overall flow stress. This is still a rough approximation to the problem, since Eqn (149) does not include strain–rate dependence, temperature dependence other than the modulus, as well as the details for dislocation generation from grain boundaries. However, Mishra et al. use this argument to emphasize that the microcrystalline behavior apparently cannot be extrapolated to nanomaterials. Instead, there may be a transition between both kinds of behavior. The large strain hardening found during superplasticity of nanocrystalline materials has still not been thoroughly explained.

A classification of nanomaterials according to the processing route has been made by the same investigators [523,526]. Nanomaterials processed by mechanical deformation (such as ECAP) are denoted by "D" and nanomaterials processed by sintering of powders are denoted by "S." A large amount of dislocations are already generated during D processing that can contribute to deformation by an "exhaustion plasticity" mechanism. Thus, at the initial stage of deformation, the applied stress is not sufficient to generate new dislocations for slip accommodation; rather, existing dislocations move. As the easy paths of grain boundary sliding become exhausted, the flow stress increases until it is high enough to generate new dislocations. S nanomaterials may not be suitable for obtaining large tensile-strains, due to the absence of preexisting dislocations.

A significant amount of grain growth occurs during deformation even with superplasticity at lower temperatures. In fact, the transition from low plasticity to superplasticity in nanomaterials is often accompanied by the onset of grain growth. This seems unavoidable, since both grain growth and grain boundary sliding are thermally activated processes. It has been found that a reduction of the superplastic temperature is usually offset by a reduction of grain growth temperature [522]. As the grain size decreases, the surface area of grain boundaries increases, and thus the reduction of grain boundary energy emerges as a new driving force for grain growth. This force is much less significant for coarser grain sizes (which, in turn, render higher superplastic temperatures).

Tensile-loading molecular dynamic simulations on nanocrystalline SiC suggests superplastic behavior [783] through forming a thin amorphous layer at the grain boundaries.

CHAPTER 7

Recrystallization

Contents

1. INTRODUCTION

The earlier chapters have described creep as a process where dislocation hardening is accompanied by dynamic recovery. It should be discussed at this point that dynamic recovery is not the only (dynamic) restoration mechanism that may occur with dislocation hardening. Recrystallization can also occur and this process can also "restore" the metal and reduce the flow stress. Often, recrystallization during deformation (dynamic recrystallization (DRX)) is observed at relatively high strain rates which is outside the common creep realm. However, any complete discussion of elevated temperature creep, and particularly, a discussion of high temperature plasticity must include this restoration mechanism. An understanding of the hot working (high strain rates and high temperatures) requires an appreciation of both dynamic recovery and recrystallization processes. Some definitions are probably useful, and we will use those definitions adopted by Doherty et al. [222].

During deformation, energy is stored in the material mainly in the form of dislocations. This energy is released in three main processes, those of recovery, recrystallization, and grain coarsening (subsequent to recrystallization). The usual definition of recrystallization [222] is the formation and migration of high-angle boundaries, driven by the stored energy of deformation. The definition of recovery includes all processes releasing stored energy that do not require the movement of high-angle boundaries. In the context of the processes discussed, creep is deformation accompanied only by dynamic recovery. Typical recovery processes involve the

Fundamentals of Creep in Metals and Alloys
ISBN 978-0-08-099427-7
http://dx.doi.org/10.1016/B978-0-08-099427-7.00007-4

rearrangement of dislocations to lower their energy, for example by the formation of low-angle subgrain boundaries, and annihilation of dislocation line length in the subgrain interior, such as by Frank network coarsening. Grain coarsening is the growth of the mean grain size driven by the reduction in grain boundary area.

It is now recognized that recrystallization is not a Gibbs I transformation that occurs by classic nucleation and growth process as described by Turnbull [530] and Christian [531]. $\Delta G^{\star}$ and $r^{\star}$, the critical Gibb's free energy and critical-sized embryos, are unrealistically large if the proper thermodynamic variables are used. As a result of this disagreement, it is now universally accepted [532], as first proposed by Cahn [533], that the new grains do not nucleate as totally new grains by the atom by atom construction assumed in the classic kinetic models. Rather, new grains grow from small regions, such as subgrains, that are already present in the deformed microstructure. Special grains do not have to form. These embryos are present in the starting structure. Only subgrains with a high misorientation angle to the adjacent deformed material appear to have the necessary mobility to evolve into new, recrystallized, grains. Typical nucleation sites include pre-existing high-angle boundaries, shear bands, and highly misoriented deformation zones around hard particles. Misoriented "transition" bands (or geometric necessary boundaries (GNBs)) inside grains are a result of different parts of the grain having undergone different lattice rotations due to different slip systems being activated and can also be included as nucleation sites. Figure 78 (from Ref. [222]) illustrates an example of recrystallization in 40% compressed pure aluminum. New grains 3 and 17 are only growing into the deformed regions A and B, respectively, with which they are strongly misoriented and not into regions with which they share a common misorientation; 17 has a low-angle misorientation with A and 3 with B. It should be mentioned that recrystallization often leads to a characteristic texture(s), usually different that any texture developed as a consequence of the prior deformation that is the driving force for any recrystallization.

2. DISCONTINUOUS DRX

Recrystallization can occur under two broad conditions: static and dynamic. Basically, static occurs in the absence of plasticity during the recrystallization. The most common case for static is heating cold-worked metal leading to a recrystallized microstructure. DRX occurs with

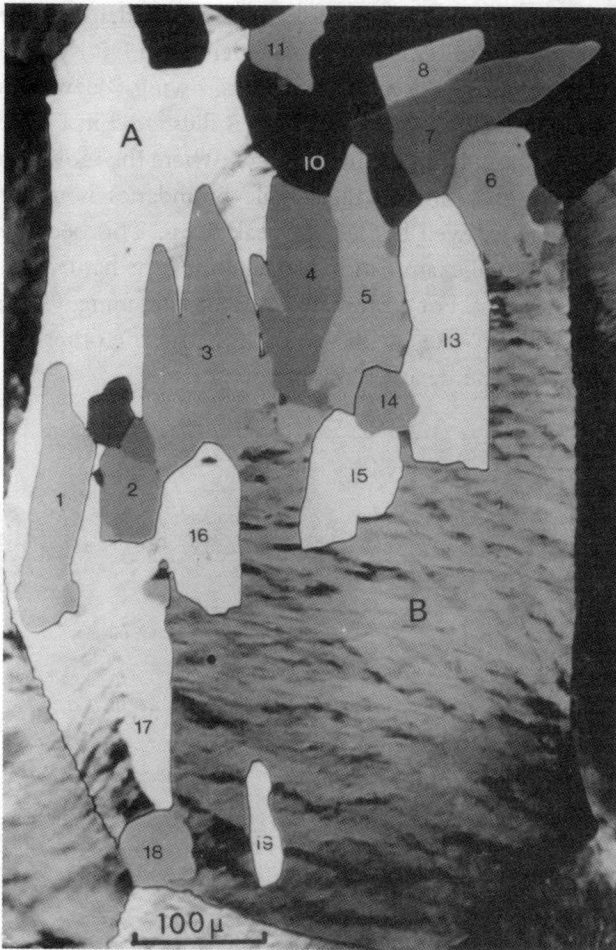

Figure 78 Static recrystallization in aluminum cold worked 40%. A large grain has fragmented into two regions, A and B. *From Ref. [222].*

concomitant plasticity. This distinction is complicated, somewhat, by the more recent suggestion of meta-dynamic recrystallization (MDRX) [534] that can follow DRX, generally at elevated temperature. Although it occurs without external plasticity, it can occur, quickly. It is distinguished from static recrystallization (SRX) in that MDRX is relatively sensitive to prior strain rate but insensitive to prestrain and temperature. SRX depends on prestrain and temperature, but only slightly on strain rate. The recrystallization remarks in the previous section are equally valid for these two cases (although Figure 78 was an SRX example) but differences are apparent. DRX is more important to discuss in the context of creep plasticity.

A single broad stress peak, where the material hardens to a peak stress, followed by significant softening is often evidenced in DRX. The softening is largely attributable to the nucleation of growing, "new," grains that annihilate dislocations during growth. This is illustrated in Figure 79. The restoration is contrasted by dynamic recovery, where the movement of, and annihilation of, dislocations at, high-angle boundaries is not important. DRX may commence well before the peak stress. This becomes evident without microstructural examination by examining the hardening rate, θ, as a function of flow stress. For customary Stage III hardening, θ decreases at a constant or decreasing "rate" with stress. DRX, on the other hand, causes an "acceleration" of the decrease in hardening rate.

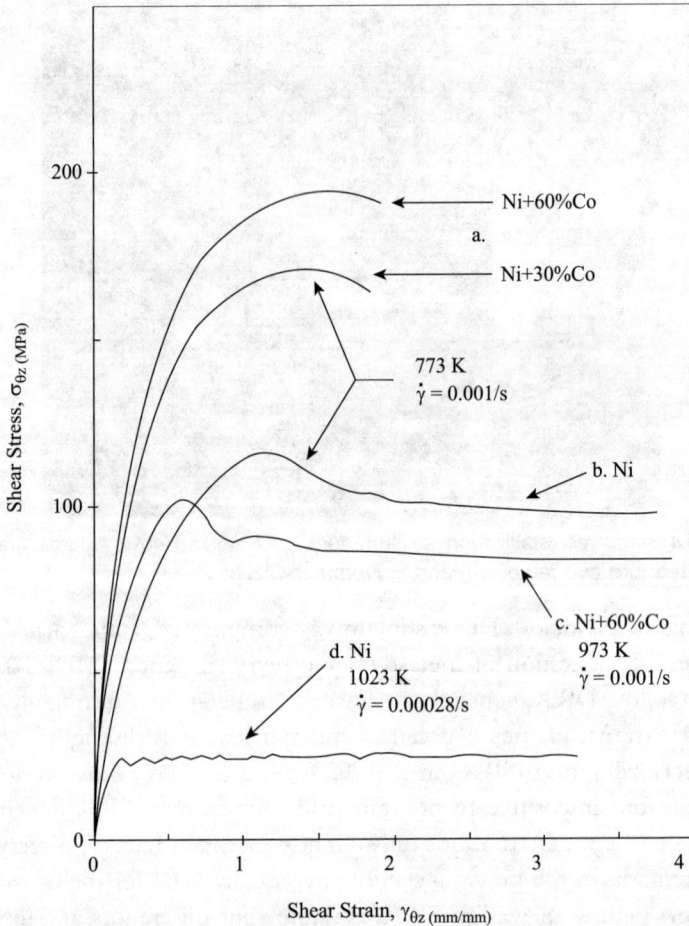

Figure 79 Dynamic recrystallization in Ni and Ni–Co alloys in torsion [13].

Sometimes the single peak in the stress versus strain behavior in DRX is not observed, rather multiple peaks may be evident leading to the appearance of undulations in the stress versus strain behavior that "dampen" into an effective "steady state." This is also illustrated in Figure 79. It has been suggested that the cyclic behavior indicates that grain coarsening is occurring while a single peak is associated with grain refinement [535].

Although DRX is frequently associated with commercial metal forming strain rates (e.g., 1 s^{-1} and higher), Figure 79 illustrates that DRX can occur at more modest rates that approach those in ordinary creep conditions. This explains why some ambiguity has been experienced in interpreting creep deformation where both dynamic recovery and DRX are occurring. As some recent analysis indicated, some of the creep data of Figure 16 of Zr may include some data for which some DRX may be occurring. Metals such as pure Ni and Cu frequently exhibit DRX [7,78]. A substantially greater discussion of DRX, particularly in Al and Al-alloys, is available in [223].

3. GEOMETRIC DRX

The starting grains of the polycrystalline aggregate distort with relatively large strain deformation. These boundaries may thin to the dimensions of the subgrain diameter with strain approaching 2–10 (depending on the starting grain size), achievable in torsion or compression. In the case of Al, the starting high-angle grain boundaries (HABs) (typically 35° misorientation) are serrated as a result of subgrain-boundary formation, in association with dynamic recovery (DRV), where the typical misorientations are about a degree, or so. As the grains thin to about twice the subgrain diameter, nearly one-third to one-half of the subgrain facets have been replaced by high-angle boundaries, which have ancestry to the starting polycrystal. The remaining two-thirds are still of low misorientation polygonized boundaries typically of a degree or so. As deformation continues, "pinching off" may occur which annihilates HABs and the high-angle boundary area remains constant. Thus, with geometric dynamic recrystallization (GDX), the HAB area can dramatically increase but not in the same discontinuous way as DRX. GDX has been confused with DRX, as well and continuous recrystallization (CR) (discussed in a later section), but has been confirmed in Al and Al–Mg alloys [146,156], and may occur in other alloys as well, including Fe-based and Zr [221,536]. Figure 44 showed the progression of GDX in Al at elevated temperature in torsion [18].

4. PARTICLE-STIMULATED NUCLEATION

As pointed out by Humphreys [222], an understanding of the effects of second-phase particles on recrystallization is important since most industrial alloys contain second-phase particles and such particles have a strong influence on the recrystallization kinetics, microstructure, and texture. Particles are often known for their ability to impede the motion of high-angle boundaries during high temperature annealing or deformation (Zener pinning). During the deformation of a particle-containing alloy, the enforced strain gradient in the vicinity of a nondeforming particle creates a region of high dislocation density and large orientation gradients, which is an ideal site for the development of a recrystallization nucleus. The mechanisms of recrystallization in two-phase alloys do not differ from those in single-phase alloys. There are not a great deal of systematic measurements of these zones, but it appears that the deformation zone may extend a diameter, or so, from the particle into the matrix and lead to misorientations of tens of degrees from the adjacent matrix.

5. CONTINUOUS REACTIONS

According to McNelley [222], it is now recognized that refined grain structures may evolve homogeneously and gradually during the annealing of deformed metals, either with or without concurrent straining. This can occur even when the heterogeneous nucleation and growth stages of primary recrystallization do not occur. "Continuous reactions" is a term that is sometimes used in place of others that imply at least similar process such as "continuous recrystallization," "in situ recrystallization," and "extended recovery." It is commonly observed that deformation textures sharpen and components related to the stable orientations within the prior deformation textures are retained [537]. These observations are consistent with recovery as the sole restoration mechanism, suggesting that the term "continuous reactions" may be more meaningful a description than "continuous recrystallization."

Mechanisms proposed to explain the role of recovery in high-angle boundary formation include subgrain growth via dislocation motion [535], the development of higher-angle boundaries by the merging of lower-angle boundaries during subgrain coalescence [537], and the increase of boundary misorientation though the accumulation of dislocations into the subgrain boundaries [535]. These processes have been envisioned to result in a progressive buildup of boundary misorientation during (static or dynamic) annealing, resulting in a gradual transition in boundary character and formation of high-angle grain boundaries. Of course, one must consider that some of the HABs may be formed as a result of GNBs, discussed earlier.

CHAPTER 8

Creep Behavior
of Particle-Strengthened Alloys

Contents

1. INTRODUCTION

This chapter will discuss the behavior of two types of materials that have creep properties enhanced by second phases. These materials contain particles with square or spherical aspect ratios that are both coherent and incoherent with the matrix, and are of relatively low volume fractions. This chapter is a review of work in this area, but it must be recognized that other reviews have been published and this chapter reflects the particularly high-quality reviews by Reppich and coworkers [538–540], Arzt [541,542], and others [543–549]. It should also be mentioned that this chapter will emphasize those cases where there are relatively wide separations between the particles, or stated another way, the volume fractions of the precipitate discussed here are nearly always less than 30% and usually less than 10%. This contrasts with the case of some γ/γ' alloys where the precipitate, γ', occupies a substantial volume fraction of the alloys; these are discussed in Chapter 11.

Fundamentals of Creep in Metals and Alloys
ISBN 978-0-08-099427-7
http://dx.doi.org/10.1016/B978-0-08-099427-7.00008-6

2. SMALL VOLUME-FRACTION PARTICLES COHERENT AND INCOHERENT WITH THE MATRIX WITH SMALL ASPECT RATIOS

2.1 Introduction and Theory

It is well known that second-phase particles provide enhanced strength at lower temperatures, and there have been numerous discussions on the source of this strength. A review of low-temperature strengthening by second-phase particles was published by Reppich [539]. Although a discussion of the mechanisms of lower-temperature second-phase strengthening is outside the scope of this chapter, it should be mentioned that the strength by particles has been believed to be provided in two somewhat broad categories of strengthening, Friedel cutting or Orowan bypassing. Basically, the former involves coherent particles and the flow stress of the alloy is governed by the stress required for the passage of the dislocation through a particle. The Orowan stress is determined by the bypass stress based on an Orowan loop mechanism. In the case of oxide dispersion strengthened (ODS) alloys, in which the particles are incoherent, the low temperature yield stress is reasonably predicted by the Orowan loop mechanism [539,550]. The Orowan bowing stress is approximated by the classic equation:

$$\tau_{\mathrm{or}} = Gb/L = \left(\frac{2T_{\mathrm{d}}}{bL}\right) \qquad (150)$$

where τ_{or} is the bowing stress, L is the average separation between particles, and T_{d} is the dislocation line tension. This equation, of course, assumes that the elastic strain energy of a dislocation can be estimated by $Gb^2/2$, which, though reasonable, is not firmly established [88].

This chapter discusses how the addition of second phases leads to enhanced strength (creep resistance) at elevated temperatures. This discussion is important for at least two reasons. First, as Figure 80 illustrates, the situation at elevated temperature is different than at lower temperatures. This figure illustrates that the yield stress of the single-phase matrix is temperature-dependent, of course, but there is a superimposed strengthening (suggested in the figure to be approximately athermal) by Orowan bowing [550]. It is generally assumed that the bowing process cannot be thermally activated, but the non-shearable particles can be negotiated by climb. At higher temperatures, it is suggested that the flow stress becomes only a fraction of this superimposed stress and an understanding of the

(a)

(b)

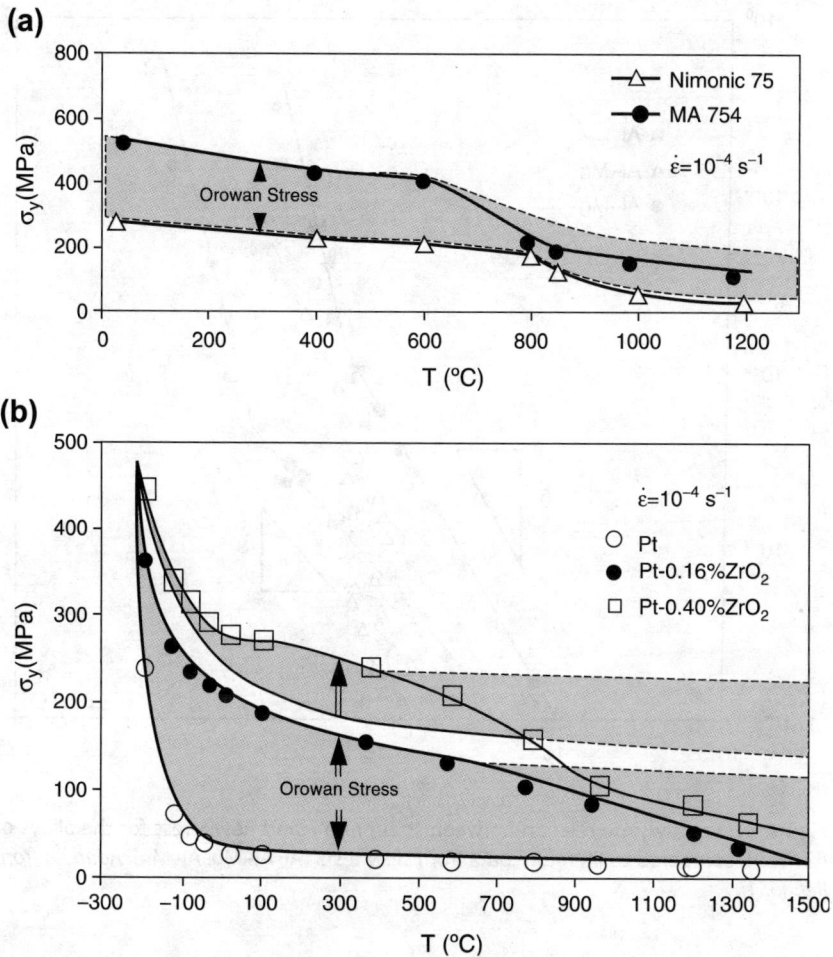

Figure 80 Compressive 0.2% yield stress versus temperature. (Shaded: Orowan stress given as low-temperature yield-stress increment due to oxide dispersoids.) (a) ODS Superalloy MA 754. (b) Pt-based ODS alloys. *From Ref. [539].*

origin is a significant focus of this chapter. Second, in the previous chapters it was illustrated how solute additions, basically obstacles, lead to increased creep strength. There is, essentially, a roughly uniform shifting of the power law, power-law breakdown, and low-stress-exponent regimes to higher stresses. This is evident in Figure 81, where the additions of Mg to Al are described. In the case of alloys with second-phase particles, however, there is often a lack of this uniform shift and sometimes the appearance that many investigators have termed a "threshold stress," σ_{th}. The intent of this

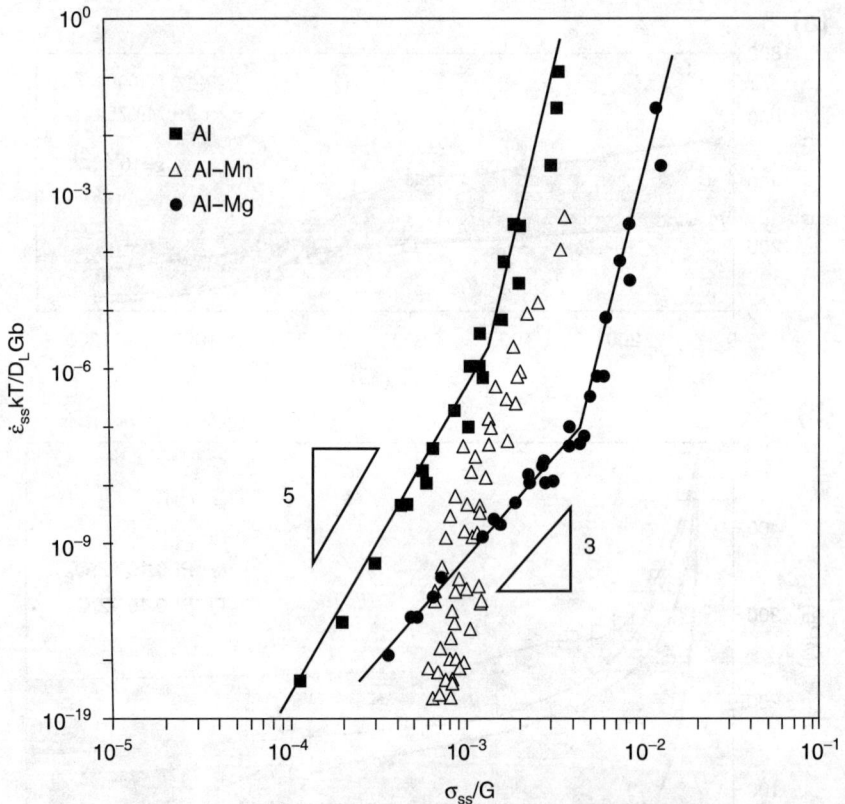

Figure 81 Steady-state relation between strain rate $\dot{\varepsilon}$ and flow stress for the alloys of this work compared to literature data from slow tests (Al, Al–Mg, Al–Mn). *Adapted from Ref. [551].*

term is illustrated in Figure 82, based on the data of Lund and Nix and additional interpretations by Pharr and Nix [552,553]. Figure 82 reflects "classic" particle strengthening by oxide dispersoids (ThO_2) in a Ni–Cr solid solution matrix. These particles, of course, are incoherent with the matrix. The "pure" solid solution alloy behavior is also indicated. Particle strengthening is evident at all steady-state stress levels, but at lower stresses there appears a modulus-compensated stress below which creep does not appear to occur or is at least very slow. Again, this has been termed the threshold stress, σ_{th}. Said another way, there is not a uniform shift of the strengthening on logarithmic axes; there appears a larger fraction of the strength provided by the particles at lower stress (high temperatures) than at higher stresses (lower temperatures). In fact, the concept of a

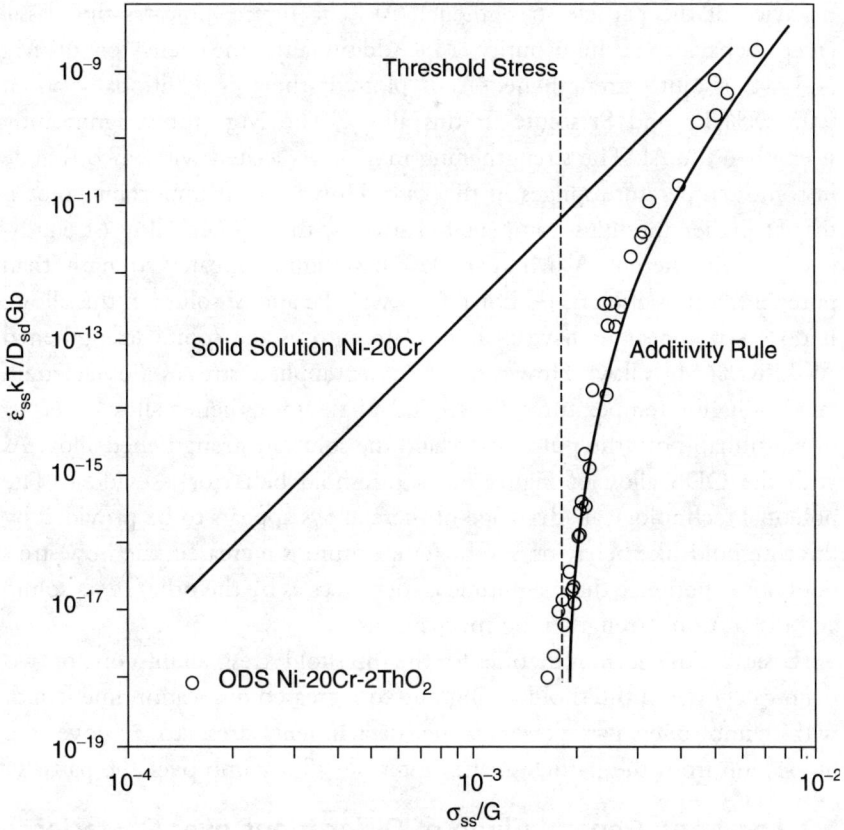

Figure 82 The normalized steady-state creep-rate versus modulus compensated steady-state stress. *Adapted from Refs [552,553].*

threshold stress was probably originally considered one of athermal strengthening. This will be discussed more later, but this "coarse" description identifies an aspect of particle strengthening that appears generally different from dislocation–substructure strengthening and solution strengthening (although the latter, in certain temperature regimes, may have a nearly athermal strengthening character). The threshold stress of Figure 82 is only about half the Orowan bowing stress, suggesting that Orowan bowing may not be the basis of the threshold. Activation energies appear relatively high (greater than lattice self-diffusion of the matrix) as well as the stress exponents being relatively high ($n \gg 5$) in the region where a threshold is apparent.

Particle strengthening is also illustrated in Figure 81, based on a figure from reference [551], where there is, again, a nonuniform shift in the behavior of the particle-strengthened Al. The figure indicates the classic creep behavior of high-purity Al. Additionally, the behavior of Mg (4.8 wt%)-solute strengthened Al is plotted (there is additionally about 0.05 wt% Fe and Si solute in this alloy). The Mg atoms significantly strengthen the Al. The strengthening may be associated with viscous glide in some temperature ranges in this case. However, an important point is that at higher modulus-compensated stresses, the Al–Mn alloy (strengthened by incoherent Al_6Mn particles) has slightly greater strength than pure Al (there is also an additional 0.05 wt% Fe and Si solute in this alloy). It does not appear to have as high a strength as the solute-strengthened Al-4.8wt% Mg alloy. However, at lower applied stresses (lower strain rates) or higher temperatures, the second-phase strengthened alloy has *higher* strength than both the pure matrix and the solution strengthened alloy. As with the ODS alloy of Figure 82, a threshold behavior is evident. The potential technological advantage of these alloys appears to be provided by this threshold-like behavior. As the temperature is increased, the flow stress does not experience the magnitude of decreases as by the other (e.g., solute and dislocation) strengthening mechanisms.

Basically, the current theories for the threshold stress fall into one of two main categories: a threshold arising due to increased dislocation line length with climb over particles and the detachment stress to remove the dislocation from the particle matrix interface after climb over the particle.

2.2 Local and General Climb of Dislocations over Obstacles

It was presumed long ago, by Ansell and Weertman [554], that dislocation climb allowed for passage at these elevated temperatures and relatively low stresses. The problem with this early climb approach is that the creep rate is expected to have a low stress-dependence with an activation energy equivalent to that of lattice self-diffusion. As indicated in the figures just presented in this chapter, the stress dependence in the vicinity of the threshold is relatively high and the activation energy in this threshold regime can be much higher than that of lattice self-diffusion. More recent analysis has attempted to rationalize the apparent threshold. One of the earlier approaches suggested that for stresses below the cutting, σ_{ct} (relevant for some cases of coherent precipitates) or Orowan bowing stress, σ_{or}, the dislocation must—as Ansell and Weertman originally suggested—climb over the obstacle. This climbing process could imply an increase in dislocation line length and hence total elastic strain energy, which would act as an impediment to plastic flow [555–560]. The schemes by which this has

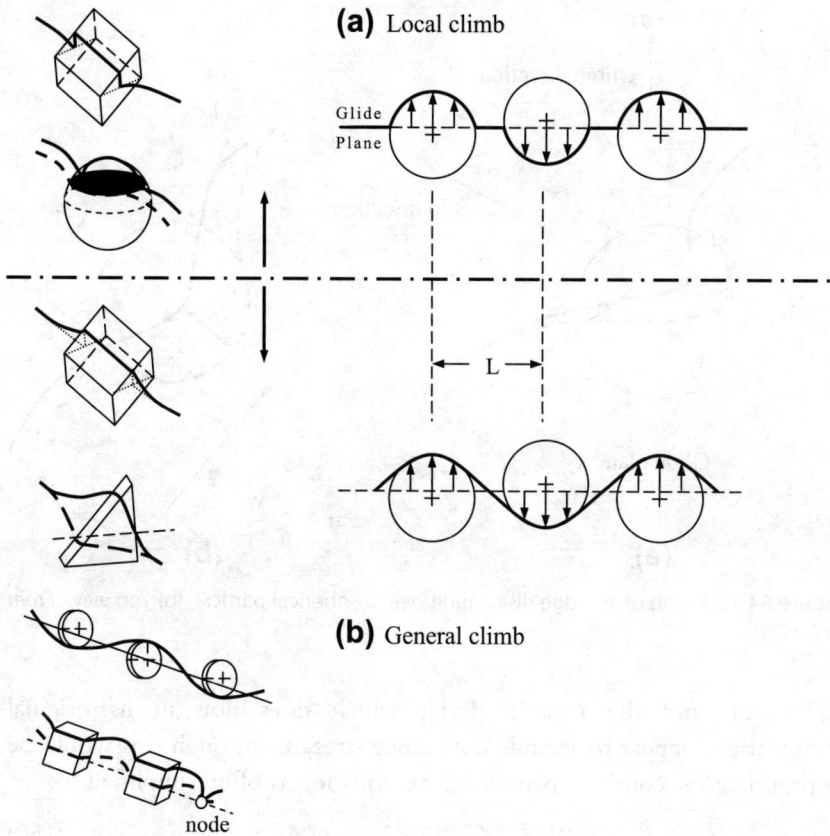

(a) Local climb

(b) General climb

node

Figure 83 Compilation by Blum and Reppich [538] of models for dislocation climb over second-phase particles.

been suggested are illustrated in Figure 83. Figure 84 shows an edge dislocation climbing, with concomitant slip, over a spherical particle. As the dislocation climbs, work is performed by the applied shear. The total energy change can be described by:

$$dE \cong (Gb^2/2)dL - \tau bLdx - \sigma_n bLdy - dE_{el} \tag{151}$$

The first term is the increase in elastic strain energy associated with the increase in dislocation line length. This is generally the principal term giving rise to the (so-called or apparent) threshold stress. The second term is the work done by the applied stress as the dislocation glides. The third term is the work done by the normal component of the stress as the dislocation climbs. The fourth term accounts for any elastic interaction between the

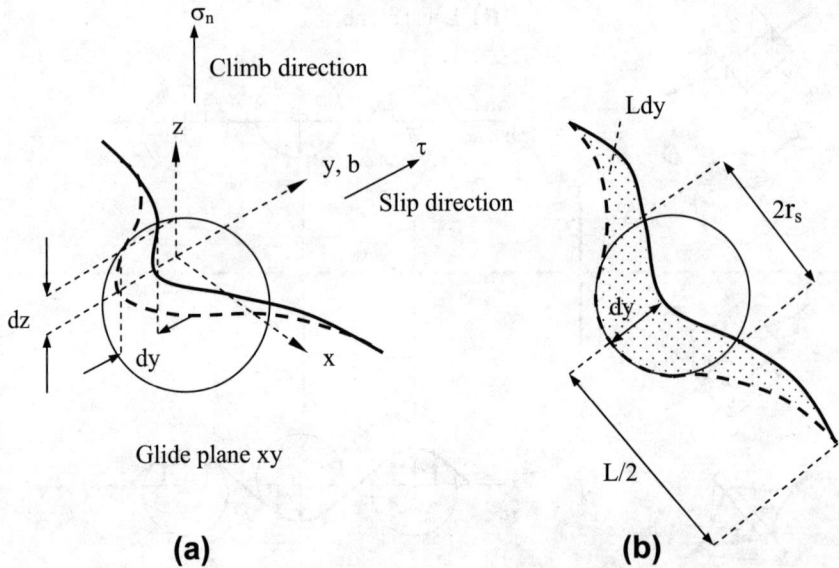

Figure 84 (a) Climb of an edge dislocation over a spherical particle. (b) Top view. *From Ref. [538].*

dislocation and the particle [561], which does not, in its original formulation, appear to include coherency stresses, although this would be appropriate for coherent particles. This equation is often simplified to:

$$dE = (Gb^2/2)dL - \tau bLdx \tag{152}$$

The critical stress for climb of the dislocation over the particle is defined under the condition where:

$$(dE/dx) = 0$$

or:

$$\tau_c = Gb/L(0.5\alpha') \tag{153}$$

Arzt and Ashby defined the α' parameter $= (dL/dx)_{max}$ as the climb resistance, and τ_c can be regarded as the apparent threshold stress. Estimates have been made of α' by relating the volume fraction of the particles, and the particle diameter to the value of L in Eqn (153). Furthermore, there is a statistical distribution of particle spacings, and Arzt and Ashby suggest:

$$\tau_c/(Gb/L) = \alpha'/(1.68 + \alpha') \tag{154}$$

while Blum and Reppich use a similar relationship that includes the so-called Friedel correction:

$$\tau_c/(Gb/L) = \alpha'^{1.5}\Big/\left(2\sqrt{2} + \sqrt{\alpha'^3}\right) \qquad (155)$$

Equations (153)–(155), together with the values of α', allow a determination of the threshold stress. Note that for general climb there is the suggestion that τ_c is particle-size independent.

The determination of α' will depend on whether climb is local or general; both cases are illustrated in Figure 83. The portion of the dislocation that climbs can be either confined to the particle matrix interfacial region (local), or the climbing region can extend beyond the interfacial region, well into the matrix (general). This significantly affects the α' calculation.

First, Eqn (153) suggests a maximum value for α' that corresponds to the Orowan bowing stress. It has been suggested that the Orowan stress can be altered based on randomness and elastic interaction considerations [556,559], giving values of $0.5 < \alpha' < 1.0$. For local climb, the value of α' depends on the shape of the particle, with $0.77 < \alpha' < 1.41$, from spherical to square shapes [555,556,559]. For extended or general climb, which is a more realistic configuration in the absence of any particular attraction to the particle [562], the α' is one order of magnitude or so smaller. Additionally, the value of α' will be dependent on the volume fraction, f, as $f^{1/2}$ and values of α' range from 0.047 to 0.14 from $0.01 < f < 0.10$ [538]. Blum and Reppich suggested that for these circumstances:

$$\text{for local climb,} \quad \tau_c = 0.19[Gb/L]$$

$$\text{and for general climb,} \quad \tau_c = 0.004 \text{ to } 0.02[Gb/L]$$

This implies that there is a threshold associated with the simple climbing of a dislocation over particle obstacles, without substantial interaction. This threshold stress is a relatively small fraction of the Orowan stress. There is the implicit suggestion in all of this analysis that the stress calculated from the above equations is athermal in nature, and this will be discussed subsequently.

2.3 Detachment Model

In connection with the above, however, there has been evidence that dislocations may interact with incoherent particles. This was observed by Nardone and Tien [563] and later by Arzt and Schroder [564] and others

[565] using transmission electron microscopy (TEM) of creep-deformed ODS alloys. Figures 85 and 86 illustrate this. The dislocations must undergo local climb over the precipitate and then the dislocation must undergo "detachment." Srolovitz et al. [566] suggested that incoherent particles have interfaces that may slip and can attract dislocations by reducing the total elastic strain energy. Thus, there is a detachment stress that reflects the increase in strain energy of the dislocation on leaving the interface. Basically, Arzt and coworkers suggest that the incoherent dispersoids strengthen by acting as, essentially, voids. Arzt and coworkers [557,568–570] analyzed the detachment process in some detail and estimated τ_d as:

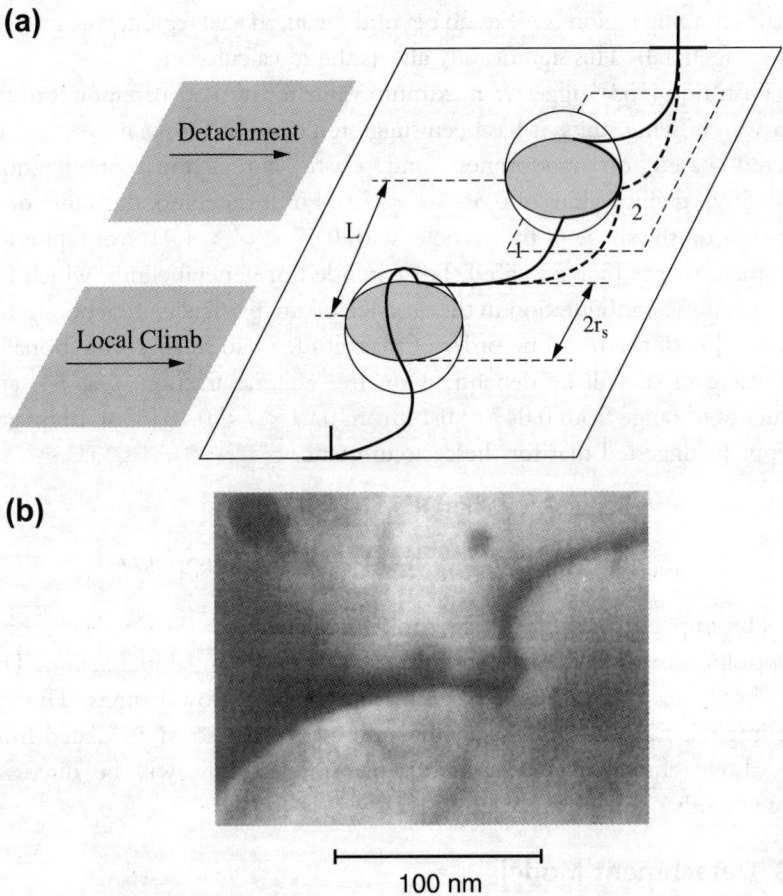

Figure 85 The mechanism of interfacial pinning. (a) Perspective view illustrating serial local climb over spherical particles of mean (planar) radius r_s and spacing l and subsequent detachment. (b) Circumstantial TEM evidence in the creep-exposed ferritic ODS superalloy PM 2000. *From Ref. [565].*

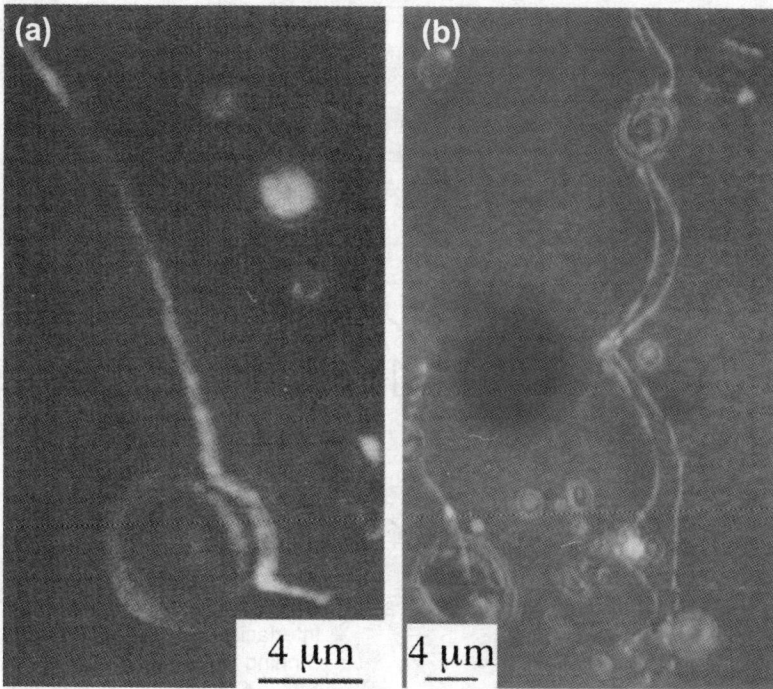

Figure 86 TEM evidence of an attractive interaction between dislocation and dispersoid particles. (a) Dislocation detachment from a dispersoid particle in a Ni alloy. (b) Dissociated superdislocation detaching from dispersoid particles in the intermetallic compound Ni_3Al. *From Refs [542,564,567].*

$$\tau_d = \left[1 - k_R^2\right]^{1/2}(Gb/L) \tag{156}$$

where k_R is the relaxation factor described by:

$$\left(Gb^2\right)_p = k_R\left(Gb^2\right)_m \tag{157}$$

where "p" refers to the particle interface and "m" the matrix. In the limit that $k_R = 1$, there is no detachment process.

Reppich [572] modified the Arzt et al. analysis slightly, using Fleischer–Friedel obstacle approximation, and suggested that:

$$\tau_d = 0.9\left(1 - k_R^2\right)^{3/4}\Big/\left[1 + \left(1 - k_R^2\right)^{3/4}\right](Gb/L) \tag{158}$$

This decreases the values of Eqn (156) by roughly a factor of 2. Note that as with general climb, τ_d and τ_c are independent of the particle size. It was suggested by both of the above groups that this detachment process

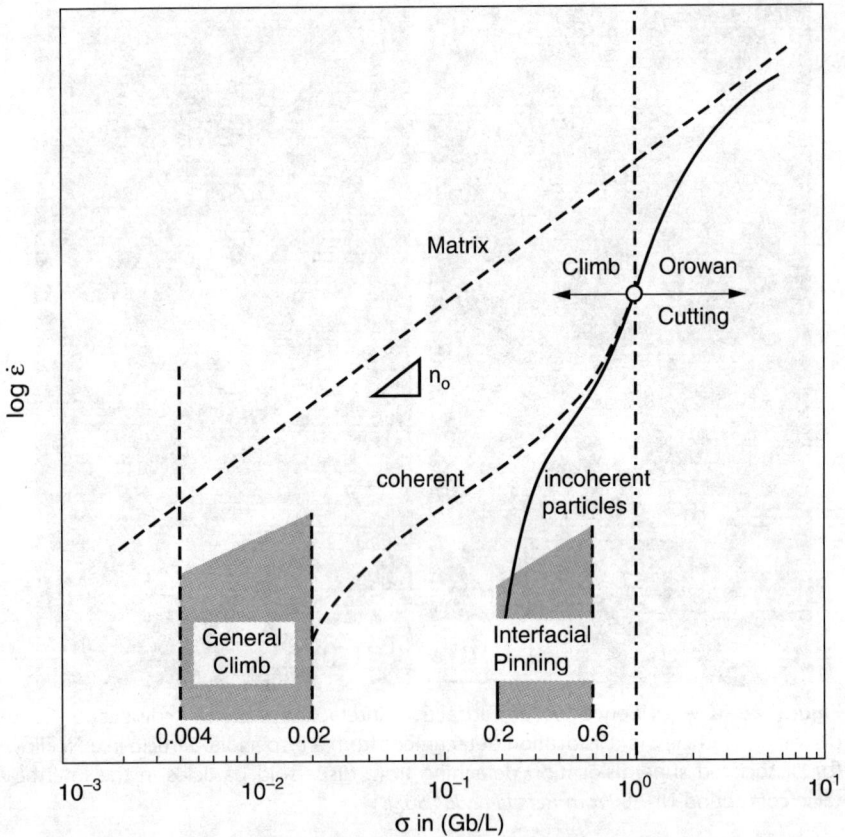

Figure 87 Creep behavior of particle-strengthened materials (schematic). The stress is given in units of the classical Orowan stress. *From Ref. [538].*

could be thermally activated. This equation suggests the τ_d is roughly $Gb/3L$, substantially higher than τ_c for general climb, as illustrated in Figure 87. Arzt and Wilkinson [557] showed that if k_R is such that there is just a 6%, or less, reduction in the elastic strain energy, then local climb become the basis of the threshold stress instead of detachment. For general climb, the transition point is k_R about equal to 1.

Rosler and Arzt [570] extended the detachment analysis to a "full kinetic model" and suggested a constitutive equation for "detachment-controlled" creep:

$$\dot{\varepsilon} = \dot{\varepsilon}_o \exp\left[\left(-Gbr_S^2/kT\right)\left(1 - k_R\right)^{3/2}\left(1 - \sigma/\sigma_d\right)^{3/2}\right] \qquad (159)$$

where $\dot{\varepsilon}_o = C_2 D_v L \rho_m / 2b$, and r_S is the particle radius. This was shown to be valid for random arrays of particles. Figure 88 plots this equation for

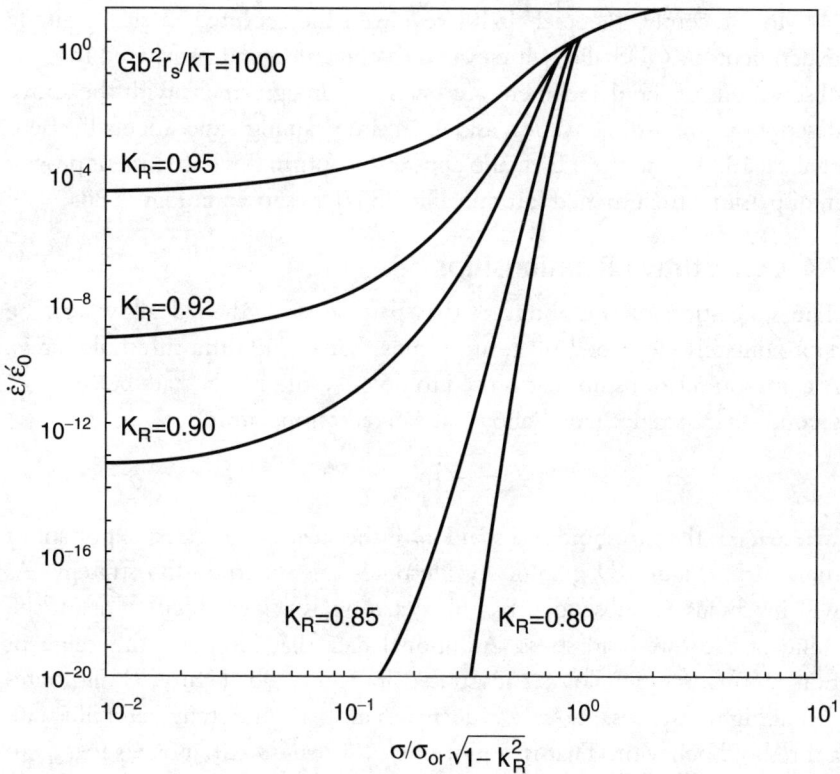

Figure 88 Theoretical prediction of the creep rate (normalized) as a function of stress (normalized) on the basis of thermally activated dislocation detachment from attractive dispersoids, as a function of interaction parameter k. The change of curvature at high strain rates (broken line) indicates the transition to the creep behavior of dispersoid-free material and does not follow from the equation. *From Ref. [569].*

several values of k_R as a function of strain rate. Threshold behavior is apparent for modest values of k_R. This model appears to reasonably predict the creep behavior of various dispersion–strengthened Al alloys [541,571] with reasonable k_R values (0.75–0.95). As Arzt points out, this model, however, does not include the effects of dislocation substructure. Arzt noted from this equation that an optimum particle size is predicted. This results from the probability of thermally activated detachment being raised for small dispersoids and that large particles (for a given volume fraction) have a low Orowan stress and, hence, small detachment stress [541]. It should be noted that Figure 88 does not suggest a "pure" threshold stress (below which plasticity does not occur). Rather, thermally activated detachment is suggested, and this will be discussed more later.

More recently, Reppich [540] reviewed the reported in situ straining experiments in ODS alloys at elevated temperatures and concluded that the observations of the detachment are essentially in agreement with the above description (thermal activation aside). In situ straining experiments by Behr et al., at 1000 °C in the TEM, also appear to confirm this detachment process in dispersion-strengthened intermetallics [567], as shown in Figure 86.

2.4 Constitutive Relationships

The suggestion of the above is that particle-strengthened alloys can be approximately described by relationships that include the threshold stress. A common relationship that is used to describe the steady-state behavior of second-phase strengthened alloys (at a fixed temperature) is:

$$\dot{\varepsilon}_{ss} = A'(\sigma - \sigma_{th})^{n_m} \tag{160}$$

where σ_{th} is the threshold stress and n_m is the steady-state stress exponent of the matrix. Figure 89 graphically illustrates this superposition strategy. As will be discussed subsequently, this equation is widely used to assess the value of the threshold stress. Additional data that illustrates the value of Eqn (160) for superalloys is illustrated in Figure 90. Figure 91 illustrates that at higher stresses, above τ_{or}, decreases in stress (and strain-rate) illustrate a threshold behavior. That is, a plot of $\dot{\varepsilon}^{1/n}$ versus σ extrapolates to τ_{or}, an "apparent σ_{th}." However, as σ decreases below τ_{or}, a new threshold appears, and this then is the σ_{th} relevant to creep plasticity. A σ_{th} can be estimated with low stress plots such as Figure 91 [575,576] (also see Figure 87). These give rise to estimates of σ_{th} and allow plots such as Figure 90. However, Figure 89(b) and (c) illustrate that the high-temperature σ_{th} is not a true threshold and creep occurs below σ_{th}. This is why σ_{th} estimates based on plots such as Figure 91 decrease with increasing temperature.

An activation energy term can be included in the form of:

$$\dot{\varepsilon}_{ss} = A'' \exp(-Q/kT) \left(\frac{\sigma - \sigma_{th}}{G} \right)^{n_m} \tag{161}$$

or

$$\dot{\varepsilon}_{ss} = \frac{A'' DGb}{kT} \left(\frac{\sigma - \sigma_{th}}{G} \right)^{n_m} \tag{162}$$

where D is the diffusion coefficient. However, the use of D works in some cases while not in others for both coherent and incoherent particles.

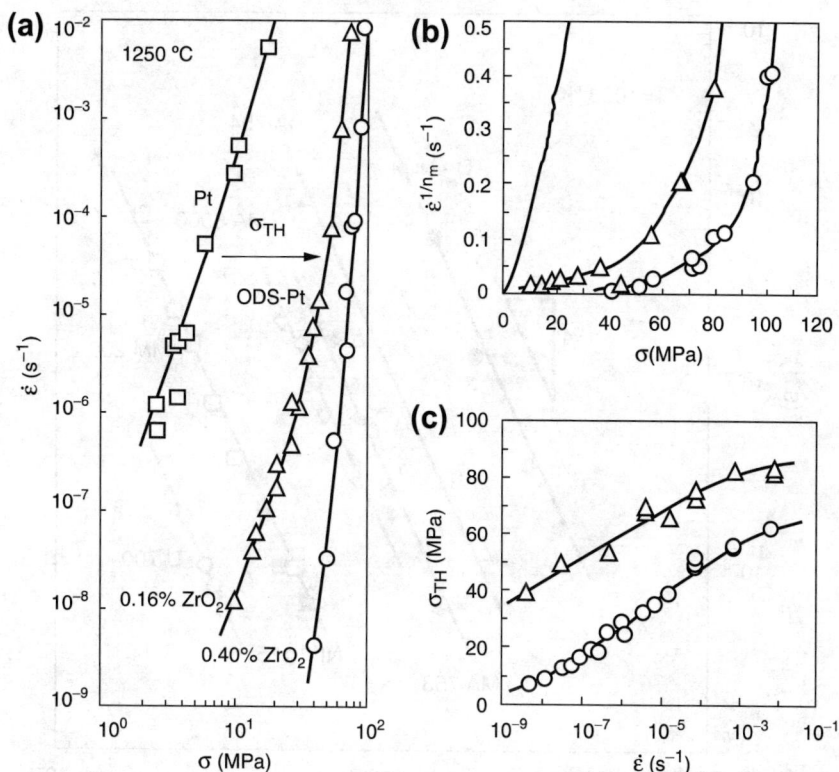

Figure 89 Comparison of the creep behavior of ZrO_2 dispersion strengthened Pt-based alloys at 1250 °C. (a) Double-logarithmic Norton plot of creep rate $\dot{\varepsilon}$ versus stress σ. (b) Lagneborg–Bergman plot. (c) Dependence of σ_{th} on $\dot{\varepsilon}$. From Ref. [567].

Figure 92 illustrates somewhat different behavior from Figure 82 in that the data do not appear to reduce to a single line when the steady-state stress is modulus-compensated and the steady-state creep rates are lattice-self-diffusion compensated. The activation energy for creep is reported to be relatively high at 537 kJ mol^{-1} (as compared to 142 kJ mol^{-1} for lattice self-diffusion). Cadek and coworkers [578,579] illustrated that for experiments on ODS Cu, the (modulus-compensated) threshold stress, determined by the extrapolation procedure described in Figure 92, is temperature-dependent. They propose that the activation energies should be determined using the usual equation but at constant $(\sigma - \sigma_{th})/G$ rather than σ/G as used in typical (especially five-power law for single-phase metals and alloys) creep activation energy calculations. The activation energies they calculated using this procedure reasonably correspond to

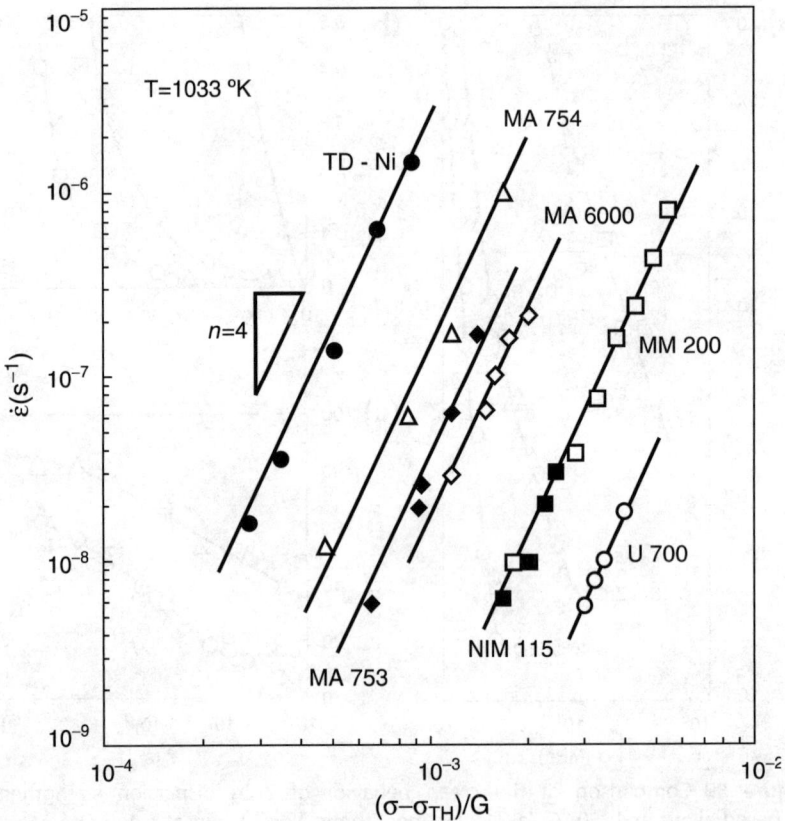

Figure 90 Double-logarithmic plot of creep rate versus reduced stress $\sigma - \sigma_{th}$ for various superalloys. *After Ajaja et al. [573].*

lattice or dislocation-core self-diffusion. Thus, the investigators argued that the activation energy for diffusion could reasonably be used as the activation energy term such as in Eqn (161).

Figure 87 is an idealized plot by Blum and Reppich that illustrates many of the features and parameters for particle strengthening. This is a classic logarithmic plot of the steady-state creep- (strain-) rate versus the steady-state stress. The value of σ_{or} is indicated and apparent threshold behavior is observed above this stress. A second threshold-like behavior is evident below the Orowan stress, one for incoherent particles and another for coherent particles at particularly low stresses (high temperature). The incoherent particles evince interfacial pinning and the more effective (or higher) threshold-like behavior is observed in the absence of a detachment

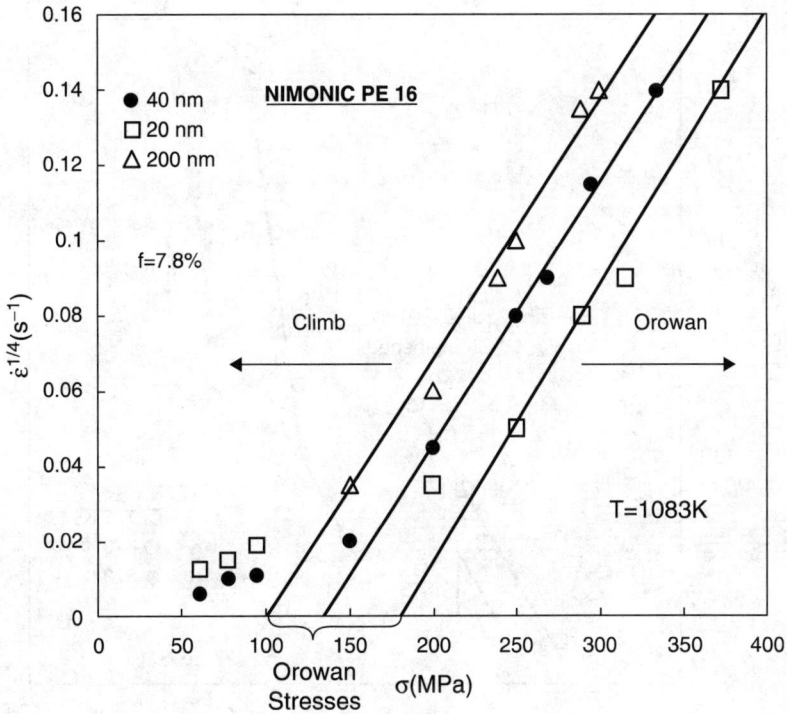

Figure 91 $\dot{\varepsilon}^{1/n}$ versus σ-plot for determination of σ_{th} in γ'-hardened Nimonic PE 16 by back extrapolation (arrows). *From Ref. [574].*

stress. There has been some discussion as to what mechanism may be applicable at stresses below the apparent threshold, and it appears that grain boundary sliding and even diffusional creep have been suggested [580,581]. These, however, are speculative, as even single crystals appear to show sub-threshold plasticity. Figure 88 was basically an attempt to explain creep below the apparent σ_{th} through thermally activated detachment. This approach has become fairly popular [582], although recently it has been applied below, but not above, an apparent threshold. A difficulty with this approach is that Figures 85 and 86 suggest that a considerable length of dislocation is trapped in the interface, which would appear to imply a very large activation energy for detachment, much larger than that for D_v in Eqn (159). In at least some instances, the plasticity below the apparent threshold is due to a change in deformation mechanism [583]. Dunand and Jansen [584,585] suggested, for larger-volume fractions of second-phase particles (e.g., 25%), dislocation pile-ups become relevant and additional stress terms

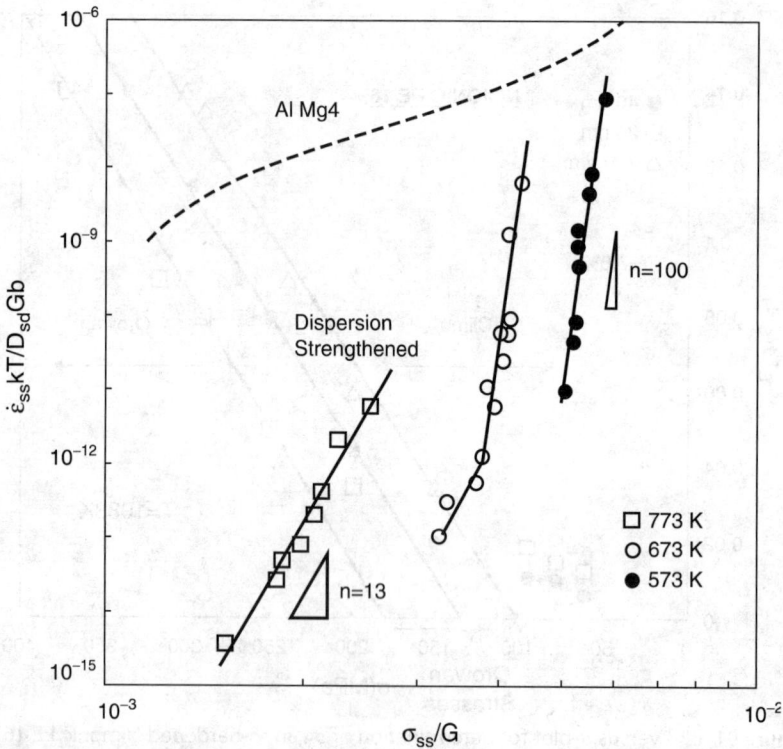

Figure 92 Lack of convergence of the different dispersion-strengthened creep curves with Q_{sd} compensation. *From Ref. [577].*

must be added to the conventional equations. However, these considerations do not appear relevant to the volume fractions being typically considered here.

2.5 Microstructural Effects

2.5.1 Transient Creep Behavior and Dislocation Structure

The strain versus time behavior of particle-strengthened alloys during primary and transient creep is similar to that of single-phase materials in terms of the strain rate versus strain trends as illustrated in Figure 93. Figure 93(a) illustrates Incoloy 800 H [544] and (b) Nimonic PE 16 [574]. Both generally evince Class M behavior although the carbide-strengthened Incoloy shows an inverted transient (such as a Class A alloy) but this was suggested to be due to particle structure changes, which must be considered with prolonged high-temperature application. The Nimonic alloy at the

(a)

(b)

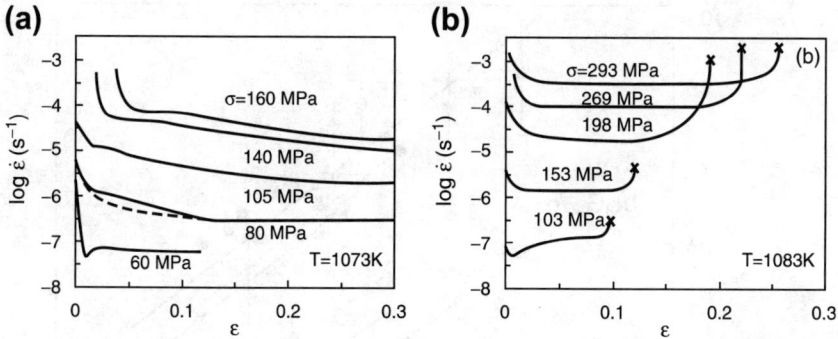

Figure 93 Half-logarithmic plot of creep rate $\dot{\varepsilon}$ versus (true) strain of a single-phase material and particle-strengthened material. (a) Incoloy 800 H. (b) Nimonic PE 16. *Adapted from Refs [538,574,586].*

lowest stress also shows such an inverted transient and this was suggested as possibly being due to particle changes in this initially coherent γ'-strengthened alloy.

There has been relatively little discussed in the literature regarding creep transients. Blum and Reppich suggest that the transients between steady states with stress drops and stress jumps are analogous to the single-phase metals both in terms of the nature of the mechanical (e.g., strain rate vs strain) trends and the final steady-state strain-rate values, as well as the final substructural dimensions.

The dislocation structure of particle-strengthened alloys has been examined, particularly by Blum and coworkers [538,551]. The subgrain size in the particle-strengthened Al–Mn alloy are essentially identical to those of high-purity Al at the same modulus-compensated stresses. Similar findings were reported for Incoloy 800H [586] and TD-Nichrome [558]. These results show that the *total* stress level affects the subgrain substructure, even if there is an interaction between the particles and the subgrain boundaries. Blum and Reppich suggest, however, that the density of dislocations within the subgrains seems to depend on $\sigma - \sigma_{\text{th}}$ rather than σ, suggesting that particle hardening diminishes the network dislocation density compared to the single-phase alloy at the same value of σ_{ss}/G. Some of these trends are additionally evident from Figure 94. One interpretation of this observation is that the subgrain size reflects the stress but does not determine the strength. In contrast, Arzt suggests that only at higher stresses, where the alloys approach the behavior of dispersoid-free matrix, have dislocation substructures been reported [577]. Blum,

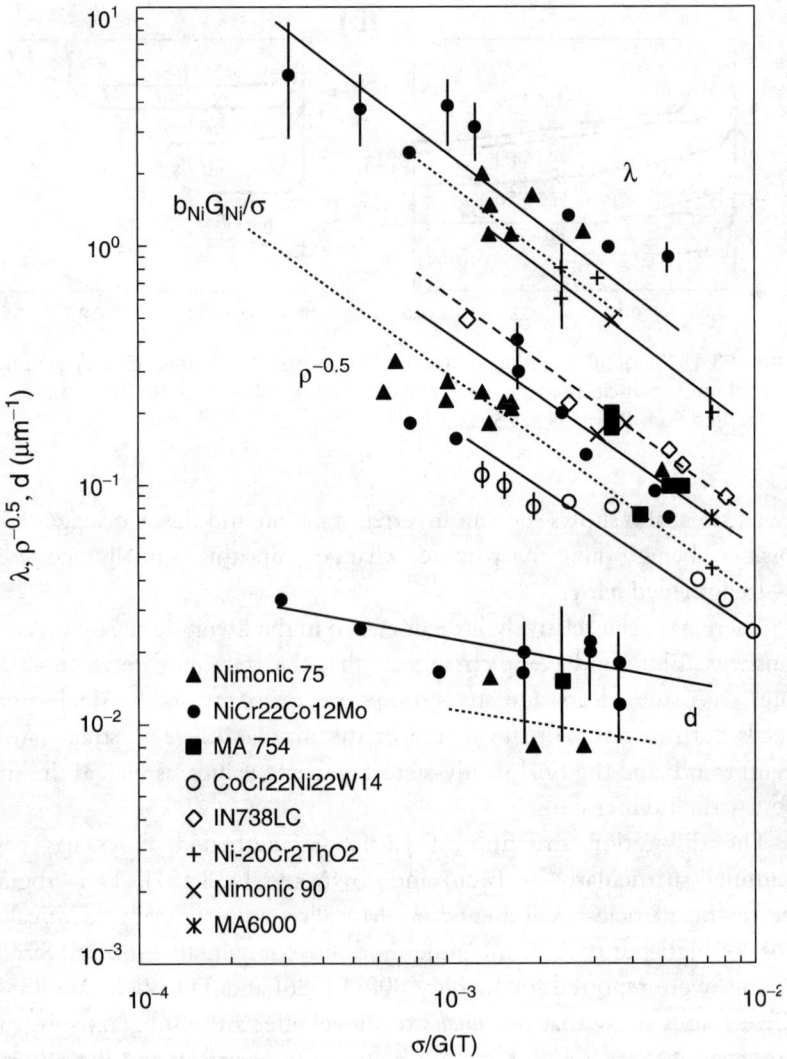

Figure 94 Steady-state dislocation spacings of Ni-base alloys. *Adapted from Ref. [90].*

however, suggests that this may be due to insufficient strain to develop the substructure that would ultimately form in the absence of interdiction by fracture [214].

2.5.2 Effect of Volume Fraction

As expected [580], higher-volume fractions, for identical particle sizes, are associated with greater strengthening and, of course, threshold behavior.

Others [587] have also suggested that the volume fraction of the second-phase particles can affect the value of the threshold stress.

2.5.3 Grain Size Effects

Lin and Sherby [548], Stephens and Nix [583], and Gregory et al. [511] examined the effects of grain size on the creep properties of dispersion strengthened Ni–Cr alloys and found that smaller grain size material may not exhibit a threshold behavior and evince stress exponents more typical of single-phase polycrystalline metals with high elongations [588]. Arzt [541] reports this sigmoidal behavior occurs in single crystals as well and the loss in strength (presumably below that of thermally activated detachment) involves other poorly understood processes including changes in the size or number of particles.

2.6 Coherent Particles

Strengthening from coherent particles can occur in a variety of ways that usually involve particle cutting. This cutting can be associated with (1) the creation of antiphase boundaries (e.g., γ-γ' superalloys), (2) the creation of a step in the particle, (3) differences in the stacking-fault energy between the particle and the matrix, (4) the presence of a stress field around the particle, and (5) other changes in the "lattice friction stress" [133].

Most of the earlier work referenced was relevant to incoherent particles. This is probably in part due to the fact that coherent particles are often precipitated from the matrix, as opposed to added by mechanical alloying, etc. Precipitates may be unstable at elevated temperatures and as a consequence, the discussion returns to that of incoherent particles. Of course, exceptions include the earlier referenced γ-γ' of superalloys and, more recently, the $AlSc_3$ precipitates in Al–Sc alloys by Seidman et al. [590,591]. In this latter work, coherent particles are precipitated. The elevated temperature strengths are much less than the Orowan bowing stress and also less than expected based on the shearing mechanism. Thus, it was presumed that the rate-controlling process is general climb over the particle, consistent with other literature suggestions. This, as discussed in the previous section, is associated with a relatively low threshold stress that is a small fraction of the Orowan bowing stress at about 0.03 σ_{or}, and is independent of the particle size. Seidman et al. found that the normalized threshold stress increases significantly with the particle size and argued that this could only be rationalized by elastic interaction effects, such as coherency strain and modulus effects. Detachment is not important. The results are illustrated in

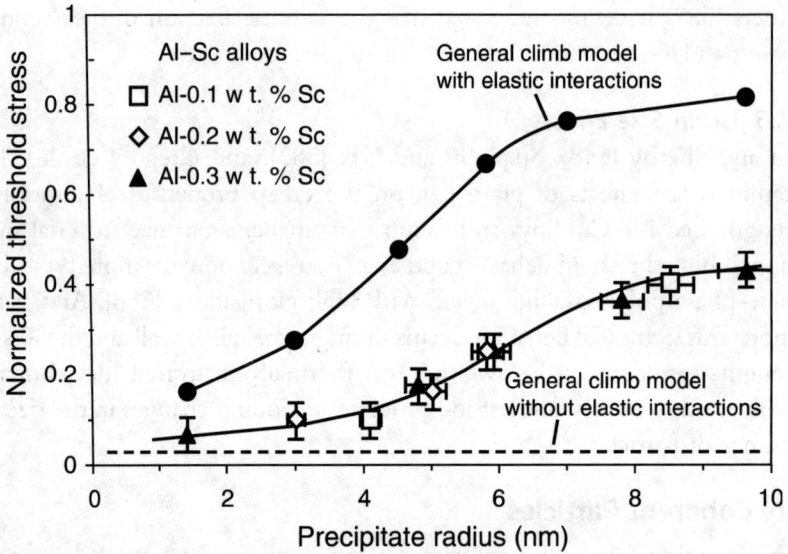

Figure 95 Normalized threshold stress versus coherent precipitate radius. *From Ref. [591].*

Figure 95. They also found that subgrains may or may not form. They do appear to obey the standard equations that relate the steady-state stress to subgrain size when they are observed. Seidman et al. do appear to suggest that steady state was achieved without the formation of subgrains.

CHAPTER 9

Creep of Intermetallics

M.-T. Perez-Prado, M.E. Kassner

Contents

Fundamentals of Creep in Metals and Alloys
ISBN 978-0-08-099427-7
http://dx.doi.org/10.1016/B978-0-08-099427-7.00009-8

189

1. INTRODUCTION

The term "intermetallics" has been used to designate the intermetallic phases and compounds that result from the combination of various metals, and which form a large class of materials [592]. There are mainly three types of superlattice structures based on the f.c.c. lattice, i.e., $L1_2$ with a variant of $L'1_2$ (in which a small interstitial atom of C or N is inserted at the cube center), $L1_0$, and $L1_2$-derivative long-period structures such as DO_{22} or DO_{23}. The b.c.c.-type structures are B2 and DO_3 or $L2_1$. The DO_{19} structure is one of the most typical superlattices based on h.c.p. symmetry. Table 6 lists the crystal structure, lattice parameter, and density of selected intermetallic compounds [593]. A comprehensive review on the physical metallurgy and processing of intermetallics can be found in Ref. [594].

Intermetallics often have high melting temperatures (usually higher than 1000 °C), due partly to the strong bonding between unlike atoms, which is, in general, a mixture between metallic, ionic, and covalent to different extents. The presence of these strong bonds is also associated with high creep resistance. Another factor that contributes to the superior strength of intermetallics at elevated temperature is the high degree of long-range order [596], which results in low diffusivity; the number of atoms per unit cell is large in a material with long-range order. Therefore, in alloys in which dislocation climb is rate-controlling, a decrease in the diffusion rate would result in a drop in the creep rate and therefore an increase of the creep resistance.

Table 6 Crystal structure, lattice parameters, and density of selected intermetallic compounds

Alloy	Structure (Bravais lattice)	Lattice parameters		Density (g cm^{-3})
		a (nm)	c (nm)	
Ni_3Al	$L1_2$ (simple cubic)	0.357	—	7.40
NiAl	B2 (simple cubic)	0.288	—	5.96
Ni_2AlTi	DO_3	0.585	—	6.38
Ti_3Al	DO_{19}	0.577	0.464	4.23
TiAl	$L1_0$	0.398	0.405	3.89
Al_3Ti	DO_{22}	0.395	0.860	3.36
FeAl	B2 (simple cubic)			5.4–6.7 [599]
Fe_3Al	DO_3			5.4–6.7 [599]
$MoSi_2$	C11b			6.3

One major disadvantage of these materials, which is limiting their industrial application, is low fracture toughness [597]. This is attributed to several factors. First, the strong atomic bonds as well as the long-range order give rise to high Peierls stresses. Second, grain boundaries are intrinsically weak. The low boundary cohesion results in part from the directionality of the distribution of the electronic charge in ordered alloys [594]. The strong atomic bonding between the two main alloy constituents is related to the p-d orbital hibridization, which leads to a strong directionality in the charge distribution. The directionality is reduced in grain boundaries and the bonding becomes much weaker. Other factors that may contribute to the brittleness in intermetallics are the limited number of operative slip systems, segregation of impurities at grain boundaries, a high-work hardening rate, planar slip, and the presence of constitutional defects. The latter may be, for example, atoms occupying sites of a sublattice other than their own sub-lattice (antisites) or vacancies of deficient atomic species (constitutional vacancies). The planar faults, dislocation dissociations, and dislocation core structures typical of intermetallics were summarized by Yamaguchi and Umakoshi [598]. Other so-called extrinsic factors that cause brittleness are the presence of segregants, interstitials, moisture in the environment, poor surface finish, and hydrogen [599]. It appears that those intermetallics with more potential as high-temperature structural materials, i.e., those that are less brittle, are compounds with high crystal symmetry and small unit cells. Thus, nickel aluminides, titanium aluminides, and iron aluminides have been most studied over the last few decades. These investigations were stimulated by both the possibility of industrial application and scientific interest [592–601].

Creep resistance is a critical property in materials used for high-temperature structural applications. Some intermetallics may have the potential to replace nickel superalloys in parts such as the rotating blades of gas turbines or jet engines [602] due to their higher melting temperatures, high oxidation and corrosion resistance, high creep resistance, and in some cases lower density. The creep behavior of intermetallics is more complicated than that of pure metals and disordered solid solution alloys due to their complex structures together with the varieties of chemical composition [23,603]. The rate-controlling mechanisms are still not fully understood despite significant efforts over the last couple of decades [592,598,604–612].

In the following, the current understanding of creep of intermetallics will be reviewed, placing special emphasis on investigations published over the last two decades and related to the compounds with potential for structural applications such as titanium aluminides, iron aluminides, and nickel aluminides.

2. TITANIUM ALUMINIDES

2.1 Introduction

Titanium aluminide alloys have potential for replacing heavier materials in high-temperature structural applications such as automotive and aerospace engine components. This is due, first, to their low density (lower than that of most other intermetallics), high melting temperature, excellent elevated temperature strength, high modulus, oxidation resistance, and favorable creep properties [613,614]. Second, they can be processed through conventional manufacturing methods such as casting, forging, and machining [610]. In fact, TiAl turbocharger turbine wheels have recently been used in automobiles [610]. Table 7 compares the properties between titanium aluminides, titanium-based conventional alloys, and superalloys (see the phase diagram for phase compositions).

Many investigations have attempted to understand the creep mechanisms in titanium aluminides over the last two decades. There are several excellent reviews in this area [601,613,615,616]. The creep behavior of

Table 7 Properties of titanium aluminides, titanium-based conventional alloys, and superalloys

Property	Ti-based alloys	Ti$_3$Al-based α_2 alloys	TiAl-based γ alloys	Superalloys
Density (g cm^{-3})	4.5	4.1−4.7	3.7−3.9	8.3
RT modulus (GPa)	96−115	120−145	160−176	206
RT yield strength (MPa)	380−1115	700−990	400−630	250−1310[a]
RT tensile strength (MPa)	480−1200	800−1140	450−700	620−1620[a]
Highest temperature with high creep strength (°C)	600	750	1000	1090
Temperature of oxidation (°C)	600	650	900−1000	1090
Ductility (%) at RT	10−20	2−7	1−3	3−5
Ductility (%) at high T	High	10−20	10−90	10−20
Structure	hcp/bcc	DO19	L1$_0$	fcc/L1$_2$

[a]Data added to the table provided in Ref. [614].

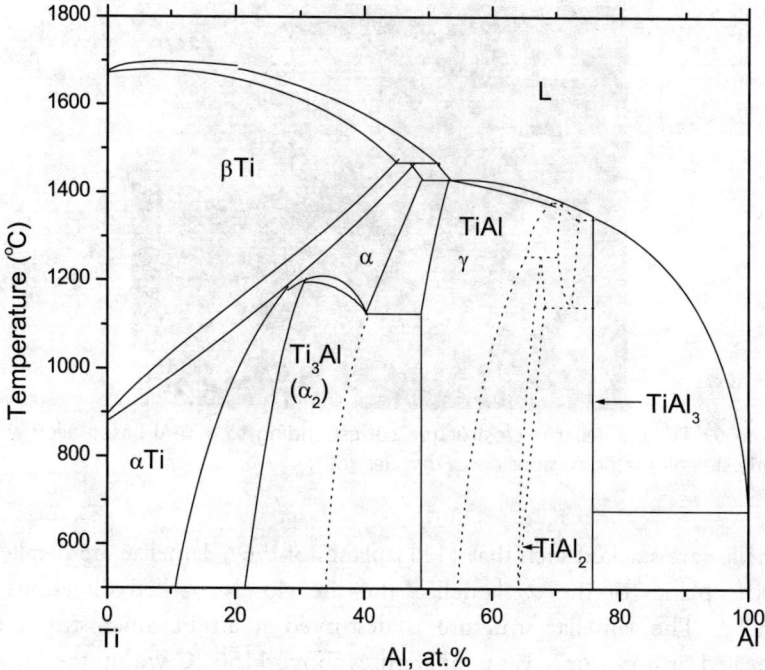

Figure 96 Ti-Al phase diagram. *From Ref. [594].*

titanium aluminides depends strongly on alloy composition and micro-structure. The different Ti-Al microstructures are briefly reviewed in the following.

Figure 96, the Ti-Al phase diagram, illustrates the following phases: γ-TiAl (ordered face-centered tetragonal, $L1_0$), α_2-Ti_3Al (ordered hexag-onal, DO_{19}), α-Ti (h.c.p., high-temperature disordered), and β-Ti (b.c.c., disordered). Gamma (γ) or near γ-TiAl alloys have compositions with 49–66 at.% Al, depending on temperature. The α_2 alloys contain from 22 at.% to approximately 35 at.% Al. Two-phase (γ-TiAl $+ \alpha_2$-Ti_3Al) alloys contain between 35 at.% and 49 at.% Al. The morphology of the two phases depends on the thermomechanical processing [610]. Alloys, for example, with nearly stoichiometric or Ti-rich compositions that are cast or cooled from the β phase, going through the α single-phase region and $\alpha \rightarrow \alpha + \gamma$ and $\alpha + \gamma \rightarrow \alpha_2 + \gamma$ reactions, have fully lamellar (FL) mi-crostructures, as illustrated in Figure 97. An FL microstructure consists of "lamellar grains," or colonies of size g_l, that are equiaxed grains composed of thin alternating lamellae of γ and α_2. The average thickness of the lamellae, termed the lamellar interface spacing, is denoted by λ_l. The γ and α_2

Figure 97 Fully lamellar microstructure corresponding to a Ti-Al based alloy with a nearly stoichiometric composition. *From Ref. [601].*

lamellae are stacked such that {111} planes of the γ lamellae are parallel to (0001) planes of the α_2 lamellae and the closely packed directions are parallel. The lamellar structure is destroyed if an FL microstructure is annealed or hot-worked at temperatures above 1150 °C within the $(\alpha + \gamma)$ phase fields. A bimodal microstructure develops, consisting of lamellar grains alternating with γ grains (or grains exclusively of γ-phase). Depending of the amount of γ-grains, the microstructure is termed "nearly lamellar" (NL) when the fraction of γ-grains is small, or duplex (DP) when the fractions of lamellar and γ-grains are comparable. Detailed studies of the microstructures of TiAl alloys are given elsewhere [617,618].

Overall, two phase γ-TiAl alloys have greater potential for high-temperature applications than α-Ti₃Al alloys due to their higher oxidation resistance and elastic modulus [614]. Simultaneously, two-phase γ-TiAl alloys have comparatively lower creep strength at high temperature than α-Ti₃Al alloys, and therefore significant efforts have been devoted to improve the creep behavior of γ-TiAl [615]. It is now well established that the optimum microstructure for creep resistance in two-phase TiAl alloys is FL [619–621]. As will be discussed in the subsequent sections, this microstructure shows the highest creep resistance, the lowest minimum creep rate, and the best primary creep behavior (i.e., longer times to attain a specified strain). Figure 98 illustrates the creep curves at 760 °C and 240 MPa corresponding to a Ti-48% Al alloy with several different microstructures. The FL microstructure shows superior creep resistance. Lamellar microstructures have also superior fracture toughness and fatigue

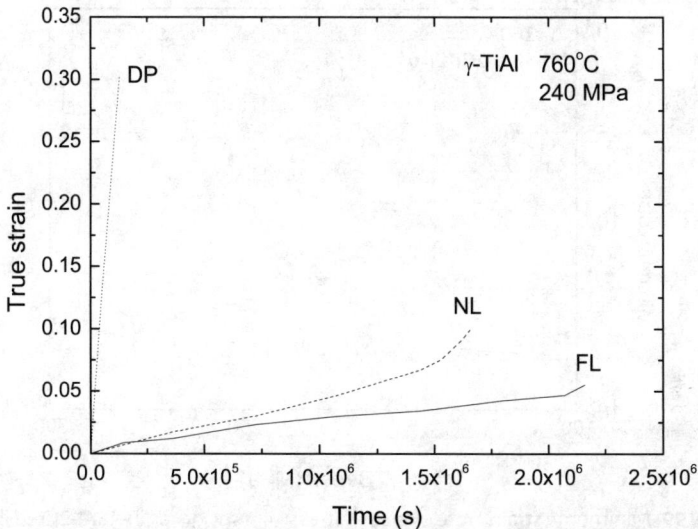

Figure 98 Creep curves at 760 °C and 240 MPa corresponding to several near γ-TiAl alloys with different microstructures. Ti-48Al alloy with a fully lamellar (FL) microstructure. Ti-48Al alloy with a nearly lamellar (NL) microstructure. Ti-48Al alloy with a duplex (DP) microstructure. *From Ref. [619].*

resistance as compared to DP structures, although the latter have, in general, better ductility [601]. This section will review the fundamentals of creep deformation in FL Ti-Al alloys. Emphasis will be placed on describing prominent recent creep models, rather than on compiling the extensive experimental data [615,616,619].

2.2 Rate-Controlling Creep Mechanisms in FL TiAl Intermetallics during Secondary Creep

Several investigations have attempted to determine the rate-controlling mechanisms during creep of FL TiAl intermetallics [396,615,616, 619–625]. Most creep studies were performed in the 676–877 °C temperature range [616] and 80–500 MPa, which are relevant to the anticipated service conditions [396]. Clarifying the rate-controlling creep mechanisms in FL TiAl alloys is difficult for several reasons. First, rationalization of creep data by conventional methods such as analysis of steady-state stress exponents is controversial, since an unambiguous secondary creep stage is not usually observed. Instead, a minimum strain rate, $\dot{\varepsilon}_{min}$, is measured, and the "secondary creep rate" or steady-state rate is presumed close to the minimum rate. Second, a continuous increase in the slope of

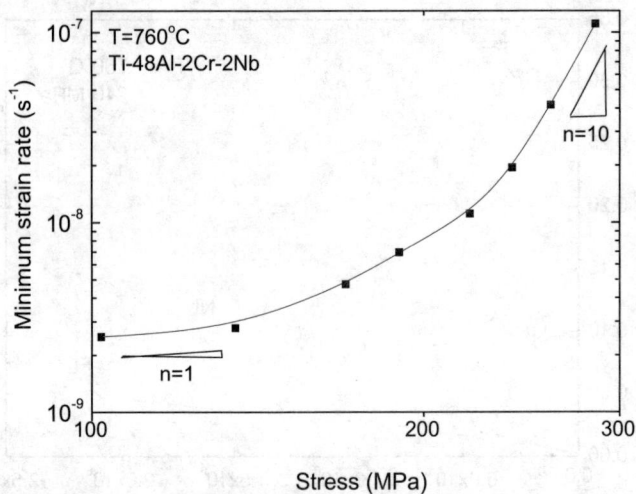

Figure 99 Minimum strain rate versus stress curve of a Ti-48Al-2Cr-2Nb alloy deformed at 760 °C. *From Ref. [619].*

the curve is observed (i.e., the stress exponent steadily increases as stress increases) when minimum strain rates are plotted versus modulus-compensated stress over a wide stress range. Figure 99 illustrates a minimum strain rate versus stress plot for an FL Ti–48Al–2Cr–2Nb at 760 °C. The stress exponent varies from $n = 1$, at low stresses, to $n = 10$ at high stresses [619]. Stress exponents as high as 20 have been measured at elevated stresses [614]. Third, the analysis of creep data is a difficult task because of the complex microstructures of FL TiAl alloys. Microstructural parameters such as lamellar grain size (g_l), lamellar interface spacing (λ_l), lamellar orientation, precipitate volume fraction, and grain boundary morphology all have significant influences on the creep properties that are difficult to incorporate into the traditional creep models that have been discussed earlier. Nevertheless, a variety of rate–controlling mechanisms of FL TiAl alloys has been proposed over the last few years.

2.2.1 High Stress–High Temperature Regime

Beddoes et al. [619] suggested, based on their own results and data from other investigators, that the gradual increase in the stress exponent with increasing stress might be due to changes in the creep mechanisms from diffusional creep at low stresses, to dislocation climb as the stress increases, and finally to power–law breakdown at very high stresses. (A note should be made that the creep data analyzed by Beddoes et al. originated from strain-rate change tests, rather than from independent creep tests.) Thus, these investigators claim that dislocation climb would most likely be rate-controlling during creep of FL microstructures at stresses higher than about 200 MPa and temperatures higher than about 700 °C. They suggested that this argument is consistent with the previous work on the creep of single-phase γ-TiAl alloys by Wolfenstine and González-Doncel [626]. These investigators analyzed the creep data of Ti-50 at.%Al, Ti-53.4 at.%Al and Ti-49 at.%Al tested from 700–900 °C, and concluded that the creep behavior of these materials could be described by a single mechanism by incorporating a threshold stress. The stress exponent was found to be close to 5 and the activation energy equal to 313 kJ mol^{-1}, a value close to that for lattice diffusion of Ti in γ-TiAl (291 kJ mol^{-1}) [627]. Additionally, several other studies on creep of FL TiAl alloys reported activation energies of roughly 300 kJ mol^{-1} [627]. Therefore, the creep of FL TiAl alloys appears to be controlled by lattice diffusion of Ti. The activation energy for lattice diffusion of Al in γ-TiAl has not been measured but it is believed to be significantly higher than that of Ti [628]. Es-Souni et al. [620] also suggested the predominance of a recovery-type dislocation-climb mechanism based on microstructural observations of the formation of dislocation arrangements (similar to subgrains) during creep. Several possible explanations have been suggested to reconcile the proposition of dislocation climb and the observation of high stress exponents ($n > 5$). First, it has been suggested [615] that backstresses may arise within lamellar microstructures due to the trapping of dislocation segments at the lamellar interfaces, which leads to bowing of dislocations between interfaces. The shear stress required to cause bowing, which was suggested as a source of backstresses during creep, is inversely proportional to the lamellar interface spacing (λ_l). Second, it has been proposed [629] that the occurrence of microstructural instabilities such as dynamic recrystallization during deformation may contribute to a rise in the strain rate, thus rendering stress exponents with less physical meaning in terms of a single, rate-controlling restoration mechanism. Finally, it has been suggested [615] that the subgrain size corresponding to a specific creep stress if dislocation climb were rate-controlling could be larger than the lamellar spacing, λ_l, which remains constant with stress. In fact, subgrains are not observed, particularly at low strains, where minimum creep rates are measured. Thus, it was suggested that λ_l may actually become the "effective subgrain size." These circumstances are similar to constant structure creep, which is associated with a relatively high stress exponent of 8 or more.

Beddoes et al. [619] later more precisely delineated the stress range in which dislocation climb was rate-controlling by performing stress-reduction tests on FL TiAl alloys. Figure 100 illustrates the results of the reduction tests on a Ti–48Al–2Cr–2Nb alloy at 760 °C, with an initial stress of 277 MPa. Data are illustrated for two different FL microstructures, both with a (lamellar) grain size of 300 μm, but with different lamellar interface spacing (120 and 450 nm). Deformation at a lower rate was observed upon reduction of the stress. An incubation period was observed for reduced stresses lower than a given stress (indicated with a dotted line) before deformation would continue. This was attributed [619] to the predominance of dislocation climb in the low-stress regime (below the dotted line), and to the predominance of dislocation glide in the high-stress regime (above the dotted line). The stress at which the change in mechanism occurs depends on the lamellar interface spacing. It was suggested that an increase in the lamellar interface spacing results in an increase of the stress below which dislocation climb becomes rate-controlling. Beddoes et al. [619] proposed an explanation for the decrease in the minimum creep rate with decreasing lamellar interface spacing for a given stress. First, for two

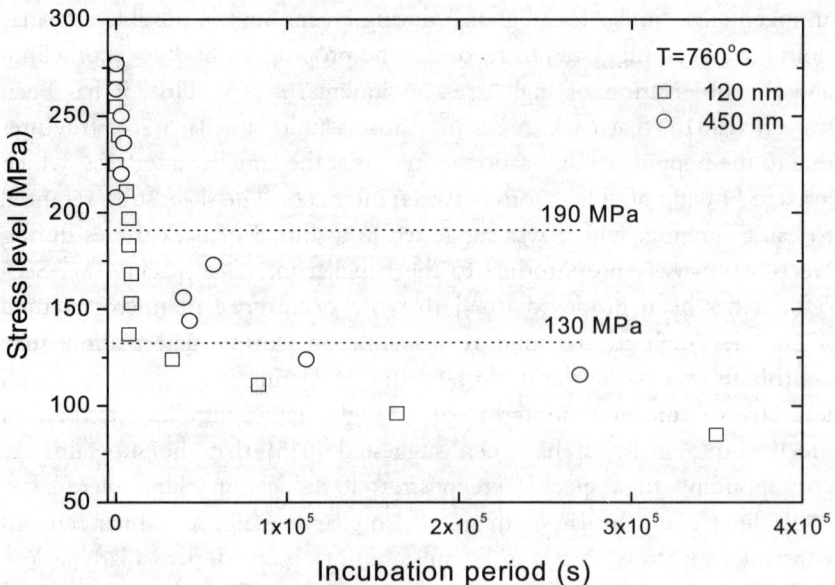

Figure 100 Incubation period following stress reduction tests to different final stresses. Tests performed at 760 °C in two Ti-48Al-2Cr-2Nb alloys with the same lamellar grain size (300 μm) but two different lamellar interface spacings (120 and 450 nm). Initial stress of 280 MPa. *From Ref. [619].*

microstructures deforming in the glide-controlled creep regime (for example, for stresses higher than 190 MPa in Figure 100), narrower spacings would increase the creep resistance, since the mean free path for dislocations would significantly decrease. The lamellae interfaces would thus act as obstacles for gliding dislocations. In fact, dislocation pile-ups have been observed at interfaces in FL structures [621]. Second, there is a stress range (for example, stresses between 130 and 190 MPa in Figure 100) for which the rate-controlling creep mechanism in microstructures with very narrow lamellae would be dislocation climb (associated with lower strain rates) whereas in others with wider lamellae it would be dislocation glide. This was attributed to the different backstresses originating at lamellae of different thicknesses. A larger Orowan stress is necessary to bow dislocations in narrow lamellae than in wider lamellae. Thus, an applied stress of 130–190 MPa would be high enough to cause dislocation bowing in the material with wider lamellae. Dislocation glide would be controlled by the interaction between dislocations and interfaces rather than climb. However, in narrower lamellae, the applied stress is not sufficient to cause dislocation bowing and dislocation movement is then controlled by climb.

Mills et al. [79,931] studied the creep properties of an FL Ti-48Al-2Cr-2Nb alloy at high stresses (207 MPa) and high temperatures (around 800 °C), particularly examining the dislocation structures developed during deformation. They mainly observed unit a/2 [110] dislocations with jogs pinning the screw segments. They found that a distribution of lamellae spacings exists in an FL microstructure. A higher dislocation density was observed in the wider lamellae, suggesting, to these investigators, that wider lamellae contribute more to creep strain than thinner lamellae. Additionally, no subgrains were observed at the minimum creep rate (typically 1.5–2% plastic strains). The absence of subgrain formation during secondary creep of single-phase TiAl alloys under conditions with an activation energy similar to that of self-diffusion was observed (thus suggesting the predominance of dislocation climb). In order to rationalize this apparent discrepancy between the behavior of single-phase Ti-Al alloys and pure metals, Mills and coworkers proposed a modification of the jogged-screw creep model discussed in a previous chapter. The original model [190,633,634] suggests that the nonconservative motion of pinned jogs along screw dislocations is the rate-controlling process. Using this model, the "natural" stress exponent is derived. The conventional jogged-screw creep model predicts strain rates that are several orders of magnitude higher than the measured values in TiAl [632]. The modification proposed by Mills et al. incorporated the presence of tall jogs instead of assuming the jog heights as equal to the Burgers vector. Additionally, it is proposed that there should be an upper bound for the jog height, above which the jog becomes a source of dislocations. This maximum jog height, h_d, depends on the applied stress and can be approximated by:

$$h_d = \left(\frac{Gb}{\{8\pi(1-\nu)\tau\}} \right) \tag{163}$$

This is suggested to reasonably predict the strain rates in single-phase TiAl alloys and could account for the absence of subgrain formation during secondary creep. At the same time, by introducing this additional stress dependence in the equation for the strain rate, the phenomenological stress exponent of 5 is obtained at intermediate stresses. This exponent increases with increasing stress. Mills et al. [930,931] claim that the same model can be applied to creep of FL microstructures, where deformation mainly occurs within the wider γ-laths by jogged a/2 [110] unit-dislocation slip. Evidence of the presence of jogged screw dislocations has been extensively reported [630,632,930–933].

Wang et al. [635] observed that, together with dislocation activity and some twinning, thinning and dissolution of α_2 lamellae and coarsening of γ-lamellae occurred during creep of two FL TiAl alloys at high stresses (e.g., >200 MPa at 800 °C and >400 MPa at 650 °C). They proposed a creep model based on the movement of ledges (or steps) at lamellar interfaces to rationalize these observations. Wang et al. [635] observed the presence of ledges at the lamellar interfaces already before deformation. Two such ledges of height h_L are illustrated in Figure 101. Growth of the γ

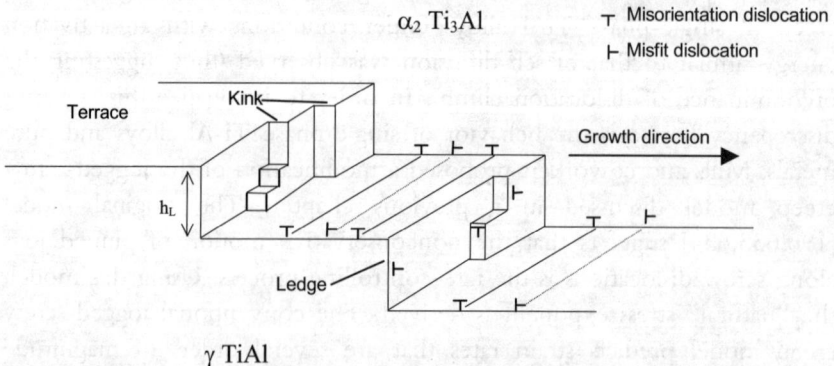

Figure 101 Interface separating γ and α_2 lamellae. Ledge size is denoted by h_L. From Ref. [635].

phase at the expense of the α_2 phase could occur by ledge movement as a consequence of the applied stress. Ledge motion was suggested to involve glide of misfit dislocations and climb of misorientation dislocations. Ledge motion leading to the transformation from α_2 to γ may account for a significant amount of the creep deformation, since (1) as mentioned above, it requires dislocation movement; and (2) it also involves a volume change from α_2 to γ. At high stresses, multiple ledges (i.e., ledges that are several {111} planes in thickness) are suggested to be able to form and dissolve, and thus deformation may occur. Diffusion of atoms is needed since the climb of misorientation dislocations is necessary for a ledge to move. Also, diffusion is needed for the composition change associated with the phase transformation from α_2 to γ. Thus, lattice self-diffusion becomes rate-controlling, in agreement with previous observations of activation energies close to Q_{SD}.

Modeling of the creep behavior of FL microstructures has also been undertaken by Clemens et al. [936,937], who found that, in the climb-dominated regime, the strain rate can be related to the applied stress by a conventional power law equation with an additional factor (a so-called structure factor), which is a function of the lamellar orientation and the mean lamellar interface spacing, λ_l. This micromechanical model reasonably predicts the decrease in the minimum creep rate with decreasing λ_l.

Alloy additions are another factor that may influence the creep rate. It is well known that additions of W greatly improve creep resistance [616]. It has been suggested that solute hardening by W occurs within the glide-controlled creep regime, whereas the addition of W may lower the diffusion rate, thus reducing the dislocation climb rate in the climb-controlled creep regime. The effect of ternary or quaternary additions on the creep resistance may be more important than that of the lamellar interface spacing [619]. Additions of W, O, Si, C, and N favor precipitation hardening [636–639,939], which may hinder dislocation motion and stabilize the lamellar microstructure. In particular, carbide and silicide particles have been observed to precipitate preferentially during creep testing or previous aging in places originally occupied by α_2 laths. In essence, the precipitates replace the α_2 lamellae, thereby maintaining restricted dislocation motion [940]. Other suggested hardening elements include Nb and Ta [616]. The addition of B does not seem to have any effect on the minimum strain rate of FL microstructures [616].

The lamellar orientation also has a significant influence on the creep properties of FL TiAl alloys [602,640,641,941,942]. Hard orientations (i.e., those in which the lamellae are parallel or perpendicular to the tensile axis) show improved creep resistance and low strain to failure; soft orientations (those in which the lamellae form an angle of 30°–60° with the tensile axis) are weaker but are more ductile [602]. The different behaviors were rationalized by considering changes in the Taylor factors and Hall–Petch strengthening [602]. Basically, in soft orientations, the shear occurs parallel to the lamellar boundaries. In hard orientations, however, the resolved shear stress in the planes parallel to the lamellae is very low, and therefore other systems are activated. Thus it was suggested that the mean free path for dislocations is larger in soft orientations than in hard orientations [640,641,943].

2.2.2 Low-Stress Regime

Hsiung and Nieh [396] investigated the rate-controlling creep mechanisms during secondary creep (or minimum creep rate) at low stresses in an FL Ti-47Al-2Cr-2Nb alloy. In particular, they studied the stress/temperature range where stress exponents between 1 and 1.5 were observed. They reported an activation energy equal to $160 \, \text{kJ mol}^{-1}$ within this range, which is significantly lower than the activation energy for lattice diffusion of Ti in γ-TiAl ($291 \, \text{kJ mol}^{-1}$) [627] and much lower than the activation energy for lattice diffusion of Al in γ-TiAl. They suggested that dislocation climb is less important at low stresses. They also discarded grain boundary sliding as a possible deformation mechanism due to the presence of inter-locking grain boundaries such as those shown in Figure 97. These are boundaries in which there is not a unique boundary plane. Instead, the lamellae from adjacent (lamellar) grains are interpenetrating at the bound-ary, thus creating steps and preventing easy sliding [619]. TEM examination revealed both lattice dislocations (including those that are free within the γ-laths and threading dislocations that have their line ends within the lamellar interfaces) and interfacial (Shockley) dislocations, the density of the latter being much larger. They proposed that, due to the fine lamellar interface spacing ($\lambda_l < 300 \, \text{nm}$), the operation and multiplication of lattice dislocations at low stresses are very sluggish. Dislocations can only move small distances ($\approx \lambda_l$) and the critical stress to bow threading dislocation lines (which is inversely proportional to the lamellar interface spacing) is, on average, higher than the applied stress. Thus, Hsiung and Nieh [396] concluded that dislocation slip by threading dislocations could not ratio-nalize the observed creep strain in alloys with thin laths. They proposed that the predominant deformation mechanism was interfacial sliding at lamellae interfaces caused by the viscous glide of interfacial (Shockley) dislocation arrays. These arrays might eventually be constituted by an odd number of partials, in which case a stacking fault is created at the interface. Stacking faults are indeed observed by TEM [396]. Segregation of solute atoms may cause Suzuki locking. Thus, according to Hsiung and Nieh, the viscous glide of interfacial dislocations (dragged by solute atoms) is the rate-controlling mechanism. It was suggested that further reduction of the lamellar interface spacing (in the range of $\lambda_l \leq 300$ nm) would not significantly affect the creep rate once the γ-laths are thin enough for interfacial sliding to occur.

Zhang and Deevi [616] recently analyzed creep data of several TiAl alloys with Al concentrations ranging from 46 to 48 at.% compiled from numerous other studies. They proposed expressions relating the minimum creep rate and the stress that could reasonably predict most of the data. They recognized that using the classical constitutive equations and power-law models could be misleading, due to the large and gradual variations of the stress exponent with stress. They utilized:

$$\dot{\varepsilon}_{min} = \dot{\varepsilon}_0 \sinh\left(\frac{\sigma}{\sigma_{int}}\right) \tag{164}$$

where σ is the applied stress, and $\dot{\varepsilon}_0$ and σ_{int} are both temperature and material-dependent constants. Additionally:

$$\dot{\varepsilon}_0 \propto \rho_{sr} D_0 \exp\left(\frac{-Q_{sd}}{kT}\right) \tag{165}$$

where ρ_{sr} is the dislocation-source density. The physical meaning of Eqn (165) is based on a viscous glide process. They suggested [616] that σ_{int} and $\dot{\varepsilon}_0$ are independent of the lamellar interface spacing for FL microstructures with $\lambda_l > 0.3$ μm, and that σ_{int} and $\dot{\varepsilon}_0$ increase with decreasing λ when $\lambda_l < 0.3$ μm. $\dot{\varepsilon}_0$ is temperature dependent and Q_{sd} is about 375 kJ mol^{-1} for microstructures with different lamellar interface spacing. This value is slightly higher than the activation energy for diffusion of Ti in TiAl (291 kJ mol^{-1}). Zhang and Deevi [616] attributed this discrepancy to the fact that the dislocation source density, ρ_s, may not be constant as assumed in Eqn (165), and that the creep of TiAl may be controlled by diffusion of both Ti and Al in TiAl. Since the activation energy for self-diffusion of Al is higher than that of Ti, a combination of diffusion of both species could justify the higher Q values measured.

The above equations do not predict the creep data of FL TiAl alloys obtained at both stresses lower than about 150 MPa and low temperatures. This was suggested to be due to grain boundary sliding being the dominant mechanism [396]. In this stress–temperature range, Zhang and Deevi found that most of the creep data could be described by:

$$\dot{\varepsilon}_{min}(GB) = 63.4 \exp(-2.18 \times 10^5/kT)g^{-2}\sigma^2 \qquad (166)$$

2.3 Primary Creep in FL Microstructures

γ-TiAl alloys are characterized by a pronounced primary creep regime. Depending on the temperature, the primary creep strain may exceed the acceptable limits for certain industrial applications. Thus, several investigations have focused on understanding the microstructural evolution during primary creep [619,623,629,640,642,643].

Figure 102 illustrates the creep curves corresponding to a TiAl binary alloy deformed at 760 °C and at an applied stress of 240 MPa. The creep curves correspond to a DP microstructure and a FL microstructure. The FL microstructure shows lower strain rates during primary creep. It has been suggested [629] that the pronounced primary creep regime in FL microstructures is due to the presence of a high density of interfaces and dislocations, since both may act as sources of dislocations. Careful TEM examination by Chen et al. [640] showed that dislocations formed loops

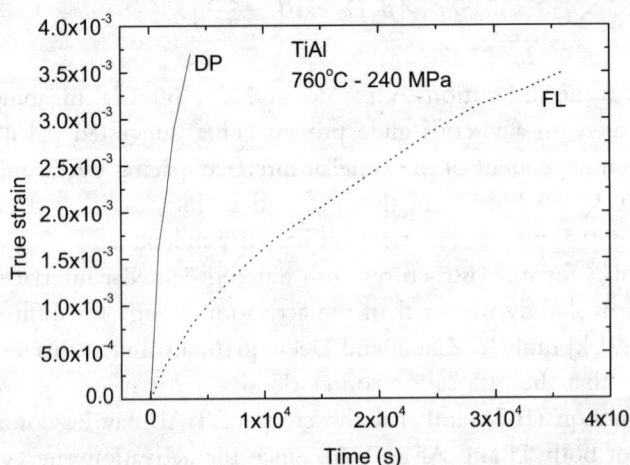

Figure 102 Primary creep behavior of a binary TiAl alloy deformed at 760 °C at an applied stress of 240 MPa. The creep curves corresponding to duplex (DP) and fully lamellar (FL) microstructures are illustrated. *From Ref. [619].*

that expand from the interface to the next lamellar interface. Other processes that may occur during primary creep of TiAl alloys are twinning and stress-induced phase transformations (SIPT) ($\alpha_2 \rightarrow \gamma$ or $\gamma \rightarrow \alpha_2$). SIPT consists of the transformation of α_2 laths into γ laths (or vice versa). This transformation, which is aided by the applied stress, has been suggested to occur by the movement of ledge dislocations at the γ/α_2 interfaces, as illustrated in Figure 101 [638]. The SIPT may be associated with a relatively large creep strain. The finding that the primary strain in microstructures with narrow lamellae (FLn) is not higher than the primary strain in alloys with wider lamellae (FLw) suggests that the contribution of interface boundary sliding by the motion of pre-existing interfacial dislocations is less important [623]. It is possible that a sufficient number of interfacial dislocations need to be generated during primary creep before the onset of sliding.

Zhang and Deevi recently analyzed primary creep of TiAl-based alloys [623] and concluded that the primary creep strain depends dramatically on stress. At stresses lower than a critical value, σ_{cr}, the primary strain is low (about 0.1–0.2%), independent of the microstructure, temperature, and composition. The relevant stresses anticipated for industrial applications are usually below σ_{cr} and therefore primary creep strain would be less important [623]. The value of σ_{cr} seems to be mainly related to the critical stress to activate dislocation sources, twinning, and stress-induced phase transformations. This value increases with W additions, with lamellar refinement, and with precipitation of fine particles along lamellar interfaces [623]. For example, the value of σ_{cr} at 760 °C for an FL Ti-47 at.% Al-2 at.%Nb-2 at.%Cr alloy with a lamellar spacing of 0.1 μm is close to 440 MPa, whereas the same alloy with a lamellar spacing larger than 0.3 μm has a σ_{cr} value of 180 MPa [623]. Primary creep strain increases significantly above the threshold stress. In order to investigate additional factors influencing the primary creep strain, Zhang and Deevi modeled primary creep of various TiAl alloys using the following expression, also utilized previously in other works [615,642]:

$$\varepsilon_p = \varepsilon_0' + A'(1 - \exp(-\alpha't)) \tag{167}$$

This expression reflects that primary creep strain consists of an "instantaneous" strain (ε_0') that occurs immediately upon loading and a transient strain that is time dependent. Zhang and Deevi suggested that the influence of temperature on ε_0', A', and α' could be modeled using the relation $X = X_0 \exp(-Q/RT)$, where X represents any of the three parameters. By fitting a large amount of data, they obtained that $Q = 190$ kJ mol^{-1} for A$'$, and $Q = 70$ kJ mol^{-1} for ε_0' and α'. The physical basis for the temperature dependence is unclear. The effect of composition and microstructure on these parameters is also complex [623]. Aging treatments before creep deformation appear to have a beneficial effect in increasing the primary creep resistance [623,638,643]. Precipitation at lamellar interfaces has been suggested to hinder dislocation generation [34] and reduces the instantaneous strain and the strain-hardening constant, A$'$. Additionally, the presence of fine precipitates may inhibit interface sliding and even twinning. Finally, the contribution of stress-induced phase transformation to the primary creep strain decreases in samples heat treated before creep deformation, since metastable phases are eliminated [623].

2.4 Tertiary Creep in FL Microstructures

Several investigations have studied the effect of the microstructure on tertiary creep of FL TiAl alloys [615,619,640]. It has been suggested that tertiary creep is initiated due to strain incompatibilities between lamellar grains with soft and hard orientations leading to particularly elevated stresses [619]. These incompatibilities may lead to intergranular and interlamellar crack formation. Crack growth is retarded when lamellar grains are smaller than 200 μm (lamellar grain sizes are typically 500 μm in diameter), since cracks can be arrested by grain boundaries or by a triple points. Grain boundary morphology also significantly influences tertiary creep behavior. In FL microstructures with wide lamellae, a well-interlocked lamellae network forms [619]. However, narrow lamellae are more planar. Grain boundaries in which lamellae are well interlocked offer greater resistance to cracking and allow larger strains to accumulate within the grains.

Tertiary creep in lamellar structures has been connected with the breakdown of the structure by coarsening and spheroidization [943]. Interfaces of γ–γ are highly unstable and they migrate and coalesce with other γ–γ boundaries, while α$_2$ laths dissolve during high-temperature exposure. Other processes that lead to tertiary creep are the onset of discontinuous dynamic recrystallization, i.e., the nucleation and growth of equiaxed gamma grains [943,944]. Coarsening and spheroidization can be prevented by designing a lamellar structure with a large number of α$_2$–γ boundaries, which have been shown to have a higher thermal stability [943].

In summary, the creep behavior of FL TiAl alloys is influenced by many different microstructural features, and it is difficult to formulate a model that incorporates all of the relevant variables. Optimal creep behavior may require [619]:

1. A lamellar grain size smaller than about 200 μm that helps to improve creep life by preventing early fracture. Some studies have suggested that grain size has only a small effect [16,17].
2. A narrow interlamellar spacing which reduces the minimum strain rate during secondary creep.
3. Interlocked lamellar grain boundaries.
4. Alloy chemistry: W, Nb, Mo, and V have a large strengthening effect as solutes; C, B, and N significantly influence properties via precipitation.
5. Stabilized microstructure, or presence of a large fraction of $\gamma-\alpha_I$ interfaces.

3. IRON ALUMINIDES

3.1 Introduction

Fe_3Al- and FeAl-based ordered intermetallic compounds have been extensively studied due to their excellent oxidation and corrosion resistance as well as other favorable properties such as low density, favorable wear resistance, and potentially lower cost than many other structural materials. Fe_3Al has a DO_3 structure. FeAl is a B2-ordered intermetallic phase with a simple cubic lattice with two atoms per lattice site, an Al atom at position (x,y,z) and a Fe atom at position $(x + 1/2, y + 1/2, z + 1/2)$. Several recent reviews summarizing the physical, mechanical and corrosion properties of these intermetallics are available [592,594–596,599–601,644–646]. Iron aluminides are especially attractive for applications at intermediate temperatures in the automotive and aerospace industry due to their high specific strength and stiffness. Additionally, they may replace stainless steels and nickel alloys to build long-lasting furnace coils and heat exchangers due to superior corrosion properties. The principal limitations of Fe–Al intermetallic compounds are low ambient-temperature ductility (due mainly to the presence of weak grain boundaries and environmental embrittlement) and only moderate creep resistance at high temperatures [646]. Many efforts have been devoted in recent years to overcoming these difficulties. The present review will discuss the strengthening mechanisms of iron aluminides as well as other high-temperature mechanical properties of these materials.

Figure 103 illustrates the Fe–Al phase diagram. The Fe_3Al phase, with a DO_3 ordered structure, corresponds to Al concentrations ranging from

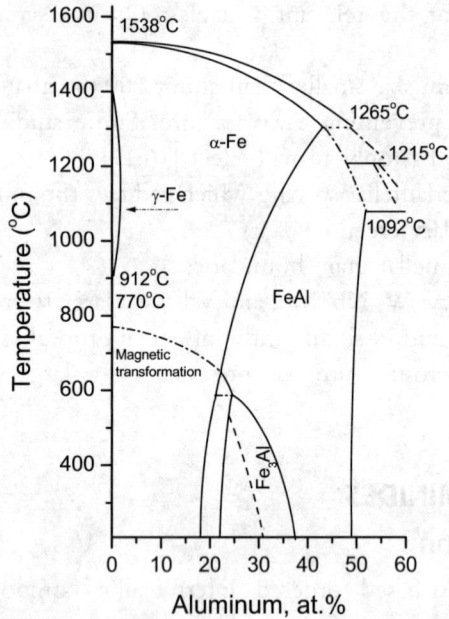

Figure 103 Fe-Al phase diagram. *From Ref. [647].*

approximately 22 at.% and 35 at.%. A phase transformation to an imperfect B2 structure takes place above 550 °C. The latter ultimately transforms to a disordered solid solution with increasing temperature. This, in turn, leads to the degradation of creep and tensile resistance at high temperatures. The FeAl phase, which has a B2 lattice, is formed when the amount of Al in the alloy is between 35 at.% and 50 at.%.

3.2 Anomalous Yield Point Phenomenon

An anomalous peak in the variation of the yield stress with temperature has been observed in Fe-Al alloys with an Al concentration ranging from 25 at.% up to 45 at.%. The peak appears usually at intermediate temperatures, between 400 °C and 600 °C [601,645,646,648–667]. This phenomenon is depicted in Figure 104, which illustrates the dependence of the yield strength with temperature for several large-grain Fe-Al alloys in tension at a strain rate of 10^{-4} s^{-1}. Several mechanisms have been proposed to rationalize the yield–strength peak, but the origin of this phenomenon is still not well understood. The different models are briefly described in the following. More comprehensive reviews on this topic as well as critical analyses of the validity of the different strengthening mechanisms are available [645,648].

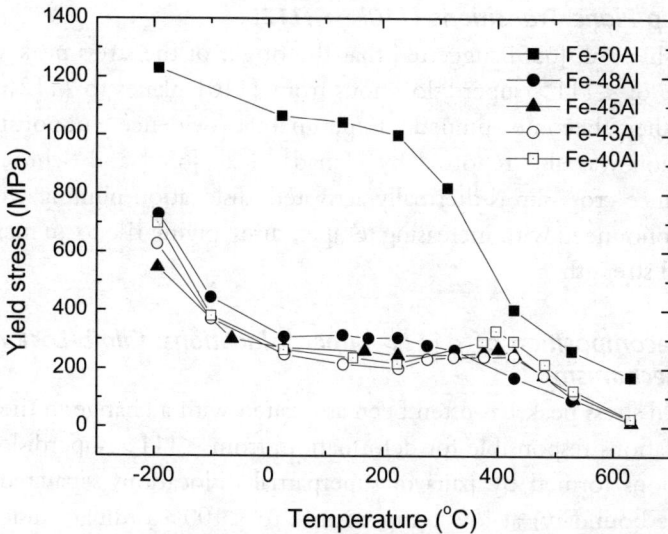

Figure 104 Variation of the yield strength with temperature for several large-grain FeAl alloys strained in tension at a strain rate of $10^{-4}\,s^{-1}$. *From Ref. [653].*

3.2.1 Transition from Superdislocations to Single Dislocations

Stoloff and Davies [654] suggested that the stress peak was related to the loss of order that occurs in Fe_3Al alloys at intermediate temperatures (transition between the DO_3 to the B2 structure). According to their model, at temperatures below the peak, superdislocations would lead to easy deformation, whereas single dislocations would, in turn, lead to easy deformation at high temperatures. At intermediate temperatures, both superdislocations and single dislocations would move sluggishly, giving rise to the strengthening observed. It has been suggested, however, that this model cannot explain the stress peak observed in FeAl alloys, where no disordering occurs at intermediate temperatures and the B2 structure is retained over a large temperature interval. Recently, Morris et al. also questioned the validity of this model [651]. They observed that the stress peak occurred close to the disordering temperature at low strain rates in two Fe_3Al alloys (Fe-28Al-5Cr-1Si-1Nb-2B and Fe-25Al, all atomic percent). The stress peak occurred at higher temperatures at higher strain rates than those corresponding to the transition. Thus, they concluded that the disordering temperature being about equal to the peak stress temperature at low strain rates was coincidental. Stein et al. [667] also did not find a correlation between these temperatures in several binary, ternary, and quaternary DO_3-ordered Fe-26Al alloys.

3.2.2 Slip Plane Transitions {110} → {112}

Umakoshi et al. [655] suggested that the origin of the stress peak was the cross-slip of <111> superdislocations from {110} planes to {112} planes, where they become pinned. Experimental evidence supporting this observation was also reported by Hanada et al. [656] and Schroer et al. [657]. Since cross-slip is thermally activated, dislocation pinning would be more pronounced with increasing temperatures, giving rise to an increase in the yield strength.

3.2.3 Decomposition of <111> Superdislocations: Climb-Locking Mechanism

The yield stress peak has often been associated with a change in the nature of dislocations responsible for deformation, from <111> superdislocations (dislocations formed by pairs of superpartial dislocations separated by an antiphase boundary) at low temperatures, to <100> ordinary dislocations at temperatures above the stress peak [658]. The <100> ordinary dislocations are sessile at temperatures below those corresponding to the stress peak and thus may act as pinning points for <111> superdislocations. The origin of the <100> dislocations has been attributed to the combination of two a/2 [111] superdislocations or to the decomposition of <111> superdislocations on {110} planes into <110> and <100> segments on the same {110} planes. As the temperature increases, decomposition may take place more easily, and thus the number of pinning points would increase leading to a stress peak. Experimental evidence consistent with this mechanism has been reported by Morris et al. [650]. However, this mechanism was also later questioned by Morris et al. [651], where detailed TEM microstructural analysis suggested that anomalous strengthening is possible without the <111> to <100> transition in some Fe₃Al alloys.

3.2.4 Pinning of <111> Superdislocations by Antiphase Boundary Order Relaxation

An alternative mechanism for the appearance of the anomalous yield stress peak is the loss of order within antiphase boundaries (APBs) of mobile <111> superdislocations with increasing temperature [660]. This order relaxation may consist of structural changes as well as variations in the chemical composition. Thus, the trailing partial of the superdislocation would no longer be able to restore perfect order. A frictional force would therefore be created that will hinder superdislocation movement. With increasing temperature, APB relaxation would be more favored, and thus increasing superdislocation pinning would take place, leading to the observed stress peak.

3.2.5 Vacancy Hardening Mechanism

The concentration of vacancies in FeAl is relatively high and increases in Al-rich alloys. Constitutional vacancies are those required to maintain the B2 structure in Al-rich non-stoichiometric FeAl alloys. Thermal vacancies are those excess vacancies generated during annealing at high temperature and retained upon quenching. For example, the vacancy concentration is 40 times larger at 800 °C than that corresponding to a conventional pure metal at the melting temperature. Constitutional vacancies may occupy up to 10% of the lattice sites (mainly located in the Fe sublattice) for Fe-Al compositions with high Al content (>50 at.%) [661]. The high vacancy concentration is due to the low value of the enthalpy of formation of a vacancy as well as to the high value of the entropy of formation (around 6 k) [662]. Vacancies have a substantial influence on the mechanical properties of iron aluminides [663].

It has been suggested that the anomalous stress peak is related to vacancy hardening in FeAl intermetallics [664]. According to this model, a larger number of vacancies are created with rising temperatures. These defects pin superdislocation movement and lead to an increase in yield strength. At temperatures higher than those corresponding to the stress peak the concentration of thermal vacancies is very large and vacancies are highly mobile. Thus, they may aid dislocation climb processes instead of acting as pinning obstacles for dislocations [665] and softening occurs.

The vacancy hardening model is consistent with many experimental observations. Recently Morris et al. [650] reported additional evidence for this mechanism in a Fe-40 at.%Al alloy. First, they observed that some time is required at high temperature for strengthening to be achieved. This may be consistent with the requirement of some time at temperature to create the equilibrium concentration of vacancies required for hardening. Second, they noted that the stress peak is retained when the samples are quenched and tested at room temperature. They concluded that the point defects created after holding the specimen at temperature for a given amount of time are responsible for the strengthening both at high and low temperatures. Additionally, careful TEM examination suggested to these investigators that vacancies were not present in the form of clusters. Instead, the small dislocation curvature observed suggested that single vacancies were mostly present, which act as relatively weak obstacles to dislocation motion.

The concentration of thermal vacancies increases with increasing Al content, and thus the effect of vacancy hardening would be substantially influenced by alloy composition. Additionally, the vacancy-hardening mechanism implies that the hardening should be independent of the strain rate, since it only depends on the amount of point defects present. In a recent investigation, Morris et al. [651] analyzed the effect of strain rate on the flow stress of two Fe_3Al alloys with compositions Fe-28Al-5Cr-1 Si-1Nb-2B and Fe-25Al-5Cr-1Si-1Nb-2B (at.%). The variation of the flow stress with temperature and strain rate (ranging from $4 \times 10^{-6}\,s^{-1}$ to $1\,s^{-1}$) for the Fe-28 at.%Al alloy is illustrated in Figure 105. It can be observed that the "strengthening" part of the peak is rather insensitive to strain rate, consistent with the predictions of the vacancy-hardening model. However, the investigators were skeptical regarding the effectiveness of this mechanism in Fe_3Al alloys, where the vacancy concentration is much lower than in FeAl alloys, and moreover where the vacancy mobility is higher. Highly mobile vacancies are not as effective obstacles to dislocation motion. Another limitation of the vacancy model is that it fails to explain the orientation dependence of the stress anomaly as well as the tension-compression asymmetry in single crystals [649]. Thus, the explanation of the yield stress peak remains uncertain. On the other hand, it is evident in Figure 105 that the softening part of the peak is indeed highly dependent on strain rate. This is attributed to the onset of diffusional processes at high

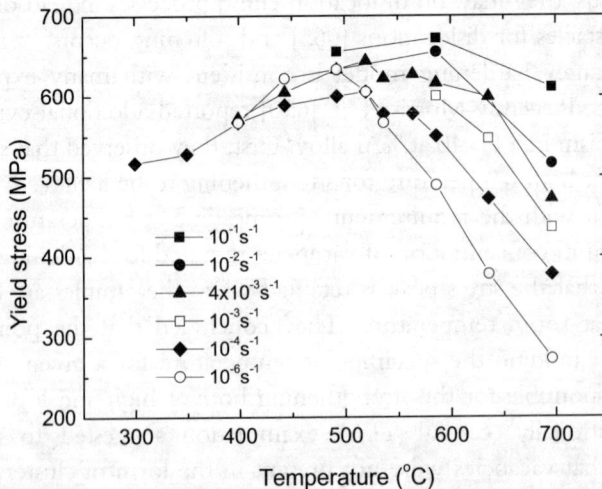

Figure 105 Variation of the yield stress with temperature and strain rate corresponding to the cast and homogeneized Fe-28 at.%Al alloy. *From Ref. [651].*

temperatures, where creep models may be applied and rate dependence may be more substantial [651].

3.3 Creep Mechanisms

The creep behavior of iron aluminides is still not well understood despite the large amount of creep data available on these materials [399,646,668–679]. The values of the stress exponents and activation energies corresponding to several creep studies are summarized in Table 8 (based on [646] with additional data). There are several factors that complicate the formulation of a general creep behavior of Fe-Al alloys. First, creep properties are significantly influenced by composition. Second, as discussed elsewhere [594,680], both the stress exponent and the activation energy have been observed to depend on temperature, in some cases. This suggests that several mechanisms may control creep of Fe-Al and Fe$_3$Al alloys. Other reasons may be the frequent absence of genuine steady-state conditions as well as the simultaneous occurrence of grain growth and discontinuous dynamic recrystallization. Nevertheless, in general, it can be inferred from Table 8 that a lower stress-exponent creep mechanism may dominate at very low stresses and high temperatures. At intermediate temperatures and stresses, diffusion-controlled dislocation climb and viscous drag have been suggested [399,646,669,760].

3.3.1 Superplasticity in Iron Aluminides

Superplasticity has been observed in both FeAl and Fe$_3$Al with coarse grains ranging from 100 to 350 μm [398,681–684]. Elongations as high as 620% were achieved in a Fe-28 at.%Al-2 at.%Ti alloy deformed at 850 °C and at a strain rate of 1.26×10^{-3} s^{-1}. The corresponding n value was equal to 2.5. Also, a maximum elongation of 297% was reported for a Fe-36.5 at.%Al-2 at.%Ti alloy with an n value close to 3. Moreover, Lin et al. [684] have reported an increasing number of boundaries misoriented between 3° and 6° with deformation. They suggest that these could be formed as a consequence of dislocation interaction, by a process of continuous recrystallization (or continuous reactions). The unusually large starting grain sizes as well as the values of the stress exponents (close to 3) would be consistent with a viscous drag deformation mechanism. However, significant grain refinement has been observed during deformation [683,684]. The correlation of grain refinement and large ductilities may, in turn, be indicative of the occurrence of grain boundary sliding.

Table 8 Stress exponents, activation energies, and suggested deformation mechanisms from various creep studies on iron aluminides

Alloy	T (°C)	Q (kJ mol⁻¹)	n	Mechanism suggested	Reference
Fe-19.4Al	500–600	305	4.6–6^a	Diffusion controlled	[668]
Fe-27.8Al	550–615	276	—	Controlled by state of order	
	Higher T	418			
Fe-15/20Al	>500	260–305^a σ dependent		Diffusion controlled	[669]
	<500			Motion of jogged screw dislocations	
Fe-28Al	625	347	3.5 (low σ)	Viscous glide	
			7.7 (high σ)	Climb	
Fe-28Al-2Mo	650	335	1.4 (low σ)	Diffusional flow	[670]
			6.8 (high σ)	Climb	
Fe-28Al-1Nb-0.013Zr	650	335	1.8 (low σ)	Diffusional flow	
			19.0 (high σ)	Dispersion strengthening	
FA-180	593	627	7.9	Precipitation strengthening	[671]
Fe-28Al	600–675	—	3.4	Viscous glide	[646]
Fe-26Al-0.1C	600–675	305	3.0	Viscous glide	
	480–540	403	6.2	—	

Alloy	Temperature	Stress	n	Mechanism	Ref.
Fe-28Al-2Cr	600—675	325	3.7	Viscous glide	[672]
Fe-28Al-2Cr-0.04B	600—675	304	3.7	Viscous glide	[673]
Fe-28Al-4Mn	600—675	302	2.6	Viscous glide	[674]
FA-129	500—610	380—395	4—5.6		[675]
Fe-24Al-0.42 Mo-0.05B-0.09 C-0.1 Zr	650—750	–	5.5		
FA-129	900—1200	335	4.81	Dispersion strengthening	
Fe-39.7Al-0.05Zr-50 ppmB	500	260—300	11	Climb	
	700	425—445	11		[399]
Fe-27.6Al	425—625	375	2.7—3.4	Viscous glide	
Fe-28.7Al-2.5Cr	425—625	325	3.5—3.8	Viscous glide	
Fe-27.2Al-3.6Ti	425—625	375	3.4—3.7	Viscous glide	[676]
Fe-24Al-0.42Mo-0.1Zr-0.005B-0.11C-0.31O	800—1150	340—430	4—7	Diffusion-controlled (climb)	
	1150 (Strain rate <0.1 s^{-1})	365	3.3	Diffusion-controlled (superplasticity)	
Fe-30.2Al-3.9Cr-0.94Ti-1.9B-0.20Mn-0.16C	600—900	280	3.3	Viscous glide	[677]
Fe-47.5Al	827—1127 ($g = 6$ μm)	487	6.3—7.2		[678]
Fe-43.2Al	927—1127 ($g = 20$ μm)	368	5.6—9.7		[678]

a Dependent on Al concentration.

3.4 Strengthening Mechanisms

The rather low creep strength of iron aluminides is a subject that has received particular attention. Several strategies to increase the creep resistance have been suggested, which are reviewed in Ref. [680]. The reduction of the high-diffusion coefficient of the rate-controlling mechanism, by micro- and macroalloying, was attempted with limited success. Another way to achieve strengthening is to add alloying elements that may hinder dislocation motion by forming solute atmospheres around dislocations or by modifying lattice order. For example, additions of Mn, Co, Ti, and Cr moderately increase the creep resistance due to solid solution strengthening. Finally, the most promising strengthening mechanism seems to be the introduction of dispersions of second phases, such as carbides, intermetallic particles, or oxide dispersions [685–687]. A sufficient volume fraction of precipitate phases (around 1–3%) must be present in order for this mechanism to be effective and the precipitates should be stable at the service temperatures. Alloying elements such as Zr, Hf, Nb, Ta, and B have been effective in improving creep resistance of FeAl by precipitation hardening.

Baligidad et al. [687,688] reported that improved creep strength is obtained in a Fe–16 wt.%Al–0.5 wt.%C possibly due to combined carbon solid-solution strengthening and mechanical constraint from the $Fe_3AlC_{0.5}$ precipitates. They claim that creep is recovery-controlled and that climb assists the recovery. Morris-Muñoz [675] analyzed the creep mechanisms in an oxide-dispersion-strengthened Fe–40 at.%Al intermetallic containing Y_2O_3 particles at 500 °C and 700 °C. The absence of substructure formation at either temperature suggested, to the investigators, constant-structure creep with a temperature-dependent threshold stress. Particle–dislocation interactions were also apparent. It was concluded that the threshold stress, based on particle–dislocation interactions, operates at 500 °C (where dislocations are predominantly <111> superdislocations) and that climb-controlled processes occur at 700 °C, where <100> dislocations are mainly present. The decrease in creep resistance observed between 500 °C and 700 °C was attributed to the rapid increase in diffusivity at high temperatures. Recently Sundar et al. [689] reported creep resistance values for two Fe–40 at.%Al alloys (with additions of Mo, Zr, and Ti for solute strengthening and additions of C and B for particle strengthening) that were comparable to, if not better than, those of many conventional Fe-based alloys. According to Sundar et al. [689], a combination of strengthening mechanisms is perhaps the best way to improve creep resistance of iron aluminides.

4. NICKEL ALUMINIDES

4.1 Ni₃Al

The Ni-Al binary phase diagram is illustrated in Figure 106. Ni$_3$Al forms at Al concentrations between 25 at.% and 27 at.%. This compound has a simple cubic Bravais lattice with four atoms per lattice site: one Al atom, located in the (x,y,z) position; and three Ni atoms, located, respectively, at the $(x + 1/2,y,z)$, $(x,y + 1/2,z)$, and $(x,y,z + 1/2)$ positions. This intermetallic received substantial attention since it is the main strengthening phase in superalloys. Furthermore it has been considered to be a technologically important structural intermetallic alloy system especially after its successful ductilization by microalloying with boron [690]. Additionally, it exhibits a flow stress anomaly, i.e., the yield stress increases with increasing temperature over intermediate temperatures (roughly ambient temperatures

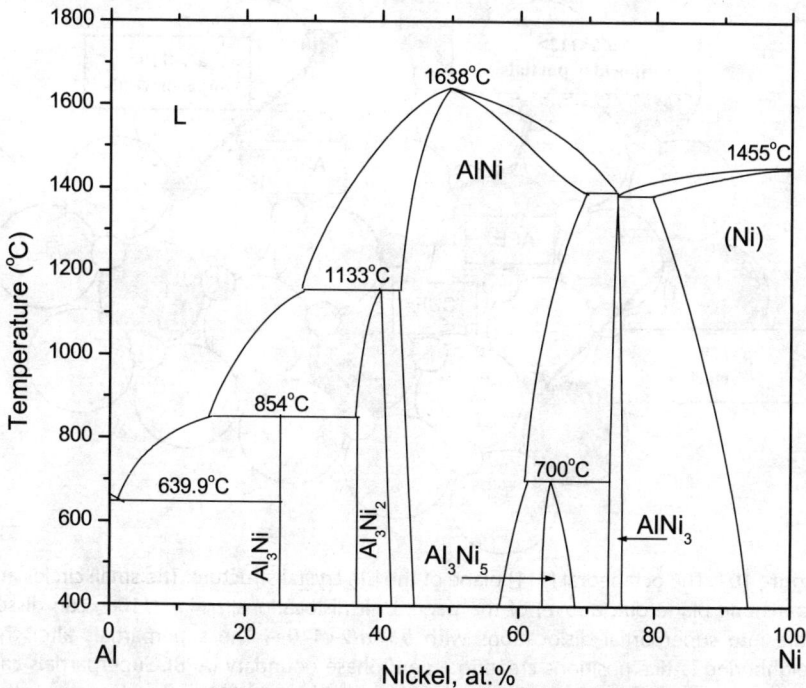

Figure 106 Binary Ni-Al phase diagram. *From Ref. [647].*

to 700 °C) as with iron aluminides. Thus, it has often been used as a model material for understanding intermetallic compounds in general.

The crystal structure of Ni_3Al is an ordered $L1_2$ (f.c.c.) structure having Al atoms at the unit cell corners and Ni atoms at the face centers. Similar to pure f.c.c. metals, the planes of easy glide are the octahedral planes $\{111\}$. Slip along $\{001\}$ planes is more difficult, since they are less compact, but it may occur by thermal activation [691]. Figure 107 illustrates an octahedral plane of this ordered structure. A perfect dislocation associated with primary octahedral glide has a Burgers vector, $b = a<110>$, that is twice as large as that corresponding to a unit dislocation in the disordered f.c.c. lattice. These dissociate into "super-partial" dislocations, with $b = a/2<110>$, and the latter may, in turn, dissociate into Shockley dislocations, with $b = a/6<112>$, as depicted in Figure 107.

The creep behavior of Ni_3Al will be briefly reviewed in the following sections.

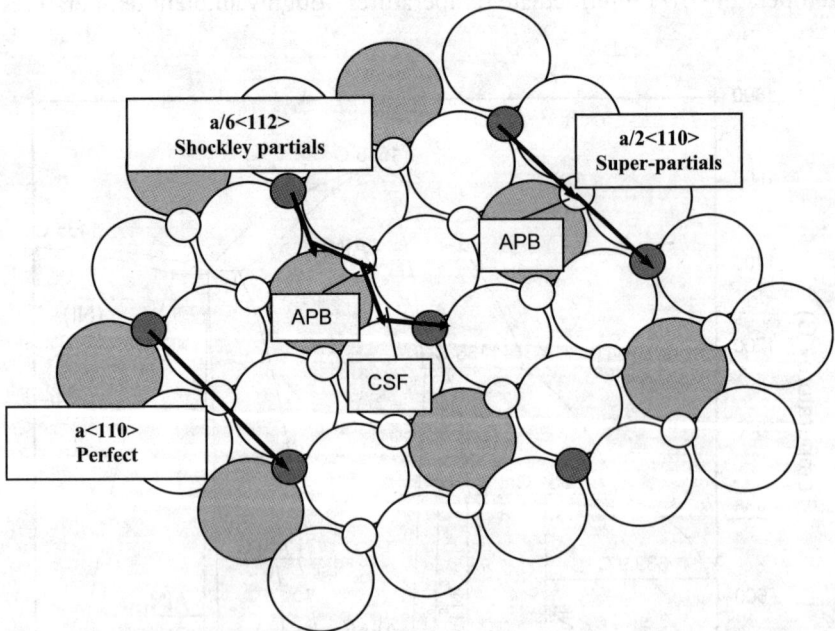

Figure 107 The octahedral $\{111\}$ plane of the $L1_2$ crystal structure. The small circles are atoms one plane out (above) of the page. Unit dislocations, $b = a<110>$, can dissociate into superpartial dislocations with $b = a/2<110>$. The superpartials alter the neighboring lattice positions creating an antiphase boundary (APB). Superpartials can dissociate into Shockley partial dislocations, with $b = a/6<112>$, that are connected by a complex stacking fault (CSF) that includes both an APB and an ordinary stacking fault. *From Ref. [691].*

4.1.1 Creep Curves

Creep tests have been performed on both single-crystal [691–706] and polycrystalline [703–705,707–720] Ni_3Al alloys in tension and compression. Most creep curves exhibit a normal shape, which consists of the conventional three stages. However, some [707–709] show sigmoidal creep, where the creep rate decreases quickly to a minimum and this is followed by a continuous increase in the creep rate with strain. A steady state may or may not be achieved before reaching tertiary creep after sigmoidal creep. This creep behavior is also frequently termed "inverse creep" among the intermetallics community. Primary creep is often limited to very small strains [605,697,706]. Figure 108 illustrates the creep curve of a Ni_3Al alloy (with 1 at.% Hf and 0.24 at.% B) deformed at 643 °C at a constant stress of 745 MPa [691]. Initially, the creep rate decreases with increasing strain and normal primary creep occurs. This is followed by an extended region where the strain rate continually increases with strain. Steady state may or may not be reached afterwards, as will be explained later.

High-temperature creep refers to creep deformation at temperatures higher than T_p, the temperature at which the peak yield stress is observed. This temperature varies with alloy composition [720] and crystal orientation

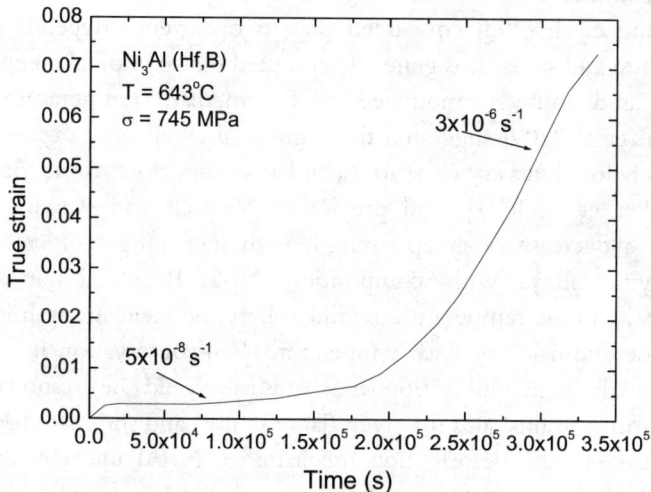

Figure 108 Sigmoidal creep curve corresponding to a Ni_3Al alloy (with 1 at.% Hf and 0.24 at.% B) deformed at 643 °C at a constant stress of 745 MPa [691]. Normal primary creep is followed by a continuous increase in the strain rate. This creep behavior has been also termed "inverse creep."

[692] but is typically observed from 0.5 to 0.6 T_m. Intermediate temperature creep usually refers to creep deformation at temperatures lower than (but close to) T_p. A steady-state regime is usually observed during high-temperature creep. The relationship between the strain rate and the stress in the high-temperature range usually follows a power-law relationship with a stress exponent of about 3. This may suggest that the viscous glide of dislocations is the rate-controlling mechanism [594]. However, at intermediate temperatures, from about 0.3 to 0.6 T_m, sigmoidal creep may occur depending on both the temperature and the stress [708,709]. Nicholls and Rawlings [712] suggested that different creep mechanisms should operate below and above T_p.

4.1.2 Sigmoidal (or Inverse) Creep

Sigmoidal creep has been observed in both single- and polycrystalline Ni_3Al alloys [691,698,700,707–709], as well as in some other intermetallics, e.g., Ni_3Ga [723] and TiAl [709]. The onset of sigmoidal creep (i.e., the increase in the creep-rate after primary creep) usually takes place at very small strains. This strain rate increase may extend over a large strain interval, leading directly to tertiary creep in the absence of a steady-state stage [691,798,707], as illustrated in Figure 108, or it may occur only for a small strain previous to the steady state [693,708].

The conditions under which sigmoidal creep occurs are relatively narrow. Rong et al. [708] concluded that its occurrence depends on both temperature and stress. It is generally accepted that sigmoidal creep is more frequent and more pronounced at intermediate temperatures [709]. Smallman et al. [709] suggested that sigmoidal creep only occurs at temperatures below but very close to T_p and at stresses close to the yield stress.

Hemker et al. [691], and previously Nicholls and Rawlings [712], observed a decrease of creep strength with increasing temperature in a single-crystal alloy, with composition Ni-22.18 at.%Al-1 at.%Hf-0.24 at.%B [691], in the temperature regime where the yield strength is known to increase anomalously with temperature. This observation led them to infer that different dislocation mechanisms would be responsible for yielding (small strains) and for creep (large strains) and stimulated a detailed investigation of the deformation mechanisms. Ni_3Al intermetallic single crystals (LI_2) show a positive temperature dependence from ambient to about 700°. A recent review of this subject by Choi et al. [595] describes the various experimental, theoretical and analytical developments over the past several decades. They proposed a constitutive model. They pointed out that experimental evidence shows that screw-superdislocations cross-slip from the octahedral planes to the cubic planes, where the dislocation core does not completely reside leading to immobilization by Kear-Wilsdorf locks (KWLs). Later, macro (or super) kink models (MKs) were developed based on microscopy studies. The Choi et al. models reasonable replicated the anomalous yield behavior. The mobilization of MKs is based on the distribution of MK heights. Hemker et al. [691] proposed a model to explain the sigmoidal creep of Ni_3Al based on careful microstructural examination. They suggested that octahedral slip during primary creep is exhausted by the formation of KW locks, due to thermally activated cube cross-slip of the screw segments. Thus, the strain rate is progressively reduced until the cross-slipped segments become thermally activated and are able to bow out and glide on the cube cross-slip plane. The KW locks act as Frank-Read type dislocation sources for glide on the {001} cube planes. The dislocation generation and subsequent glide on the cube planes leads to an increasing mobile dislocation density and thus to a larger strain rate and sigmoidal creep occurs. An alternative dislocation model for sigmoidal creep was proposed by Hazzledine and Schneibel [724]. They suggested that two highly stressed octahedral slip systems that share a common cube cross-slip plane may interact "symbiotically" and unlock each other's superdislocations, giving rise to an increasing number of <001> dislocations that are glissile on the cube plane. Thus, sigmoidal creep occurs.

Smallman et al. [709] pointed out that cube cross-slip is a necessary but not sufficient condition for sigmoidal creep. The operation of this mechanism, which leads to a strain rate increase under some conditions, is compensated by the strain rate decrease due to the exhaustion of dislocations on the octahedral slip systems. In fact, Smallman et al. [709] observed cube cross-slip in a polycrystalline Ni_3Al alloy creep deformed at 380 °C ($T << T_p$), where sigmoidal creep was not apparent. Zhu et al. [692] also reported cube cross-slip in the absence of sigmoidal creep in single crystals of Ni_3Al with different orientations. In order to rationalize these observations, Rong et al. [708] suggested that sigmoidal creep would occur only when the length of a significant number of screw segments cross-slipped onto cube planes from octahedral planes, as suggested by Hemker et al. [691], is larger than a critical value. In this case, the density of mobile dislocations on the cube cross-slip planes would increase significantly, leading to an increase in the creep rate. Rong et al. [708] also observed an anomalous temperature dependence of the creep strength in a polycrystalline Ni_3Al alloy, contrary to what Hemker et al. [691] had reported for their single-crystal alloy. Rong et al. [708] found this anomalous dependence consistent with their TEM observations of a larger density of dislocations on cube cross-slip planes at the lower temperatures. They suggested that the average length of the screw segments on cube cross-slip planes would increase with decreasing temperature. Thus, at low temperatures, there would be more dislocations with lengths larger than the critical value or a higher mobile dislocation density and this would lead to a lower creep resistance.

The occurrence of sigmoidal creep has not only been found to depend on temperature and stress, but also on the prior deformation and processing history. For example, sigmoidal creep disappears in a Ni_3Al alloy prestrained 3% at ambient temperature [707]. A recent investigation of a single crystal of $Ni_3Al(0.5\%Ta)$ indicated that the temperature of pre-creep deformation also affects the subsequent creep behavior [702]. The implications of these observations in terms of the nature of the creep mechanisms were not discussed.

4.1.3 Steady-State Creep

Steady-state creep can start very early in Ni_3Al alloys and extend over a considerable strain range (up to 20%) at high [696] as well as at intermediate temperatures [692,706]. Occasionally this stage may be delayed or even absent at intermediate temperatures if sigmoidal creep occurs, as described above. As in many other intermetallic systems, the minimum creep rate is used to calculate the stress exponent and the activation energy for creep, using the well-established power-law relations described elsewhere in this book, when clear steady state is not observed.

Table 9 summarizes some of the creep data obtained in various investigations on Ni_3Al-based alloys, mostly at high temperatures [694–699, 703–706,710–713,715–719,725–729]. This section will mainly focus on single-phase alloys. The stress exponent, n, ranges mostly between 3.2 and 4.4 in both single-crystal and polycrystalline alloys. A lower value of about 1 was reported at low stresses in a polycrystalline $Ni_3Al(Hf, B)$ [713,716]. A few studies have reported higher values of 6.7 [706], 8 [718], and 9 [730]. The values of the activation energy, Q_c, for creep ranged from 263 to 530 kJ mol^{-1}, but are generally between 320 and 380 kJ mol^{-1}. It is not possible to normalize all the creep data of various Ni_3Al alloys in a single plot, such as in earlier chapters, due to the lack of diffusion coefficient and modulus of elasticity values at various temperatures over the large range of compositions investigated.

Table 9 Creep data of Ni$_3$Al alloys

Alloy	Structure	T (°C)	n	Q$_c$ (kJ mol^{-1})	Reference
Single-phase					
Ni$_3$Al(10Fe)	P	871–1177	3.2	327	[711]
Ni$_3$Al(11Fe)	P	680–930	2.6	355	[712]
Ni$_3$Al(Zr, B)	P	860–965	4.4	406	[710]
Ni$_3$Al(Hf, B)	P	760	2–3 (HS)	–	[713]
			1 (LS)	–	
Ni$_3$Al(Zr, B)	P	760–860	2.9	339–346	[704]
Ni$_3$Al(8Cr, Zr, B)	P	760–860	3.3	391–400	[704]
Ni$_3$Al(5V)	P	850–950	2.89–3.37	–	[715]
Ni$_3$Al(Hf, B)	P	760–867	1 (LS)	313	[716]
Ni$_3$Al(Ta)	P	950–1100	3.3	383	[717]
Ni$_3$(Al, 4Ti)	P	750	8	–	[718]
Ni$_3$Al(8Cr, Hf, Ta, Mo...)	P	650–900	4.7	327	[719]
Ni$_3$Al(Hf, B)	S	924–1075	4.3	378	[696]
Ni-23.5Al	S	982	3.5	–	[695]
Ni$_3$Al(Cr, Ta, Ti, W, Co)	S	900–1000	3.5	380	[697]
Ni$_3$(Al, 4Ti)	S	852–902	3.3	282	[699]
Ni$_3$Al(Ta, B)	S	810–915	3.2	320	[698]
Ni$_3$Al(Ta, B)		1015–1115	3.2 (HS)	360	[698]
			4.3 (LS)	530	
Ni$_3$Al(Ta, B)	S	850–1000	3.5	420	[694]
Ni$_3$Al(X), X = Ti, Hf, Cr, Si	S	850–950	3.01–4.67	263–437	[703]
Ni$_3$Al(4Cr)	S	760–860	–	362–466	[705]
Ni$_3$Al(Ti, 2Ta)	S	850	6.7	383	[706]
		1150	3.3	–	
Multiphase (precipitation strengthened)					
Ni$_3$(Al, 4Ti)	γ/γ′	650, 750	31, 22	–	[718]
Ni-20.2Al-8.2Cr-2.44Fe	γ/γ′-α(Cr)	777–877	4.1	301	[725]
Oxide-dispersion strengthened Ni$_3$Al					
Ni$_3$Al(5Cr, B)	2 vol.% Y$_2$O$_3$	1000–1200	7.2, 7.8	650, 697	[726,727]
Ni$_3$Al(5Cr, B)	2 vol.% Y$_2$O$_3$	649	13.5	–	[728]
		732, 816	5.1 (LS)	239	[729]
			22, 13 (HS)		
		982	9.1		

HS, high stress; LS, low stress; P, polycrystalline; S, single crystal.

The rate-controlling mechanism during creep of Ni_3Al intermetallics is still unclear. Based on the analysis of the stress exponents, several studies suggested dislocation glide ($n \approx 3$) [697,698,711,715,717], while others propose that dislocation climb predominates ($n \approx 4$–5) [698,702,710,719], and yet others point toward Coble or Nabarro-Herring creep ($n = 1$) [713,716]. However, others have questioned the predominance of a single mechanism, since the stress exponents vary from 3 to 5. Also, the Q_c values have been found to be stress-dependent, in some cases [705], and values are often much higher than the activation energy for diffusion (the activation energy for diffusion of Ni in Ni_3Al varies from 273 to 301 kJ mol^{-1} [731]). The diffusivity of Al in NiAl is believed to be higher but it has still not been measured directly due to the lack of suitable radioactive tracers [732].

Several TEM studies have been performed to investigate the micro-structural evolution of Ni_3Al alloys during steady-state creep. In general, subgrains do not readily form. Wolfenstine et al. [698] observed randomly distributed, curved, dislocations in the $n = 3$ region between 810 °C and 915 °C, and a homogeneous dislocation distribution with some evidence for subgrain formation in the $n = 4.3$ region (lower stresses) but no evidence for subgrain formation in the $n = 3$ region (higher stresses) between 1015 °C and 1115 °C. Knobloch et al. [733] examined the microstructure of [001], [011], and [111] oriented Ni_3Al single crystals creep deformed at 850 °C at a stress of 350 MPa. They observed a homogeneous dislocation distribution for all orientations and creep stages. Stress exponents and activation energies were not calculated. As mentioned above, the most common slip systems operative during creep of Ni_3Al are the <110>{111} (octahedral slip) and <110>{100} (cube slip) [692,734], although dislocation glide on <100>{100} [735] and <110>{110} [733,736] systems has also been reported. This suggests that multiple slip takes place and that dislocation interactions may be important during creep [696,733].

4.1.4 Effect of Some Microstructural Parameters on Creep Behavior

Crystal orientation has a considerable influence on creep behavior [692–696,706,734–736] and this influence is highly dependent on temperature. At high temperatures, [001] is the weakest orientation, showing the highest creep rate; [111] is the strongest orientation associated with the lowest creep rate, about 1/5 to 1/2 of that of the [001] orientation. Finally, the [011] and [123] orientations show an intermediate strength and the creep rate is about 1/3 to 1/2 of the creep rate of the [001] orientation [693,695,696,706,733]. At intermediate temperatures, the [111] orientation is softer than the [001], and the [123] has, again, an intermediate creep strength [692,696]. Thus, the orientation dependence of creep strength at intermediate temperatures is opposite to that at high temperatures. Models considering the operation of octahedral slip, cube slip, and multiple slip have been proposed to explain and predict the creep anisotropy at different temperatures [599,695,737]. However, it seems that the crystal orientation has no obvious influence on the stress exponent n [694,695,706] and on the activation energy Q_c for creep [694,706].

Only a few studies of the influence of grain size on the creep of Ni_3Al have been published. Schneibel et al. [713] observed a grain size dependence of the creep rate of a cast $Ni_3Al(Hf, B)$ alloy creep deformed at 760 °C. They tested specimens with average grain sizes of 12, 50, and 120 μm. Figure 109 illustrates the strain rate versus stress data from tests performed at high stresses in the samples with larger grain sizes. The stress exponent is 3 for small grain sizes (50 μm) and significantly higher for larger grain sizes (120 μm). The increase in the stress exponents is attributed to scatter of the experimental data corresponding to the lowest strain rate. Thus, these investigators assume a stress exponent around 3 to be characteristic of the high-stress regime, which would be consistent with viscous glide over the investigated grain sizes. The shear strain rate in the samples with large grain sizes (>50 μm) was found to be proportional to $g^{-1.9}$ at

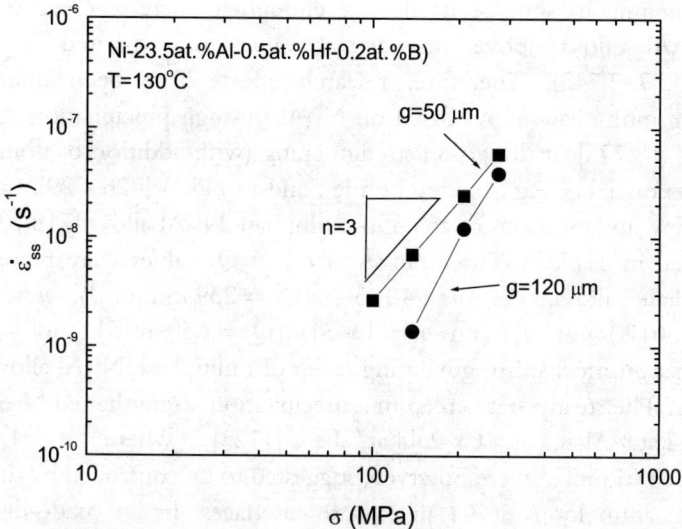

Figure 109 Stress dependence of the creep rate of Ni-23.5 at.%Al-0.5 at.%Hf-0.2 at.%B at high stresses for two grain sizes. *From Ref. [713].*

low stresses ($\sigma < 10$ MPa). This observation, together with the finding of a stress-exponent value equal to 1, may suggest Nabarro-Herring creep (if, in fact, it exists). For smaller grain sizes (12 μm), Coble (if it exists) creep may predominate, although conclusive evidence was not presented [713]. Hall–Petch strengthening was not discussed.

Hayashi et al. [715] and Miura et al. [703] investigated the effects of off-stoichiometry on the creep behavior of binary and ternary Ni$_3$Al alloys. They reported that, in both single-crystal and polycrystalline alloys, the creep resistance increases with increasing Ni concentration on both sides of the stoichiometric composition and a discontinuity exists in the variation at the stoichiometric composition. The values of the activation energy, Q_c, for creep were also found to be strongly dependent on the Ni concentration and the alloying additions [703,704,715]. The characteristic variation in creep resistance with Ni concentration was explained by the strong concentration dependence of the activation energy for creep [715]. The n values (mostly about 3–4), however, appear nearly independent of the stoichiometric composition and the alloying additions [703,715].

Attempts have been made to improve the creep strength of Ni$_3$Al by adding various alloying elements [697,700,703,711,713,715,738–741]. Several solutes have been found to be beneficial, such as Hf, Cr, Zr, and Ta [594]. In some cases the improvement in creep strength was accompanied by a non-desirable increase in density [594]. However, solid-solution strengthening has not been effective enough to increase creep resistance of Ni$_3$Al alloys above the typical values of Ni-based superalloys [697,740,742,743]. Therefore, research efforts have been directed to develop multiphase alloys based on Ni$_3$Al through precipitation strengthening [718,725] or dispersion strengthening (with addition of nonmetallic particles or fibers, e.g., oxides, borides, and carbides) [726–729].

A few investigations of creep in multiphase Ni$_3$Al alloys [718,725–729] are listed in Table 9. Three ranges of n and Q_c values were reported for multiphase alloys, i.e., $n = 4.1–5.1$ ($Q_c = 239$ kJ mol^{-1}), $n = 7.2–9.1$ ($Q_c = 301$ kJ mol^{-1}), and $n = 13–31$ ($Q_c = 650–697$ kJ mol^{-1}). The deformation mechanism governing creep of multiphase Ni$_3$Al alloys is still unclear. The steady-state creep in a precipitation-strengthened Ni$_3$Al alloy Ni-20.2 at.%Al-8.2 at.%Cr-2.44 at.%Fe [1725], where $n = 4.1$ and $Q_c = 301$ kJ mol^{-1} were observed, suggested to be controlled by the climb of dislocation loops at Cr precipitate interfaces. In an oxide-dispersion strengthened (ODS) Ni$_3$Al alloy [726–729] the n values were showed to be strongly dependent on the temperature and the stress. The stress exponents increased from 5.1 (with $Q_c = 239$ kJ mol^{-1}) at low stresses to 13 ~ 22 at high stresses at temperatures of 732 °C and 815 °C [728,729], which are typical of ODS alloys, as discussed earlier. At higher temperatures (from 1000 °C to 1273 °C) [726,727] the stress exponents were 7.2 and 7.8 (with $Q_c = 650$ and 697 kJ mol^{-1}, respectively). It was suggested that the stress exponent of 5.1 in the ODS Ni$_3$Al should not be considered indicative of dislocation climb-controlled creep as observed in pure metals and Class M alloys, and as proposed for some single-phase Ni$_3$Al alloys. Carreño et al. [726] emphasized that both Arzt and co-workers' detachment model (described in the chapter on second-phase strengthening) and incorporating a threshold stress [598] are not appropriate approaches to describe the creep behavior of ODS Ni$_3$Al at higher temperatures. There is relatively poor agreement between the data and the predictions by these models. Alternatively, they developed a "ñ-model" approach, which separates the contribution of the particles and that from the matrix. They assume the measured stress exponent is equal to the sum of the stress exponent corresponding to the matrix, termed $h_{\tilde{n}}$, and an additional stress exponent, ñ, that is necessary in order to account for the dislocation–particle interactions. The measured activation energy can be obtained by multiplying the activation energy corresponding to the matrix deformed under the same stress and temperature conditions by a factor equal to $(h_{\tilde{n}} + \tilde{n}/h)$. This approach satisfactorily models the data.

Figure 110 illustrates a comparison of the creep properties of an ODS Ni$_3$Al alloy, Ni-19 at.%Al-5 at.%Cr-0.1 at.%B with 2 vol.% of Y$_2$O$_3$ (filled circles) with a single crystal Ni$_3$Al and two nickel-based superalloys, NASAIR 100 and MA6000. Although the introduction of an oxide dispersion contributed to strengthening of the Ni$_3$Al alloy with respect to the single crystal alloy, the creep performance of ODS Ni$_3$Al was still poorer than that of commercial nickel-based superalloys.

4.2 NiAl

4.2.1 Introduction

NiAl is a B2 ordered intermetallic phase with a simple cubic lattice with two atoms per lattice site, an Al atom at position (x,y,z) and a Ni atom at position $(x + 1/2, y + 1/2, z + 1/2)$. This is a very stable structure that remains ordered until nearly the melting temperature. As illustrated in the phase diagram of Figure 106, NiAl forms at Al concentrations ranging from 40 to about 55 at.%. Excellent reviews of the physical and mechanical properties of NiAl are available in [594,745,746].

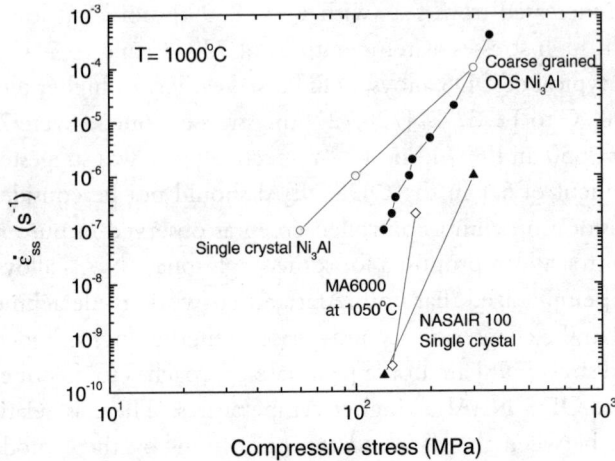

Figure 110 Comparison of the creep behavior corresponding to an oxide-dispersion strengthened (ODS) Ni$_3$Al alloy of composition Ni-19 at.%Al-5 at.%Cr-0.1 at.%B with 2 vol.% of Y$_2$O$_3$ ($g \sim 400$ μm) [727] with a single-crystal Ni$_3$Al alloy [796] and with the Ni superalloys NASAIR 100 [744] and MA6000. *From Ref. [581].*

NiAl alloys are attractive for many applications due to their favorable oxidation, carburization and nitridation resistance, as well as their high thermal and electrical conductivity. They are currently used to make electronic metallizations in advanced semiconductor heterostructures, surface catalysts, and high current vacuum circuit breakers [745]. Additionally, these alloys are attractive for aerospace structural applications due to their low density (5.98 g cm^{-3}) and high melting temperature [599]. However, two major limitations of single-phase NiAl alloys are precluding their application as structural materials, namely poor creep strength at high temperatures and brittleness below about 400 °C (brittle–ductile transition temperature). The following sections of this chapter will review the deformation mechanisms during creep of single-phase NiAl and the effects of different strengthening mechanisms.

4.2.2 Creep of Single-Phase NiAl

Most of the available creep data of NiAl were obtained from compression tests at constant strain rate or constant load [747–767]. Only limited data from tensile tests are available [755,769]. It is generally accepted that creep in single-phase NiAl is diffusion controlled. This has been inferred from the analysis of the stress exponents and activation energies. The values for these parameters are listed in Table 10. In most cases, the stress exponents range

Table 10 Creep parameters for NiAl

Al, at.%	Grain size, μm	T (°C)	n	Q, kJ mol⁻¹	Reference
48.25	5—9	727—1127	6.0—7.5	313	[757]
44—50.6	15—20	727—1127	5.75	314	[761]
50	12	927—1027	6	350	[742]
50	450	800—1045	10.2—4.6	283	[763]
50	500	900	4.7		[764]
50.4	1000	802—1474	7.0—3.3	230—290	[765]
50	Single crystal [123]	750—950	7.7—5.4		[766]
50	Single crystal	750—1055	4.0—4.5	293	[767]
49.8	39	727	5	260	[768]

From Ref. [746] with additional data from recent publications.

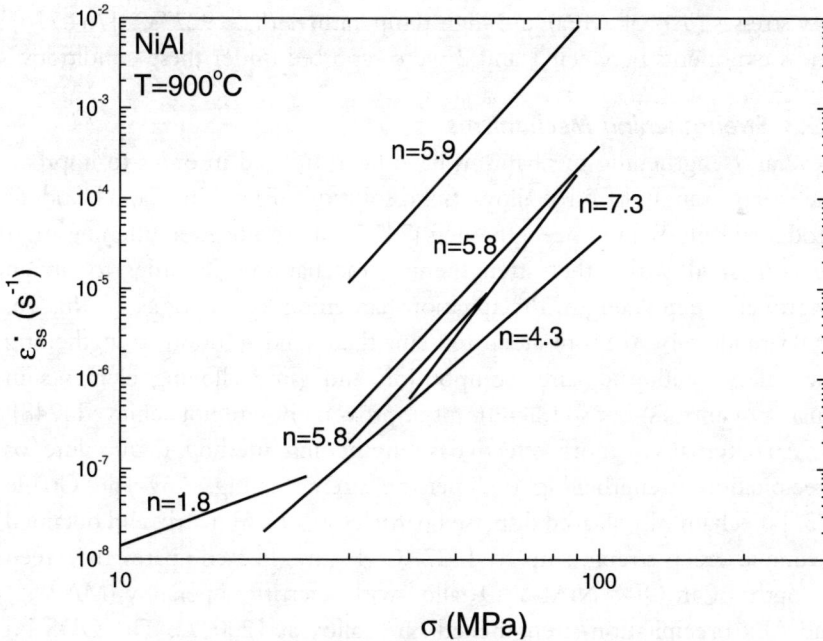

Figure 111 Creep behavior of several binary NiAl alloys. *From Ref. [594].*

from 4 to 7.5. Figure 111 illustrates the creep behavior of several binary NiAl alloys. The values of the activation energies, in many investigations, are close to 291 kJ mol⁻¹, the value of the activation energy for bulk diffusion of Ni in NiAl. Additionally, subgrain formation during deformation was observed [173], consistent with climb.

However, occasionally stress exponents different from those mentioned above have been reported. For example, values as low as 3 were measured in NiAl single crystals by Forbes et al. [768] and Vanderwoort et al. [765], who suggested that both viscous glide and dislocation climb would operate. The contributions of each mechanism would depend on texture, stress, and temperature [765]. Recently, Raj et al. [769] reported stress exponents as high as 13 in a Ni-50 at.%Al alloy tested in tension at 427 °C, 627 °C, and 727 °C and at constant stresses of 100–170 MPa, 40–80 MPa, and 35–65 MPa, respectively. Although no clear explanations for this high stress exponent value are provided, they note that the creep behavior of NiAl in tension and compression is significantly different. For example, Raj et al. observed that NiAl material creeps much faster in tension than in compression, especially at the lower temperatures.

Diffusional creep has been suggested to occur in NiAl when tested at low stresses ($\sigma < 30$ MPa) and high temperatures ($T > 927$ °C) [764,770]. Stress exponents between 1 and 2 were reported under these conditions.

4.2.3 Strengthening Mechanisms

Several strengthening mechanisms have been utilized in order to improve the creep strength of NiAl alloys. Solid solution of Fe, Nb, Ta, Ti, and Zr produced only limited strengthening [747,771]. Solute strengthening must be combined with other strengthening mechanisms in order to obtain improved creep strength. Precipitation hardening by additions of Nb, Ta, or Ti renders NiAl more creep resistant than solid solution strengthening (i.e., alloys with the same composition and same alloying elements in smaller quantities) but still significant improvements are not achieved [748].

An alternative, more effective strengthening method than solute or precipitation strengthening is dispersion strengthening. Artz and Grahle [751] mechanically alloyed dispersed particles in a NiAl matrix and obtained favorable creep strength up to 1427 °C. Figure 112 compares the creep strengths of an ODS NiAl-Y_2O_3 alloy with a ferritic superalloy (MA 956) and AlN precipitation-strengthened NiAl alloy at 1200 °C. The ODS Ni alloy is more creep resistant than the Ni superalloy MA956 at these high temperatures. The ODS NiAl alloy is also more resistant than the precipitation-strengthened alloy at low strain rates. Artz and Grahle [751] also observed that the creep behavior of ODS NiAl is significantly influenced by the grain size. In a coarse grain size ($g = 100$ μm) ODS NiAl-Y_2O_3 alloy, the creep behavior showed the usual characteristics of dispersion-strengthened systems, i.e., high stress exponents ($n = 17$) and

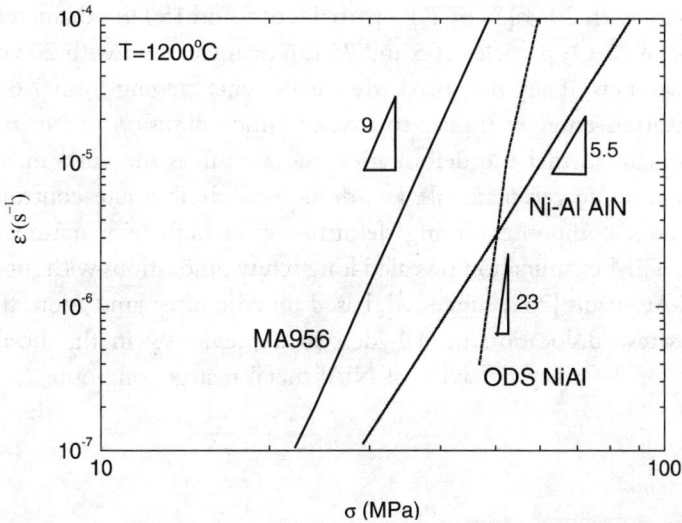

Figure 112 Comparison of the creep behavior of a coarse-grained ($g = 100 \ \mu m$) oxide-dispersion strengthened (ODS) NiAl alloy [751] with the ferritic ODS superalloy MA956 and with a precipitation strengthened NiAl-AlN alloy produced by cryomilling. *From Ref. [754].*

activation energies much higher than that of lattice self-diffusion ($Q = 576 \ kJ \ mol^{-1}$). However, intermediate stress exponents of 5 and very high activation energies ($Q = 659 \ kJ \ mol^{-1}$) were measured in an ODS NiAl-Y_2O_3 alloy with a grain size of $0.9 \ \mu m$. This creep data could not be easily modeled using established relationships for diffusional or for detachment-controlled dislocation creep [751]. Arzt and Grahle [751] suggested that the presence of particles at grain boundaries partially suppresses the sink/source action of grain boundaries by pinning the grain boundary dislocations and thus hindering Coble creep. However, the low stress exponent indicates that Coble creep (if it exists) may not be completely suppressed. Artz and Grahle [751] proposed a phenomenological model based on the coupling between Coble creep and the grain boundary dislocation–dispersoid interaction (controlled by thermally activated dislocation detachment from the particles, as described in an earlier chapter). The predictions of this model correlate with the experimental data. HfC and HfB_2 can also provide significant particle strengthening [747].

Matrix reinforcement by larger particles such as TiB_2 and Al_2O_3 or whiskers also increases significantly the strength of NiAl alloys [747,752]. Xu and Arsenault [752] investigated the creep mechanisms of NiAl matrix composites with 20 vol.% of TiB_2 particles of 5 and 150 μm diameter, with 20 vol.% of Al_2O_3 particles of 5 and 75 μm diameter, and with 20 vol.% of Al_2O_3 whiskers. They measured stress exponents ranging from 7.6 to 8.4 and activation energies similar to that of lattice diffusion of Ni in NiAl. They concluded that the deformation mechanism is the same in unreinforced and reinforced materials, i.e., dislocation climb is rate-controlling in NiAl matrix composites during deformation at high temperature. Additionally, TEM examination revealed long screw dislocations with superjogs. Xu and Arsenault [752] suggested, based on computer simulation, that the jogged screw dislocation model, described previously in this book, can account for the creep behavior of NiAl metal matrix composites.

CHAPTER 10

Creep Fracture

M.E. Kassner

Contents

1. BACKGROUND

Creep plasticity can lead to tertiary, or Stage III, creep and failure. It has been suggested that creep fracture can occur by w-type, or wedge-type, cracking, as illustrated in Figure 113(a), at grain boundary (GB) triple points. Some have suggested that w-type cracks form most easily at higher stresses (lower temperatures) and larger grain sizes [772] when GB sliding (GBS) is not accommodated. Some have suggested that the w-type cracks nucleate as a consequence of GBS. Another mode of fracture has been associated with r-type irregularities or cavities illustrated in Figure 114. The wedges may be brittle in origin or simply an accumulation of r-type voids (Figure 113(b)) [773]. These wedge cracks may propagate only by r-type void formation [774,775]. Inasmuch as w-type cracks are related to r-type voids, it is sensible to devote this chapter on creep fracture to cavitation.

Fundamentals of Creep in Metals and Alloys
ISBN 978-0-08-099427-7
http://dx.doi.org/10.1016/B978-0-08-099427-7.00010-4

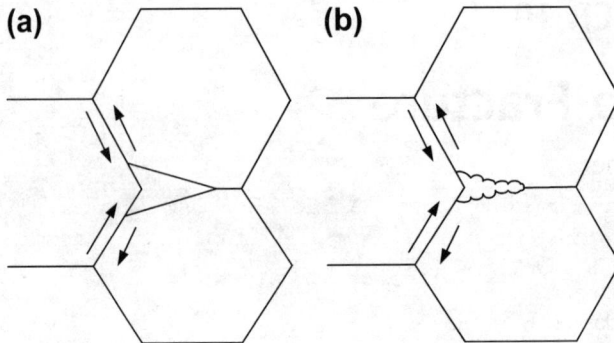

(a) **(b)**

Figure 113 (a) Wedge (or *w*-type) crack formed at the triple junctions in association with grain boundary sliding. (b) A wedge crack as an accumulation of spherical cavities.

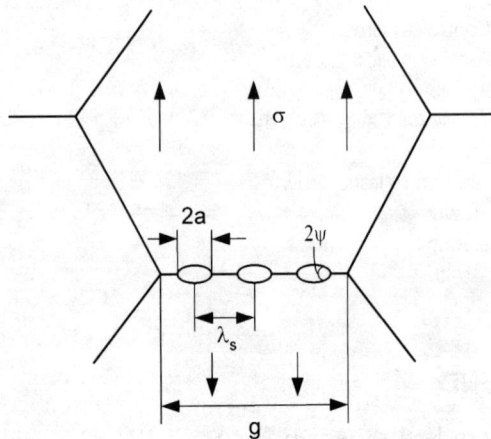

Figure 114 Cavitation (*r*-type) or voids at a transverse grain boundary. Often, ψ is assumed to be approximately 70°.

There has been, in the past, a variety of reviews of creep fracture by Cocks and Ashby [776] and Needleman and Rice [778], a series of articles in a single issue of a journal [779–781], a chapter by Cadek [20], and particularly books by Riedel [785] and Evans [30], although most of these were published 25–30 years ago. This chapter will review these and more recent works. Some of these works are compiled in recent bibliographies [786] and are quite extensive, of course, and this chapter is intended as a balanced and brief summary. The books by Riedel and Evans are considered good references for further reading. This chapter will also reference those works published subsequent to these reviews.

Creep fracture in uniaxial tension under constant stress has been described by the Monkman-Grant relationship [35], which states that the fracture of creep deforming materials is controlled by the steady-state creep rate, $\dot{\varepsilon}_{ss}$, Eqn (4),

$$\dot{\varepsilon}_{ss}^{m''} t_f = k_{MG} \tag{168}$$

where k_{MG} is sometimes referred to as the Monkman–Grant constant and m'' is a constant, typically about 1.0. Some data that illustrate the basis for this phenomenological relationship are given in Figure 115, based on previous work [30,787]. Although not extensively validated over the past 20 years, it has been shown recently to be valid for creep of dispersion-strengthened cast aluminum [788] where cavities nucleate at particles and not located at grain boundaries. Modifications have been suggested to this relationship based on fracture strain [789]. Although some more recent data on Cr-Mo steel suggest that Eqn 4 is valid [790], the same data have been interpreted to suggest the modified version. The Monkman–Grant (phenomenological) relationship, as will be discussed subsequently, places constraints on creep cavitation theories.

Another relationship to predict rupture time uses the Larson-Miller parameter [791] described by

$$LM = T[\log t_r + C_{LM}] \tag{169}$$

This equation is not derivable from the Monkman–Grant or any other relationship presented. The constant C_{LM} is phenomenologically determined as that value that permits LM to be uniquely described by the logarithm of the applied stress. This technique appears to be currently used for zirconium alloy failure time prediction [792]. C_{LM} is suggested to be about 20, independent of the material.

One difficulty with these equations is that the constants determined in a creep regime, with a given rate-controlling mechanism, may not be used for extrapolation to the rupture times within another creep regime, where the constants may change [792]. The Monkman–Grant relationship appears to be more popular.

The fracture mechanisms that will be discussed are those resulting from the nucleation of cavities followed by growth and interlinkage, leading to catastrophic failure. Figure 116 illustrates such creep cavitation in Cu, already apparent during steady state (i.e., before Stage III or tertiary creep). It will be initially convenient to discuss fracture by cavitation as consisting of two steps: nucleation and subsequent growth.

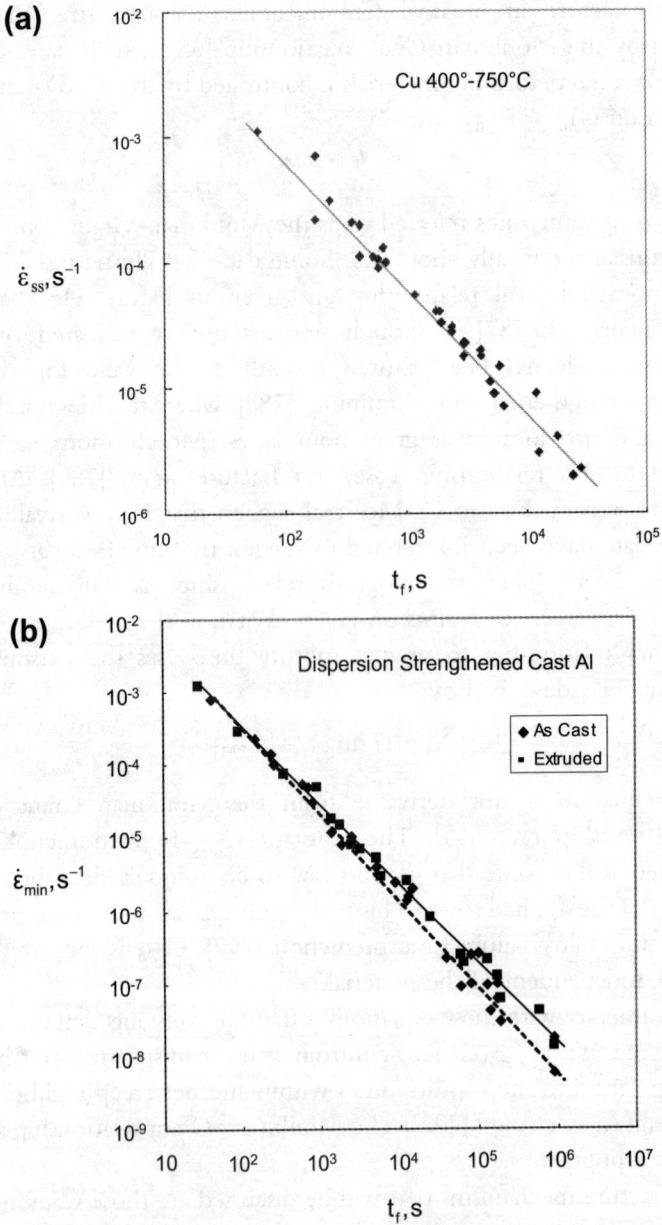

Figure 115 (a) The steady-state creep-rate (strain rate) versus time-to-rupture for Cu deformed over a range of temperatures, adapted from Evans [30], and (b) dispersion strengthened cast aluminum. *Adapted from Dunand et al. [788].*

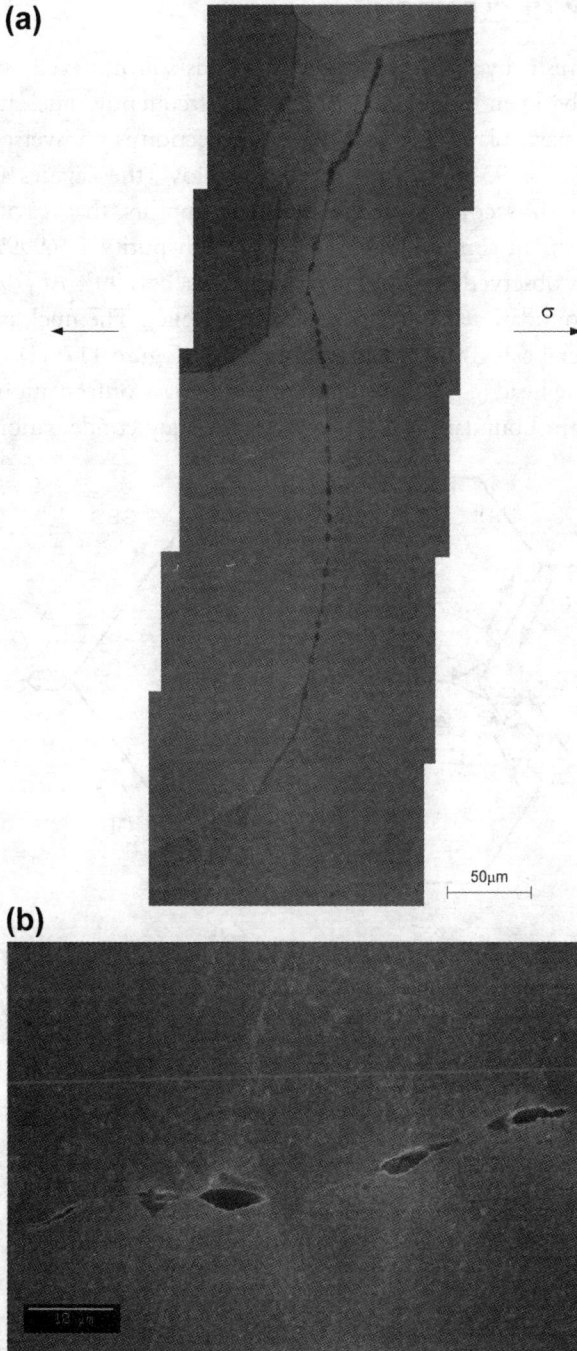

Figure 116 Micrograph of cavities in Cu deformed at 20 MPa and 550 °C to a strain of about 0.04 (within stage II, or steady state). This is shown at (a) low and (b) high magnification.

2. CAVITY NUCLEATION

The mechanism by which cavities nucleate is still not well established. It has generally been observed that cavities frequently nucleate on grain boundaries, particularly on those whose projection is transverse to a tensile stress (e.g., [775,793–797]). In commercial alloys, the cavities appear to be associated with second-phase particles. It appears that cavities do not generally form in some materials such as high–purity (99.999% pure) Al. Cavitation is observed in lower-purity metal such as 99% Al [798] (in high-purity Al, boundaries are serrated and very mobile). The nucleation theories fall into several categories that are illustrated in Figure 117: (1) GBS leading to voids at the head (e.g., triple point) of a boundary or formation of voids by "tensile" grain boundary (GB) ledges, (2) vacancy condensation, usually at

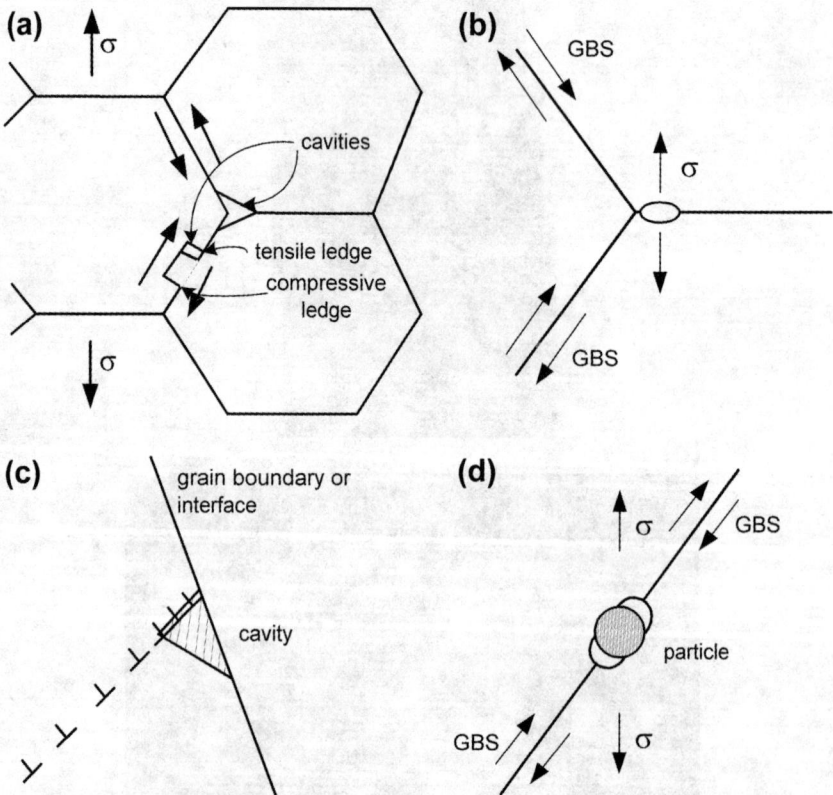

Figure 117 Cavity nucleation mechanism. (a) Sliding leading to cavitation from ledges (and triple points). (b) Cavity nucleation from vacancy condensation at a high stress region. (c) Cavity nucleation from a Zener-Stroh mechanism. (d) The formation of a cavity from a particle-obstacle in conjunction with the mechanisms described in (a–c).

grain boundaries at areas of high stress concentration, and (3) the cavity formation at the head of a dislocation pile-up such as by a Zener-Stroh mechanism (or anti–Zener-Stroh mechanism [799]). These mechanisms can involve particles as well (4).

2.1 Vacancy Accumulation

Raj and Ashby [800] developed an earlier [801] idea that vacancies can agglomerate and form stable voids (nuclei) as in Figure 117(b). Basically, the free energy terms are the work performed by the applied stress with cavity formation balanced by two surface energy terms. The change in the total free energy is given by

$$\Delta G_T = -\sigma \Omega N + A_v \gamma_m - A_{gb} \gamma_{gb} \tag{170}$$

where N is the number of vacancies, A_v and A_{gb} are the surface areas of the void and (displaced) area of GB, respectively, and γ_m and γ_{gb} are surface and interfacial energy terms of the metal and GB, respectively. (Note: All stresses and strain rates are equivalent uniaxial and normal to the GB in the equations in this chapter.)

This leads to a critical radius, $a^\star$, and free energy, ΔG_T^*, for critical-sized cavities and a nucleation rate

$$\dot{N} \cong n^* D_{gb} \tag{171}$$

where $n^* = n_o \exp(-\Delta G_T^*/kT)$, D_{gb} is the diffusion coefficient at the GB, and n_o is the density of potential nucleation sites. (The nucleation rate has the dimensions $m^2 s^{-1}$.) (Some [20,785] have included a "Zeldovich" factor in Eqn (171) to account for "dissolution" of "supercritical" nuclei $a > a^\star$.)

Some have suggested that vacancy supersaturation may be a driving force rather than the applied stress, but it has been argued that sufficient vacancy supersaturations are unlikely [785] in conventional deformation (in the absence of irradiation or Kirkendall effects).

This approach leads to expressions of nucleation rate as a function of stress (and the shape of the cavity). An effective threshold stress for nucleation is predicted. Argon et al. [802] and others [785] suggest that the cavity nucleation by vacancy accumulation (even with modifications to the Raj-Ashby nucleation analysis to include, among other things, a Zeldovich factor) requires large applied (threshold) stresses (e.g., 10^4 MPa) orders of magnitude larger than observed stresses leading to fracture, which can be lower than 10 MPa in pure metals [785]. Cavity nucleation by vacancy accumulation thus appears to require significant stress concentration. Of course, with elevated temperature plasticity, relaxation by creep plasticity and/or diffusional flow will accompany the elastic loading and relax the stress concentration. The other mechanisms illustrated in Figure 117 can involve cavity nucleation by direct "decohesion," which, of course, also requires a stress concentration.

2.2 Grain Boundary Sliding

GBS can lead to stress concentrations at triple points and hard particles on the grain boundaries, although it is unclear whether the local stresses are sufficient to nucleate cavities [20,803]. These mechanisms are illustrated in Figures 117(a), (b), and (d). Another sliding mechanism includes (tensile) ledges (Figure 117(a)) where tensile stresses generated by GBS may be sufficient to cause cavity nucleation [804], although some others [805] believed the stresses are insufficient. The formation of ledges may occur as a result of slip along planes intersecting the grain boundaries.

One difficulty with sliding mechanisms is that transverse boundaries (perpendicular to the principal tensile stress) appear to have a propensity to cavitate where sliding may be less substantial. Cavitation has been observed in bicrystals [806] where the boundary is perpendicular to the applied stress, such that there is no resolved shear and an absence of sliding. Hence, it appears that sliding is not a necessary condition for cavity nucleation. Others [780,793,807], however, still do not appear to rule out a relationship between GBS and cavitation along transverse boundaries. The ability to nucleate cavities via GBS has been demonstrated by prestraining copper bicrystals in an orientation favoring GBS, followed by subjecting the samples to a stress normal to the previously sliding GB and comparing those results to tests on bicrystals that had not been subjected to GBS [804]. Extensive cavitation was observed in the former case while no cavitation was observed in the latter. Also, as will be discussed later, GBS (and concomitant cavitation) can lead to increased stress on transverse boundaries, thereby accelerating the caviation at these locations. More recently, Ayensu and Langdon [793] found a relation between GBS and cavitation at transverse boundaries, but also note a relationship between GBS and strain. Hence, it is unclear whether GBS either nucleates or grows cavities in this case. Chen [808] suggested that transverse boundaries may slide due to compatibility requirements.

It must be remembered that so-called "transverse" boundaries may have a significant shear as they may be inclined to the photographic/image plane and only appear transverse in the particular two-dimensional projection.

2.3 Dislocation Pile-ups

As transverse boundaries may slide less, in general, perhaps the stress concentration associated with dislocation pile-ups against, particularly, hard second phase particles at transverse grain boundaries, has received significant acceptance [784,809,810] as a mechanism by which vacancy accumulation can occur. Pile-ups against hard particles within the grain interiors may nucleate cavities, but these may grow relatively slowly without short-circuit diffusion through the GB and may also be of lower (areal) density than at grain boundaries.

It is still not clear, however, whether vacancy accumulation is critical to the nucleation stage. Dyson [784] showed that tensile creep specimens that were prestrained at ambient temperature appeared to have a predisposition for creep cavitation. This suggested that the same process that nucleates voids at ambient temperature (that would not appear to include vacancy accumulation) may influence or induce void nucleation at elevated temperatures. This could include a Zener-Stroh mechanism (Figure 117(c)) against hard particles at grain boundaries. Dyson [784] showed that the nucleation process can be continuous throughout creep and that the growth and nucleation may occur together, a point also made by several other investigators [783,805,811,812]. This and the effect of prestrain are illustrated in Figure 118. The impact of cavitation rate on ductility is illustrated in Figure 119. Thus, the nucleation process may be controlled by the (e.g., steady-state) plasticity. The suggestion that cavity nucleation is associated with plastic deformation is consistent with the observation by Watanabe et al. [813], Greenwood et al. [814], and Dyson et al. [812] that the cavity spacing is consistent with regions of high dislocation activity (slip band spacing). Davanas and Solomon [801] argue that if continuous nucleation occurs, modeling of the fracture process can lead to a Monkman–Grant relationship (diffusive and plastic coupling of cavity growth and cavity interaction considered). One consideration against the slip band explanations is that in situ straining experiments with the transmission electron microscope, Dewald et al. [807] suggested that slip dislocations may easily pass through a boundary in a pure metal and that the stress concentrations from slip may be limited. This may not preclude such a mechanism in combination with second phase particles. Kassner et al. [69] performed creep fracture experiments on high-purity Ag at about 0.25 T_{m}. Cavities appeared to grow by (unstable) plasticity rather than diffusion. Nucleation was continuous, and it was noted that nucleation only occurred in the vicinity of high-angle boundaries where

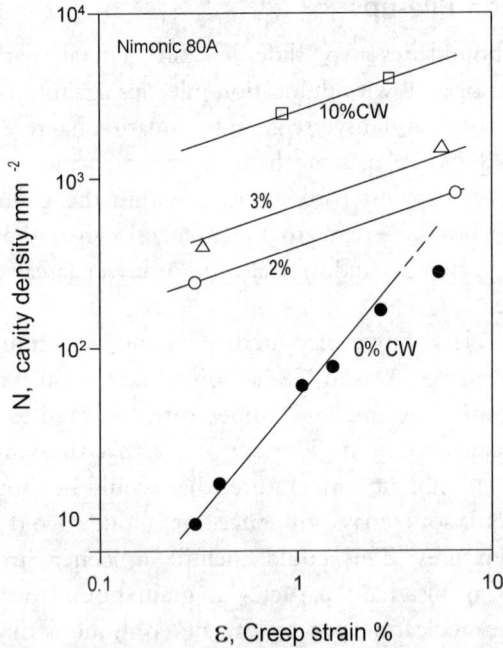

Figure 118 The variation of the cavity concentration versus creep strain in Nimonic 80A (Ni-Cr alloy with Ti and Al) for annealed and prestrained (cold-worked) alloy. *Adapted from Dyson [784]. Cavities were suggested to undergo unconstrained growth.*

obstacles existed (regions of highly twinned metal surrounded by low twin-density metal). High-angle boundaries without barriers did not appear to cavitate. Thus, nucleation (in at least transverse boundaries) appears to require obstacles and a Zener-Stroh or anti–Zener-Stroh mechanism appears most likely.

2.4 Location

It has long been suggested that (transverse) grain boundaries and second-phase particles are the common locations for cavities. Solute segregation at the boundaries may predispose boundaries to cavity nucleation [780]. This can occur due to the decrease in the surface and GB energy terms.

Some of the more recent work has found cavitation to be associated with hard second-phase particles in metals and alloys [816–824]. Second-phase particles can result in stress concentrations on application of a stress and increase cavity nucleation at a GB through vacancy condensation by increasing the GB free energy. Also, particles can be effective barriers to dislocation pile-ups.

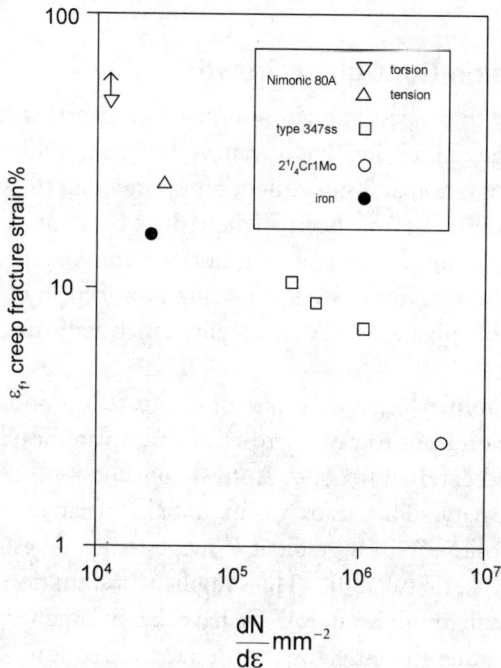

Figure 119 Creep ductility versus the "rate" of cavity production with strain. *Adapted from Dyson [784] (various elevated temperatures and stresses).*

The size of critical-sized nuclei is not well established, but the predictions based on the previous equations is about 2–5 nm [20], which are difficult to detect. Scanning electron microscopy under optimal conditions can be used to observe (stable) creep cavities as small as 20 nm [825]. It has been suggested the small-angle neutron scattering can characterize cavity distributions from less than 10 nm to almost 1 μm [20]. TEM has detected stable cavities (gas) to 3 nm [826]. Interestingly, observations of cavity nucleation not only suggest continual cavitation but also no incubation time [827] and that strain rather than time is more closely associated with nucleation [20]. Figure 119 illustrates the effect of stress states on nucleation. Torsion, for comparable equivalent uniaxial stresses in Nimonic 80 leads to fewer nucleated cavities and greater ductility than tension. Finally, another nucleation site that may be important as damage progresses in a material is the stress concentration that arises around existing cavities. The initial (elastic) stress concentration at the cavity "tip" is a factor of 3 larger than the applied stress, and even after relaxation by diffusion, the stress may still be elevated [828], leading to increased local nucleation rates.

3. GROWTH

3.1 GB Diffusion-Controlled Growth

The cavity growth process at grain boundaries at elevated temperature has long been suggested to involve vacancy diffusion. Diffusion occurs by cavity surface migration and subsequent transport along the GB, with either diffusive mechanism having been suggested to be controlling depending on the specific conditions. This contrasts creep void growth at lower temperatures where cavity growth is accepted to occur by (e.g., dislocation glide-controlled) plasticity. A carefully analyzed case for this is described [825].

Hull and Rimmer [829] were one of the first to propose a mechanism by which diffusion leads to cavity growth of an isolated cavity in a material under an applied external stress, σ. A stress concentration is established just ahead of the cavity. This leads to an initial "negative" stress gradient. However, a "positive" stress gradient is suggested to be established due to relaxation by plasticity [30]. This implicit assumption in diffusion-controlled growth models appears to have been largely ignored in later discussions by other investigators, with rare exception (e.g. [30]). The equations that Hull and Rimmer and, later, others [785,800,830] subsequently derive for diffusion-controlled cavity growth are similar. Basically

$$J_{gb} = -\frac{D_{gb}}{\Omega kT}\nabla f \tag{172}$$

where J_{gb} is the flux, Ω is the atomic volume, $f = -\sigma_{loc}\Omega$, and σ_{loc} is the local normal stress on the GB. Also

$$\nabla f \sim \frac{\Omega}{\lambda_s}\left(\sigma - \frac{2\gamma_m}{a}\right) \tag{173}$$

where "a" is the cavity radius, σ is the remote or applied normal stress to the GB, and λ_s is the cavity separation. Below a certain stress $\left[\sigma_0 = \frac{2\gamma_m}{a}\right]$, the cavity will sinter. Equations 172 and 173 give a rate of growth

$$\frac{da}{dt} \cong \frac{D_{gb}\delta\left(\sigma - \frac{2\gamma_m}{a}\right)\Omega}{2kT\lambda_s a} \tag{174}$$

where δ is the GB width. Figure 120 is a schematic that illustrates the basic concept of this approach.

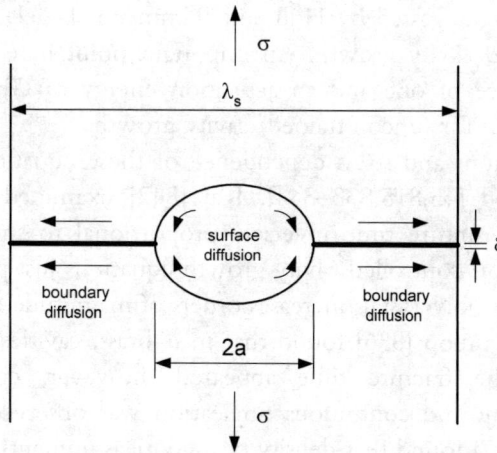

Figure 120 Cavity growth from diffusion across the cavity surface and through the grain boundaries due to a stress gradient.

By integrating between the critical radius (below which sintering occurs) and $a = \lambda_s/2$

$$t_r \cong \frac{kT\lambda_s^3}{4D_{gb}\delta\left(\sigma - \frac{2\gamma_m}{a}\right)\Omega} \qquad (175)$$

This is the first relationship between stress and rupture time for (unconstrained) diffusive cavity growth. Raj and Ashby [800,831], Speight and Beere [830], Riedel [785], and Weertman [832] later suggested improved relationships between the cavity growth rate and stress of a similar form to that of Hull and Rimmer (Eqn (174)). The subsequent improvements included modifications to the diffusion lengths (the entire GB is a vacancy source), stress redistribution (the integration of the stress over the entire boundary should equal the applied stress), cavity geometry (cavities are not perfectly spherical), and the "jacking" effect (atoms deposited on the boundary causes displacement of the grains). Riedel, in view of these limitations, suggested that the equation for unconstrained cavity growth of widely spaced voids is, approximately

$$\frac{da}{dt} = \frac{\Omega\delta D_{gb}[\sigma - \sigma''_o]}{1.22kT\,\ln(\lambda_s/4.24a)a^2} \qquad (176)$$

where σ''_o is the sintering stress. Again, integrating to determine the time for rupture shows that $t_r \propto 1/\sigma$. Despite these improvements, the basic description long suggested by Hull and Rimmer is largely representative of unconstrained cavity growth. An important point here is a predicted stress dependence of one and an activation energy of GB diffusion for Eqns (174)–(176) for (unconstrained) cavity growth.

The predictions and stress dependence of these equations have been frequently tested [45,815,833–844]. Raj [842] examined Cu bicrystals and found the rupture time inversely proportional to stress, consistent with the diffusion controlled cavity growth equations just presented. The fracture time for polycrystals increases orders of magnitude over bicrystals. Svensson and Dunlop [836] found that in α-brass, cavities grow linearly with stress. The fracture time appeared, however, consistent with Monkman–Grant and continuous nucleation was observed. Hanna and Greenwood [837] found that density change measurements in prestrained (i.e., prior cavity nucleation) and with hydrogen bubbles were consistent with the stress dependency of the earlier equations. Continuous nucleation was not assumed. Cho et al. [838] and Needham and Gladman [843,844] measured the rupture times and/or cavity growth rate and found consistency with a stress to the first power dependency if continuous nucleation was assumed. Miller and Langdon [835] analyzed the density measurements on creep-deformed Cu based on the work of others and found that the cavity volume was proportional to σ^2 (for fixed T, t, and ε). If continuous nucleation occurs with strain, which is reasonable, and the variation of the nucleation rate is "properly" stress dependent (unverified), then consistency between the density trends and unconstrained cavity growth described by Eqns (168) and (170) can be realized.

Creep cavity growth experiments have also been performed on specimens with preexisting cavities by Nix and coworkers [45,834]. Cavities, here, were created using water vapor bubbles formed from reacting dissolved hydrogen and oxygen. Cavities were uniformly "dispersed" (unconstrained growth). Curiously, the growth rate, da/dt, was found to be proportional to σ^3. This result appeared inconsistent with the theoretical predictions of diffusion-controlled cavity growth. The disparity is still not understood. Interestingly, when a dispersion of MgO particles was added to the Ag matrix, which decreased the Ag creep rate, the growth rate of cavities was *unaffected*. This supports the suggestion that the controlling factor for cavity growth is diffusion rather than plasticity or GBS. Similar findings were reported by others [845].

3.2 Surface Diffusion-Controlled Growth

Chuang and Rice [846] and later Needleman and Rice [778] suggested that surface rather than grain-boundary diffusion may actually control cavity growth (which is not necessarily reasonable) and that these assumptions can give rise to a 3-power stress relationship for cavity growth at low stresses [847].

$$\frac{da}{dt} \cong \frac{\Omega \delta D_s}{2kT\gamma_m^2}\sigma^3 \tag{177}$$

At higher stresses, the growth rate varies as $\sigma^{3/2}$. The problem with this approach is that it is not clear in the experiments, for which three-power stress dependent cavity growth is observed, that $D_s < D_{gb}$. Activation energy measurements by Nieh and Nix [834] for (assumed unconstrained) growth of cavities in Cu are inconclusive as to whether it better matches D_{gb} versus D_s. Also, the complication with all of these growth relationships (Eqns (174)–(177)) is that they are inconsistent with the Monkman–Grant phenomenology. That is, for common 5-power-law creep, the Monkman–Grant relationship suggests that the cavity growth rate $(1/t_f)$ should be proportional to the stress to the fifth power rather than 1- to 3-power. This, of course, may emphasize the importance of nucleation in the rate-controlling process for creep cavitation failure, since cavity-nucleation may be controlled by the plastic strain (steady-state creep rate). Of course, small nanometer-sized cavities (nuclei), by themselves, do not appear sufficient to cause cavitation failure. Dyson [784] suggested that the Monkman–Grant relationship may reflect the importance of both (continuous) nucleation and growth events.

3.3 Grain Boundary Sliding

Another mechanism that has been considered important for growth is GBS [848]. This is illustrated in Figure 121. Here cavities are expected to grow predominantly in the plane of the boundary. This appears to have been observed in some temperature-stress regimes. Chen appears to have invoked GBS as part of the cavity growth process [779], also suggesting that transverse boundaries may slide due to compatibility requirements. A suggested consequence of this "crack sharpening" is that the tip velocity during growth becomes limited by surface diffusion. A stress to the third power, as in Eqn (177), is thereby rationalized. Chen suggests that this phenomenon may be more applicable to higher strain rates and closely spaced cavities (later stages of creep) [808]. The observations that cavities are

(a)   **(b)**

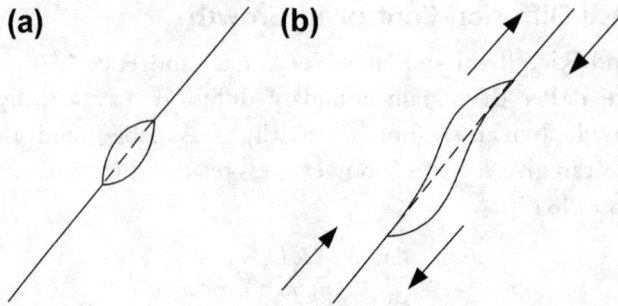

Figure 121 Cavity growth from a sliding boundary. (a) initial cavity and (b) sliding cavity. *From [780].*

often more spherical rather than plate-like or lenticular and that, of course, transverse boundaries may be less prone to slide also suggest that cavity growth does not substantially involve sliding.

Riedel [785] predicted that (constrained) diffusive cavity growth rates are expected to be a factor of $(\lambda_S/2a)^2$ larger than growth rates by (albeit, constrained) sliding. It has been suggested that sliding may affect growth in some recent work on creep cavitation of dual phase intermetallics [849].

3.4 Constrained Diffusional Cavity Growth

Cavity nucleation may be heterogeneous, inasmuch as regions of a material may be more cavitated than others. Adams [850] and Watanabe et al. [851] both suggested that different geometry (e.g., as determined by the variables necessary to characterize a planar boundary) high-angle grain boundaries have a different tendencies to cavitate, although there was no agreement as to the nature of this tendency in terms of the structural factors. A given misorientation (3 of 5 degrees of freedom) boundary may have varying orientations to the applied stresses. Another important consideration is that the zone ahead of the cavity experiences local elongating with diffusional growth, and this may cause constraint in this region by those portions of the material that are unaffected by the diffusion (outside the cavity diffusion "zone"). This may cause a "shedding" of the load from the diffusion zone ahead of the cavity. Thus, cavitation is not expected to be homogeneous and uncavitated areas may constrain those areas that are elongating under the additional influence of cavitation. This is illustrated in Figure 122. Fracture could then be controlled by the plastic *creep rate* in uncavitated regions that can also lead to cavity nucleation. This leads to consistency with the Monkman–Grant relationship [852,853].

(a)   **(b)**

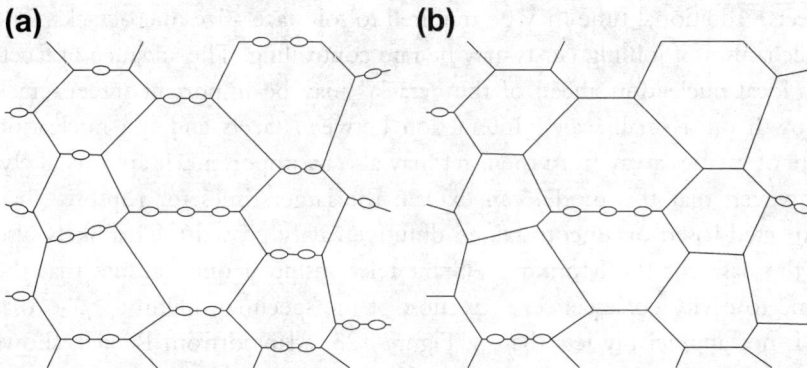

Figure 122 Uniform (a) and heterogeneous (b) cavitation at (especially transverse) boundaries. The latter condition can particularly lead to constrained cavity growth. *From [780].*

Constrained diffusional growth was originally suggested by Dyson and further developed by others [776,793,808,846,853]. This constrained cavity growth rate has been described by the relationship [785]

$$\frac{da}{dt} \cong \frac{\sigma - (1 - \omega)\sigma_o''}{\frac{a^2 kT}{\Omega \delta D_{gb}} + \frac{\sigma_{ss}\pi^2(1+3/n)^{1/2}}{\dot{\varepsilon}_{ss}\lambda_s^2 g}a^2} \tag{178}$$

where ω is the fraction of the GB cavitated.

This is the growth rate for cavities expanding by diffusion. One notes that for higher strain rates, where the increase in volume can be easily accommodated, the growth rate is primarily a function of the GB diffusion coefficient.

If only certain GB facets cavitate, then the time for coalescence, t_c, on these facets can be calculated

$$t_c \cong \frac{0.004kT\lambda^3}{\Omega \delta D_{gb}\sigma_{ss}} + \frac{0.24(1 + 3/n)^{1/2}\lambda}{\dot{\varepsilon}_{ss}g} \tag{179}$$

where, again, n is the steady-state stress exponent, g is the grain size, and σ_{ss} and $\dot{\varepsilon}_{ss}$ are the steady-state stress and strain rate, respectively, related by

$$\dot{\varepsilon}_{ss} = A_o \exp[-Q_c/kT](\sigma_{ss}/E)^n \tag{180}$$

where $n = 5$ for classic 5-power-law creep. However, it must be emphasized that failure is not expected by mere coalescence of cavities on isolated facets. Additional time may be required to join facet-size microcracks. The mechanism of joining facets may be rate controlling. The advance of facets by local nucleation ahead of the "crack" may be important (creep-crack growth on a small scale). Interaction between facets and the nucleation rate of cavities away from the facet may also be important. It appears likely, however, that this model can explain the larger times for rupture (than expected based on unconstrained diffusive cavity growth). This likely also is the basis for the Monkman–Grant relationship if one assumes that the time to cavity coalescence, t_c, is most of the specimen lifetime, t_f, so that t_c is not appreciably less than t_f. Figure 123, adapted from Riedel, shows the cavity growth rate versus stress for constrained cavity growth as solid lines. Also plotted in this figure (as the dashed lines) is the equation for unconstrained cavity growth (Eqn (176)). It is observed that the equation for unconstrained growth predicts much higher growth rates (lower t_f) than constrained growth rates. Also, the stress dependency of the growth rate for constrained growth leads to a time-to-fracture relationship that more closely matches that expected for steady-state creep as predicted by the Monkman–Grant relationship.

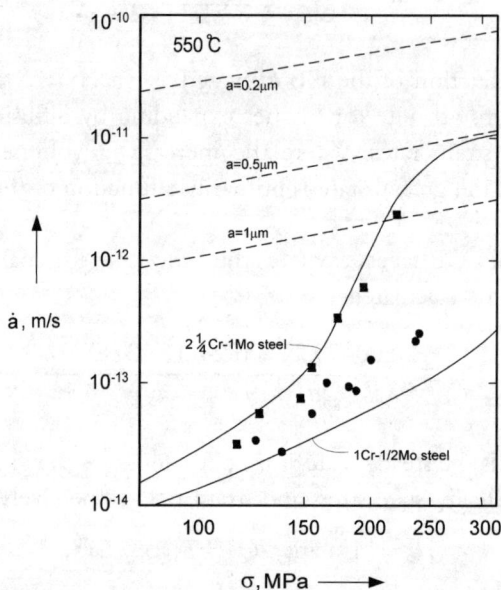

Figure 123 The cavity growth rate versus stress in steel. The dashed lines refer to unconstrained growth and solid lines to constrained growth. *Based on Riedel [785].*

One must, in addition to considering constrained cases, also consider that cavities are continuously nucleated. For continuous nucleation and unconstrained diffusive cavity growth, Riedel suggests

$$t_f = \left[\frac{kT}{5\Omega\delta D_{gb}\sigma}\right]^{2/5} \left(\frac{\omega_f}{\dot{N}}\right)^{3/5} \tag{181}$$

where ω_f is the critical cavitated area fraction and, consistent with Figure 119 from Dyson [784]

$$\dot{N} = \alpha'\dot{\varepsilon} = \alpha'\beta\sigma^n \tag{182}$$

with $\dot{\varepsilon}$ according to Eqn (180).

Equation (181) can be approximated by

$$t_f \propto \frac{1}{\sigma^{(3n+2)/5}}$$

For continuous nucleation with the constrained case, the development of reliable equations is more difficult, as discussed earlier, and Riedel suggests that the time for coalesce on isolated facets is

$$t_c = 0.38 \left[\frac{\pi\left(1 + \frac{3}{n}\right)}{\dot{N}}\right]^{1/3} \frac{\omega_f}{[\dot{\varepsilon}g]^{2/3}} \tag{183}$$

which is similar to the version by Cho et al. [838]. Figure 124, also from Riedel, illustrates the realistic additional effects of continuous nucleation, which appear to match the observed rupture times in steel. The theoretical curves in Figure 124 correspond to Eqns (175), (190), (191), and (183). One interesting aspect of this figure is that there is a very good agreement between t_c and t_f for constrained cavity growth. These data were based on the data of Cane [817] and Riedel [854], who determined the nucleation rate by apparently using an empirical value of α'. No adjustable parameters were used. Later, for NiCr steel, at 823K, Cho et al. [838] were able to reasonably predict rupture times assuming continuous nucleation and constrained cavity growth.

It should be mentioned that accommodated GBS can eliminate the constraint illustrated in Figure 122 (two-dimensional); however, in the three-dimensional case, sliding does not preclude constrained cavity

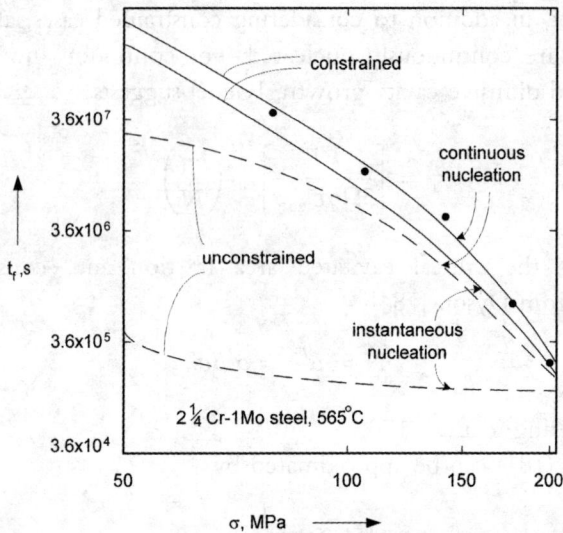

Figure 124 The time to rupture versus applied stress for unconstrained (dashed lines) cavity growth with instantaneous or continuous nucleation, and constrained cavity growth (t_c) with instantaneous and continuous nucleation. Dots refer to experimental t_f. *Based on Riedel [785].*

growth, as shown by Anderson and Rice [855]. Yousefiani et al. [857] used a calculation of the principal facet stress to predict the multiaxial creep rupture time from uniaxial stress data. Here, it is suggested that GBS is accommodated and the normal stresses on (transverse) boundaries are increased. Van der Giessen and Tvergaard [858] appear to analytically (3D) show that increased cavitation on inclined sliding boundaries may increase the normal stresses on transverse boundaries for constrained cavity growth. Thus, the Riedel solution may be nonconservative, in the sense that it overpredicts t_f. Dyson [859] suggested that within certain temperature and strain-rate regimes, there may be a transition from constrained to unconstrained cavity growth. For an aluminum alloy, it was suggested that decreased temperature and increased stress could lead to unconstrained growth. Interestingly, Dyson also pointed out that for constrained cavity growth, uncavitated regions would experience accelerated creep beyond that predicted by the decrease in load carrying area resulting from cavitation.

3.5 Plasticity

Cavities can grow, of course, exclusively by plasticity. Hancock [860] initially proposed the creep controlled cavity growth model based on the idea that cavity growth during creep should be analogous to McClintock's [818] model for a cavity growing in a plastic field. Cavity growth according to this model occurs as a result of creep deformation of the material surrounding the GB cavities in the absence of a vacancy flux. This mechanism becomes important under high strain-rate conditions, where significant strain is realized. The cavity growth rate according to this model is given as

$$\frac{da}{dt} = a\dot{\varepsilon} - \frac{\gamma}{2G} \tag{184}$$

This is fairly similar to the relationship by Riedel [785] discussed earlier. It has been suggested, on occasion, that the observed creep cavity growth rates are consistent with plasticity growth (e.g., [835]) but it is not always obvious that constrained diffusional cavity growth is not occurring, which is also controlled by plastic deformation.

3.6 Coupled Diffusion and Plastic Growth

Cocks and Ashby [776], Beere and Speight [861], Needleman and Rice [778], and others [777,862–867] suggested that there may actually be a coupling of diffusive cavity growth of cavities with creep plasticity of the surrounding material from the far-field stress. It is suggested that as material from the cavity is deposited on the GB via surface and GB diffusion, the length of the specimen increases due to the deposition of atoms over the diffusion length. This deposition distance is effectively increased (shortening the effective required diffusion-length) if there is creep plasticity in the region ahead of the diffusion zone. This was treated numerically by Needleman and Rice and later by van der Giessen et al. [868]. Analytic descriptions were performed by Chen and Argon [863]. A schematic of this coupling is illustrated in Figure 125. The diffusion length is usually described as [778]

$$\Lambda = \left(\frac{D_{gb}\Omega\delta\sigma}{kT\dot{\varepsilon}}\right)^{1/3} \tag{185}$$

Chen and Argon describe coupling by

$$\frac{dV}{dt} = \dot{\varepsilon}2\pi\Lambda^3 \bigg/ \left[\ell n\left(\frac{a+\Lambda}{a}\right) + \left(\frac{a}{a+\Lambda}\right)^2 \times \left(1 - \frac{1}{4}\left(\frac{a}{a+\Lambda}\right)^2\right) - \frac{3}{4}\right] \tag{186}$$

as illustrated in Figure 126.

Coupled Cavity Growth

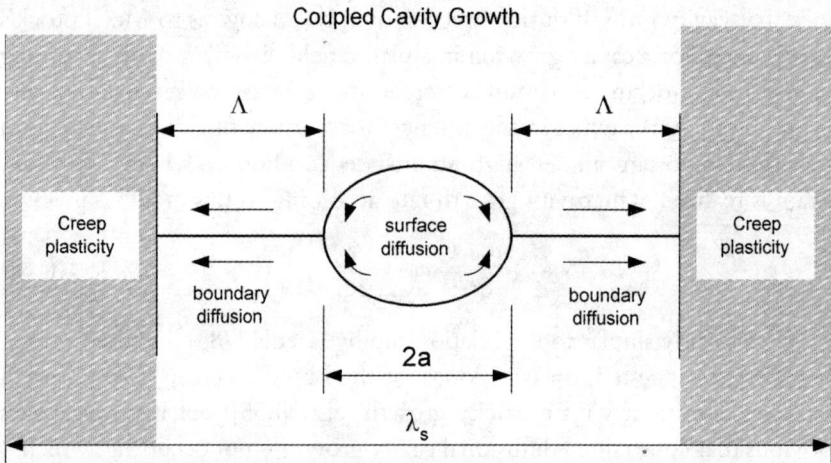

Figure 125 The model for coupled diffusive cavity growth with creep plasticity. The diffusion length is suggested to be reduced by plasticity ahead of the cavity. *Based on [777].*

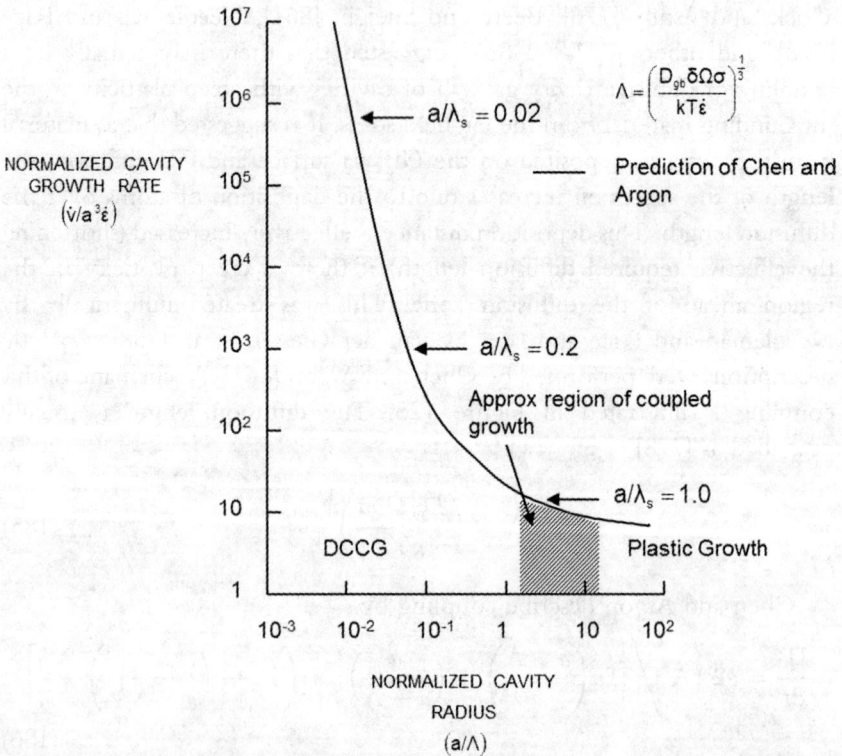

Figure 126 Prediction of growth rate for different ratios of cavity spacing λ and diffusion zone sizes Λ. *From [780].*

Similar analyses were performed by others with similar results [60,778,863,864]. It has been shown that when $\Lambda \ll a$ and λ [863,869], diffusion-controlled growth no longer applies. In the extreme, this occurs at low temperatures. Creep flow becomes important as a/Λ increases. At small creep rates, but higher temperatures, Λ approaches $\lambda_s/2$, a/Λ is relatively small, and the growth rate can be controlled by diffusion-controlled cavity growth (DCCG). Coupling, leading to "enhanced" growth rates over the individual mechanisms, occurs at "intermediate" values of a/Λ as indicated in Figure 126. Of course, the important question is whether, under "typical creep" conditions, the addition of plasticity effects (or the coupling) is important. Needleman and Rice suggest that for $T > 0.5\ T_m$, the plasticity effects are important only for $\sigma/G > 10^{-3}$ for pure metals (relatively high stress). Riedel suggests that, for pure metals, as well as creep-resistant materials, diffusive growth predominates over the whole range of creep testing. Figure 125 illustrates this coupling [777].

Even under the most relevant conditions, the cavity growth rate due to coupling is, at most, a factor of 2 different than the growth rate calculated by simply adding the growth rates due to creep and diffusion separately [870]. It has been suggested that favorable agreement between Chen and Argon's analytical treatments is fortuitous because of limitations to the analysis [865,866,869]. Of course, at lower temperatures, cavity growth occurs exclusively by plasticity [860]. It must be recognized that cavity growth by simply plasticity is not as well understood as widely perceived. In single-phase metals, for example, under uniaxial tension, a 50% increase in cavity size requires large strains, such as 50% [871]. Thus, a 1000-fold increase in size from the nucleated nanometer-sized cavities would not appear to be easily explained. Figure 115(b), interestingly, illustrates a case where plastic growth of cavities appears to be occurring. The cavities nucleate within grains at large particles in the dispersion-strengthened aluminum of this figure. Dunand et al. [872] suggest that this transgranular growth occurs by plasticity, as suggested by others [776] for growth inside grains. Perhaps the interaction between cavities explains modest ductility. One case where plasticity in a pure metal is controlling is constrained thin silver films under axisymmetric loading where $\sigma_1/\sigma_2 \ (=\sigma_3) \cong 0.82$ [2,825]. Here, unstable cavity growth [873] occurs via steady-state deformation of silver. The activation energy and stress-sensitivity appear to match those of steady-state creep of silver at ambient temperature. Cavities nucleate at high-angle boundaries where obstacles are observed (high twin-density metal) by slip-plasticity. A scanning electron photomicrograph of these cavities is illustrated in Figure 127. The cavities in Figure 127 continuously nucleate and appear to undergo plastic cavity growth. Interestingly, if a plastically deforming base metal is used (creep deformation of the constraining base metal of a few percent), the additional concomitant plastic strain (over that resulting from a perfectly elastic base metal) increases the nucleation rate and decreases the fracture time by several orders of magnitude, consistent with Figure 124. Cavity growth can also be affected by segregation of impurities, as these may affect surface and GB diffusivity. Finally, creep

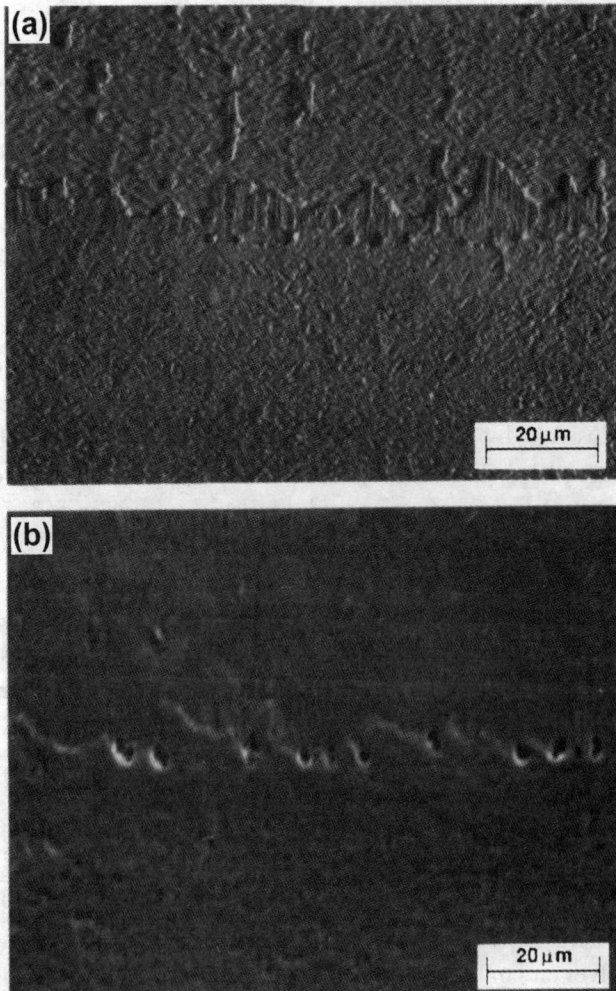

Figure 127 Creep cavitation in silver at ambient temperature. Cavities grow by un-stable cavity growth, with the rate determined by steady-state creep of silver. (a) Optical and (b) SEM micrographs. *From [2].*

fracture predictions must consider the scatter present in the data. This important, probabilistic, aspect recently has been carefully analyzed [874].

3.7 Creep Crack Growth

Cracks can occur in creeping metals from preexisting flaws, fatigue, corrosion-related processes, and porosity [875,876]. In these cases, the cracks are imagined to develop relatively early in the lifetime of the metal. This contrasts the case where cracks can form in a uniformly strained (i.e., unconstrained) cavity growth and uniform cavity nucleation metal where interlinkage of cavities leading to crack formation is the final stage of the rupture life. Crack formation by cavity interlinkage in constrained cavity growth cases may be the rate-controlling step(s) for failure. Hence, the subject of creep crack growth is relevant in the context of cavity formation. Figure 128 (from [877]) illustrates a Mode I crack. The stress/strain ahead of the crack leads to cavity nucleation and growth. The growth can be considered to be a result of plasticity-induced expansion or diffusion-controlled cavity growth. Crack growth occurs by the coalescence of cavities with each other and the crack.

Nix et al. [877] showed that plastic growth of cavities ahead of the crack tip can lead to a "steady-state" crack growth rate. Nucleation was not included in the analysis. They considered the load parameter to be the stress intensity factor for (elastic) metals with Mode I cracks, K_I

$$v_c = \frac{k_8 \lambda_s}{2(n-2)\ln(\lambda_s/2a)} \left(\frac{K_I}{n\sqrt{\lambda_s}} \right)^n \tag{187}$$

where k_8 is a constant.

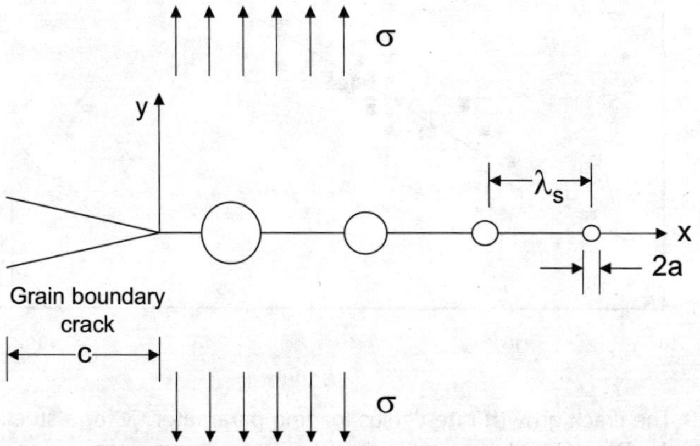

Figure 128 Grain boundary crack propagation controlled by the creep growth of cavities near a crack tip. *From [877].*

However, for cases of plasticity, and in the present case with time-dependent plasticity, the load parameters have been changed to J and $C^\star$ [878], respectively. Much of the creep cavitation work since 1990 appears to have focused on creep cracks and analysis of the propagation in terms of $C^\star$. The $C^\star$ term appears to be a reasonable loading parameter that correlates crack growth rates, although factors such as plane-stress/plane-strain (i.e., stress-state), crack branching, and extent of the damage zone from the crack tip may all be additionally important in predicting the growth rate [879–883].

Of course, another way that cracks can expand is by linking up with diffusionally growing cavities. This appears to be the mechanism favored by Cadek [20] and Wilkinson and Vitek [884,885] and others [882]. Later, Miller and Pilkington [886] and Riedel [785] suggest that strain (plasticity)-controlled growth models (with a critical strain criterion or with strain controlled nucleation) better correlate with existing crack growth data than diffusional growth models. However, Riedel indicates that the uncertainty associated with *strain*-controlled nucleation complicates the unambiguous selection of the rate-controlling growth process for cavities ahead of a crack. Figure 129 illustrates a correlation between the crack growth rate, $\dot{c}$ and the loading parameter $C^\star$. Riedel argued that the crack growth rate is best

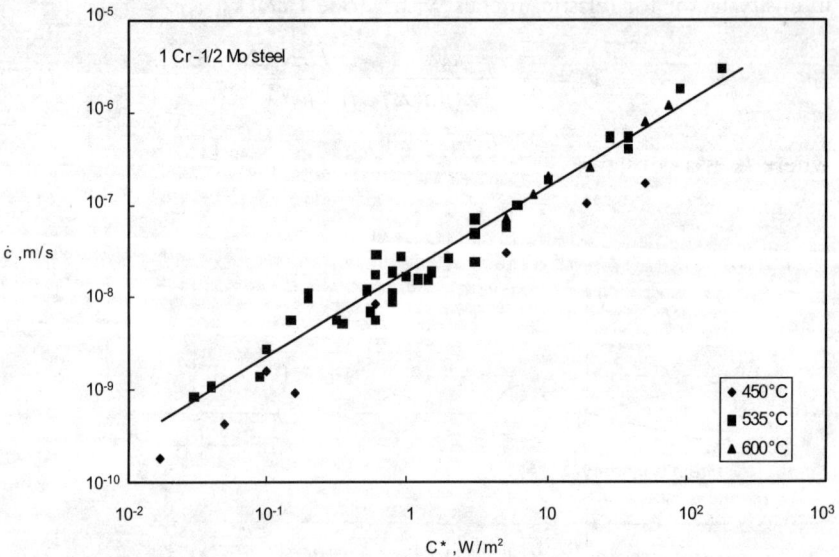

Figure 129 The crack growth rate versus loading parameter C^* for a steel. The line is represented by Eqn (150). *From [785]*.

described by the plastic cavity growth relationship, based on a local critical strain criterion

$$\dot{c} = k_9 \lambda^{1/n+1} (C^*)^{n/n+1} \left[\left(\frac{c - c_o}{\lambda_s} \right)^{1/n+1} - k_{10} \right] \tag{188}$$

Riedel similarly argued that if cavity nucleation occurs instantaneously, diffusional growth predicts

$$\dot{c} = \frac{k_{11} D_b (\Omega \delta)}{2kT\lambda_s^3} C^{*1/n+1} (c - c_o)^{n/n+1} \tag{189}$$

where c_0 is the initial crack length and c is the current crack length. These constants are combined (some material) constants from Riedel's original equation and the line in Figure 129 is based on Eqn (188) using some of these constants as adjustable parameters.

Note that Eqn (189) gives a strong temperature dependence (the "constants" of the equation are not strongly temperature dependent). Riedel also develops a relationship of strain-controlled cavity growth with strain-controlled nucleation, which also reasonably describes the data of Figure 129.

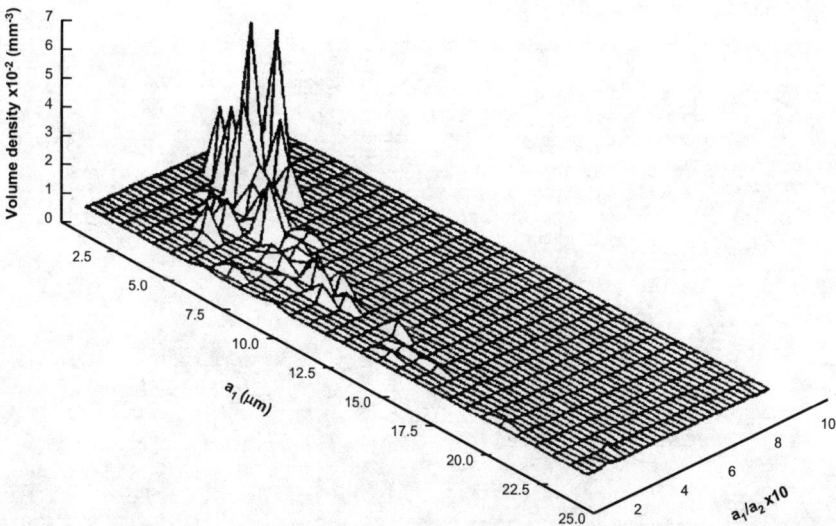

Figure 130 The cavity density versus size and aspect ratio of creep deformed 304 stainless steel. *From [866].*

3.8 Other Considerations

As discussed earlier, Nix and coworkers [811,815,833,834] produced cavities by reacting with oxygen and hydrogen to produce water-vapor bubbles (cavities). Other (unintended) gas reactions can occur. These gases can include methane, hydrogen, and carbon dioxide. A brief review of environmental effects was discussed recently by Delph [869].

The randomness (or lack or periodicity) of the metal microstructure leads to randomness in cavitation and (e.g.) failure time. Figure 130 illustrates the cavity density versus major radius a_1 and aspect ratio a_1/a_2. This was based on metallography of creep deformed AlSl 304 stainless steel. A clear distribution in sizes is evident. Creep failure times may be strongly influenced by the random nature of grain boundary cavitation.

CHAPTER 11

γ/γ' Nickel-Based Superalloys

M.E. Kassner

Contents

1. INTRODUCTION

This book deals primarily with the fundamentals of creep in crystalline materials. Systems have generally been well defined and have less ambiguous creep responses. These basic insights are intended to be useful in understanding more complex systems. For this reason, the book often deals with "simplified" materials. There have been several reviews of superalloys, and those of Pollock et al. [600,1009] are important ones. This review relied on these and other reviews [600,978,979]. A discussion of superalloys, albeit relatively brief, makes particular sense because intermetallics were discussed earlier and it appears that superalloys (that often have an ordered intermetallic phase as the majority component) have a continued important role in high-temperature materials applications.

Superalloys are an example of at least two phases: one ordered and one disordered. Industrial alloys often have additional phases to the primary disordered and ordered phases, but a discussion of the fundamentals of superalloys can be best reduced to a discussion of the common two-phase γ/γ' systems. These two-phase systems are often referred as single crystals in the absence of high-angle grain boundaries. The Ni-Al binary phase diagram, the basis of many superalloys, is illustrated in Figure 131 and shows details absent in the earlier Figure 106.

Here, the phase diagram illustrates that at compositions in the vicinity of 80 atomic weight percent Ni, two phases exist: the disordered Ni fcc solid-solution γ and the Ni_3Al ordered primitive cubic intermetallic γ', discussed earlier in Chapter 9, which has the $L1_2$ structure. Typically, the γ' occupies

Fundamentals of Creep in Metals and Alloys
ISBN 978-0-08-099427-7
http://dx.doi.org/10.1016/B978-0-08-099427-7.00011-6

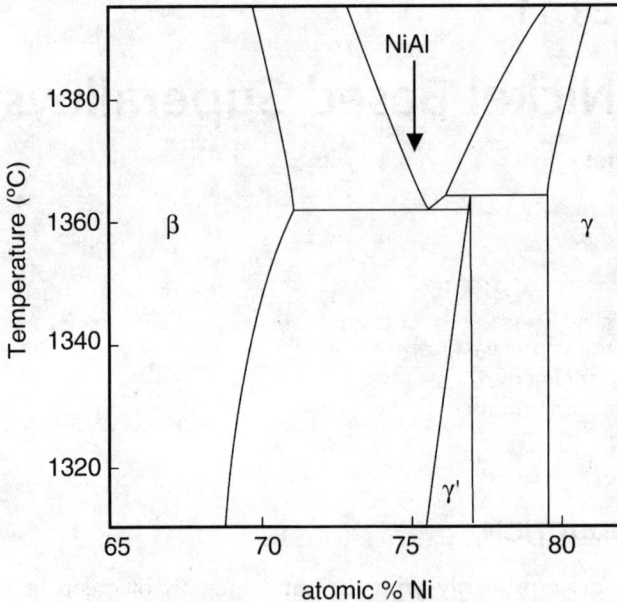

Figure 131 Binary phase diagram of Ni–Al from [1011]. For a more complete phase diagram and discussion. *See Ref. [1012]*.

60–75% of the alloy, although this fraction decreases with increasing temperature as the phase diagram illustrates. As will be discussed further, an important aspect of these alloys is coherency between the γ and γ' phases. The, ordered, cubic precipitates have faces that are parallel to <100>. Hence, we are addressing a composite material, rather than precipitation-strengthened or dispersion-strengthened material as discussed earlier, in Chapter 8, where the fraction of second phase particles was much less. Figure 132 is a micrograph of a γ/γ' microstructure. The cubic γ' phase is observed in the γ matrix.

Superalloys that are discussed here will generally be single crystals (e.g., turbine blade applications) and include NASAIR 100 and CMSX-4, -6, and -10. Typical compositions are listed in Table 11.

Obviously, elements are added to the Ni-Al "basis" for various purposes and they may partition to different phases (i.e., γ or γ'). The elements Ti and Ta are γ' hardening elements, while Re, Nb, and W are γ hardening elements [1009]. Mo may affect lattice misfit [978] and thereby affects the spacing of interfacial dislocations, which, in turn, may affect the creep properties. The role of specific additions to the creep properties of the superalloys is not always well understood [999,1013].

Figure 132 Microstructure of a single-crystal Ni–Al superalloy CMSX-10 (variant). *Courtesy of S. Tin, IIT.*

NASAIR100 and CMSX-4 (over a temperature range of about 750–950 °C) have an average elevated-temperature creep activation energy of 420 kJ/mol, which appears to be much higher than the activation energy for creep of either pure Al or Ni at roughly 280 kJ/mol. Furthermore, the creep-rate stress exponents may be higher than the pure elements as well, at comparable fractions of the melting temperature [600,1000,1001] (e.g., perhaps double). The Q and *n* values may show a trend of decreasing with increasing temperature. It should be mentioned that these values appear to be based on minimum rather than steady-state creep rates. As will be observed subsequently, genuine steady states as defined in earlier chapters may not be relevant here. Pollock and Field illustrate that the creep resistance of an Ni-Al γ/γ' superalloys is substantially better than the isolated bulk phases.

The discussion of this system, in terms of elevated-temperature creep, must consider that at different temperatures, different dislocation mechanisms are observed, in conjunction with changes in the phase coherency as well as phase morphology (e.g.,"rafting"). The approach, here, is to first divide the discussion of the creep behavior of γ/γ' coherency between the

Table 11 Typical compositions of some Ni–Al superalloys in weight percent

Alloy	Co	Cr	Mo	W	Ta	Al	Ti	Other
CMSX-4	9.2	6.6	0.6	5.9	6.2	5.6	0.9	3.3 Re
CMSX-6	5.0	10.0	3.0	–	6.0	4.8	4.7	0.1 Hf
CMSX-10	3.0	2.0	0.4	5.0	8.0	5.7	0.2	6.0 Re, 0.1 Nb, 0.3 Hf
NASAIR100	–	9.5	1.0	10.0	3.2	5.5	1.2	–
Rene 88 DT[a]	13.0	16.0	4.0	4.0	–	2.1	3.7	0.015 B, 0.3 C, 0.7 Nb

[a]Polycrystal.
From [600,978,1002,1003,1009]. Balance is Ni.

superalloys by discussing the behavior of Ni-Al superalloys into regimes where the overall behaviors are different. Stresses, composition, time, and especially temperature can affect the creep mechanisms. Different reviews separate these regimes at somewhat different temperature ranges, but here we will identify the following three regimes:

1. Creep below 800 °C (or below about 0.66 T_m, where T_m is, here, defined as the invariant temperature of the peritectic). This could be referred as low-temperature creep. Creep occurs in both phases.
2. Creep between about 800 and 950 °C, where dislocation activity is generally confined to the γ channels and only at larger strains and/or longer times, once interface dislocations form and rafting may begin, does shear in the γ' occur.
3. Creep above 950 °C, or above about 0.75 T_m, where interface dislocation networks form early in creep leading to rafting of the γ', particularly at higher temperatures in this range. This is referred to as high-temperature creep.

The meanings of "low," "intermediate," and "high" temperature have different meanings than the systems discussed in earlier chapters. The uses here are intended to be consistent with the conventions used with super-alloys. It should be mentioned that in considering creep processes in these commercial alloys, plastic strains of about 1% can become irrelevant in that dimensional consideration of components would usually consider strains above this level excessive. Therefore, creep investigations have focused on smaller strains.

As an early, general, consideration of the superalloy microstructure, it should be mentioned that there are residual stresses in the alloy because of the coherency between the γ/γ' interface. The coherency can be manipulated by adjusting the composition (adding new elements), as well as by annealing and bringing edge dislocations to the interface that relax the strains. The lattice misfit δ, is defined here, according to [978]:

$$\delta = 2[\alpha_{\gamma'} - \alpha_{\gamma}]/[\alpha_{\gamma'} + \alpha_{\gamma}] \tag{190}$$

Pollock and Argon [981] estimated these residual stresses for CMSX-3 Ni-Al superalloy with a negative misfit of about -0.3% at 850 °C. The von Mises stress in the channels (γ) was on 1 order of magnitude higher than in the precipitate (γ') at about 456 MPa vs. 58 MPa in the unloaded condition. These residual stresses can have a variety of effects on the elevated temperature behaviors, as will be discussed subsequently. This misfit, if sufficient in magnitude, leads to the cubic morphology.

2. LOW-TEMPERATURE CREEP

The early consensus of investigations of creep in superalloys is that slip occurs in <211> directions in both the γ and γ', particularly below roughly 5% strain. Slip occurs both in the matrix and the precipitates on {111} planes. This has been confirmed by Burgers vector analysis in transmission electron microscopy, as well as observations of the crystallographic rotations during single-crystal deformation. This suggestion goes back to the early work of Leverant and Kear [985]. Detailed reactions of the dislocations vary, but one recent summary is discussed [978] that captures many of the essential elements of other recent reviews and investigations.

Basically, two $a/2<110>$ dislocations react to form an $a/2<112>$ within the disordered γ matrix that consists of two partials, $a/6<112>$ and $a/3<112>$, or

$$a/2[011] + a/2[\overline{1}01] \rightarrow a/2[\overline{1}12] \rightarrow a/3[\overline{1}12] + a/6[\overline{1}12]$$

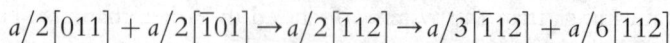

Then, an $a/3[\overline{1}12]$ could shear the γ' on a $(1\overline{1}1)$ plane, leaving a SISF (superlattice intrinsic stacking fault). The $a/6[\overline{1}12]$ could be considered, initially, to remain at the γ/γ' interface. Next, a second $a/2[\overline{1}12]$ arrives and the γ' precipitate now contains $a/6[\overline{1}12] + a/3[\overline{1}12]$ separated by an SISF. The $a/6[\overline{1}12]$ that is the leading partial of the "second" $a/2[\overline{1}12]$ is separated from the trailing $a/6[\overline{1}12]$ partial of the "first" dislocation by an antiphase boundary (APB). A superlattice extrinsic stacking fault (SESF) separates the two partials of the second dislocation. This mechanism is intended to rationalize the frequent observation of <211> type slip at lower temperatures in γ/γ' superalloys. This is a glide-type mechanism and the thermal activation anticipated for creep processes is not easily rationalized. This mechanism is illustrated in Figure 133.

More recently, Viswanathan et al. [996], as a variation of a mechanism by Condat and Decamps [997] and Mukherji [998], proposed a different explanation (Figure 134) that is applicable to polycrystals (René 88DT, a disk alloy). The $a/2[011]$ matrix dislocation is dissociated into Shockley partials (αC and Dα) in the γ matrix. As Dα crosses into the precipitate, a narrow complex stacking fault (CSF) is formed. This high-energy fault can be eliminated by the nucleation of an identical Dα Shockley partial on the $(1\overline{1}1)$ plane above or below the plane of the CSF. The *two-layer*, double

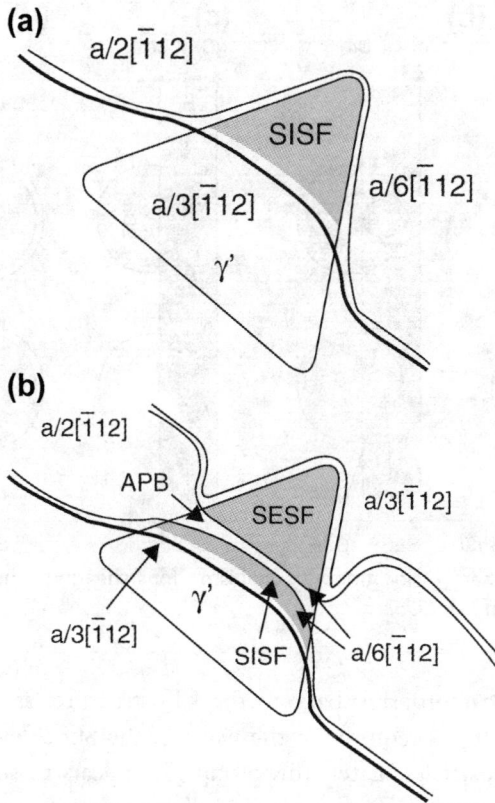

Figure 133 A mechanism for shearing γ′ at low temperatures. *From Ref. [996].*

Shockley partial 2Dα can then essentially shear the precipitate as a net super-Shockley partial dislocation. The low-energy SESF fault is actually achieved only if the (Kolbe [983]) reordering of the nearest-neighbor violations occur, which is a rate-dependent diffusional process. Viswanathn et al. suggest that, as the $a/2[110]$ moves forward in the matrix, a net Shockley partial αB wraps around the precipitate, composed of a αC and a Dα on adjacent planes. In smaller particles (in polycrystalline alloys with a small bimodal dispersion such as René88 DT), this group suggests microtwinning as a deformation mechanism within this low temperature range. It is proposed that shearing of the matrix and the precipitate, in this case, occurs by the passage of identical $a/6<11\overline{2}>$ Shockley partials on successive {111} planes, similar to the above mechanism in large particles [996]. This is a similar mechanism by which twinning may occur in traditional fcc materials (e.g., one atom per lattice point). Twinning is strongly thermally

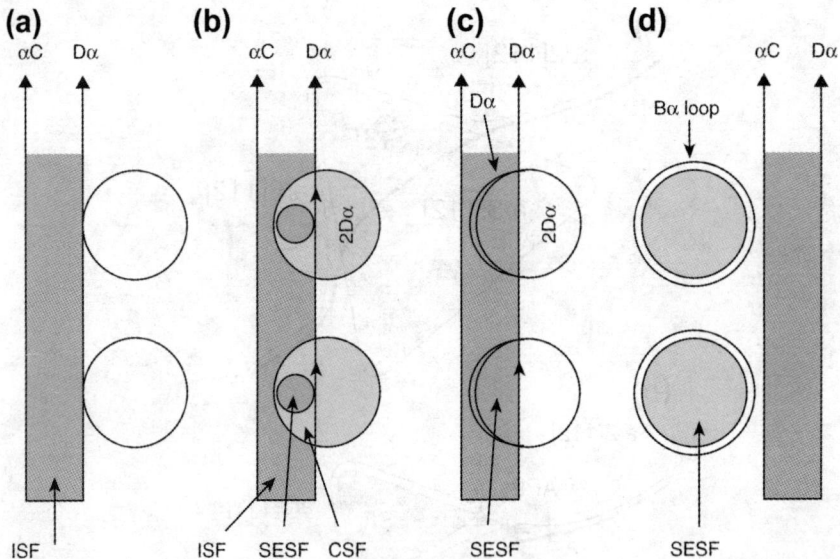

Figure 134 Another dislocation mechanism for shearing the (γ') at low temperatures. *From Ref. [996].*

activated because reordering back to the $L1_2$ structure, as originally proposed by Kolbe [6], is required in the wake of the Shockley partials when shearing the γ' particles. Later, this group [5] appears to suggest that the microtwinning mechanism can occur in larger-phase γ' as with the tertiary particles. That is, there is a suggestion that in single crystals, the "faulted regions" frequently observed in the γ' within this temperature range may be, in some instances, microtwins. The latter mechanisms by Vismanathan et al. are somewhat more satisfying than the former, more-established mechanism, in part, because thermal activation, in terms of a diffusional reordering to the $L1_2$ structure, leads to a temperature dependence that appears more consistent with observations (Figure 135).

It should also be mentioned that the creep properties in all temperature ranges have a strong orientation effect [1010]. This is consistent with both climb and glide control.

3. INTERMEDIATE-TEMPERATURE CREEP

This creep temperature range is generally regarded as consisting of slip within the γ channels. Plasticity occurs in the channels of the alloys with (typical) negative misfit, as the von Mises stress is highest. There are

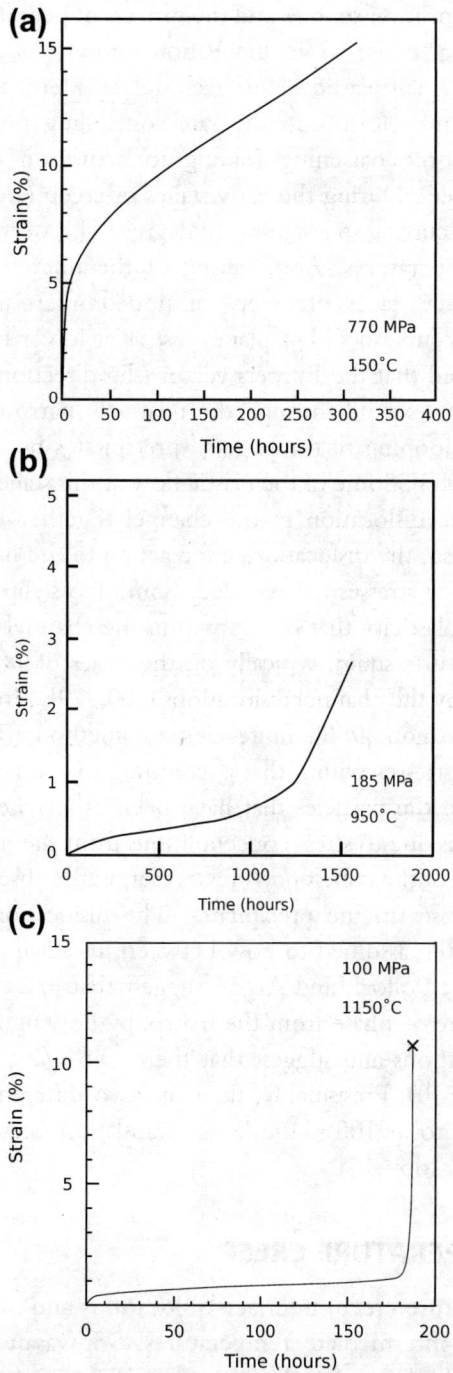

Figure 135 The creep curves of CMSX-4 superalloy at (a) 750, (b) 950, and (c) 1150 °C at a constant engineering stress and a <110> orientation. In (a), the strain rate over the first few percent strain may increase by several factors. *Taken from Ref. [978,1010].*

relatively few dislocation sources and the process of percolation throughout the matrix may give rise to an incubation period (e.g., seconds) during which only very small plastic strains accumulate. Here, three-dimensional Frank networks may form and the rate-controlling process is generally considered network coarsening leading to activation of, for example, Frank-Read sources. During the early stages of creep (e.g.,<1%), shearing of the γ' is not occurring to a significant degree. This occurs at larger strains or at higher temperatures. Also, rafting of the microstructure is absent except at the later stages of creep (of noncommercial interest). Here, dislocation slip occurs on {111} planes just as at lower temperatures. It is generally presumed that the Burgers vector (slip direction) in the matrix is $a/2<110>$. Loops expand through the relatively narrow channels of the matrix. Orowan looping of the large γ' precipitates by the dislocations is rarely observed [986]. Some of the plastic flow at this stage can be described by bowing of the dislocation in the channel together with some solute strengthening. Also, the dislocations are reacting to the interfaces in such a way the coherency stresses are reduced. Some cross-slip is generally suggested with this plasticity that occurs within the channels.

After some plastic strain, typically on the order of 1%, the γ' particles become sheared by the channel dislocations [600,979]. However, according to Pollock and Argon [986], finite element method (FEM) calculations indicate that the stresses within the γ' continue to rise to such a level that slip occurs within the particles that have been otherwise dislocation free. Presumably, this is due to stress concentrations from the matrix dislocations and/or reduction of the coherency stresses that, in the absence of relaxation, diminish the stresses in the precipitates. The dislocations that enter the precipitate are often assumed to bow between interface dislocations.

It appears that Pollock and Argon suggest that $a/2<110>$ {111} dislocations enter the γ' phase from the matrix by bowing around interfacial (network) dislocations and suggest that there two $a/2<110>$ dislocations separated by an APB. Presumably, these are two different Burgers vectors that do not sum to $a<100>$, the lattice translation vector [1007], which may be a complication.

4. HIGH-TEMPERATURE CREEP

In high-temperature creep, interface dislocations and rafting occur more quickly than at intermediate temperatures. Srinivasan et al. [993] and Eggeler et al. [994] suggest slip occurs in the matrix but also in the

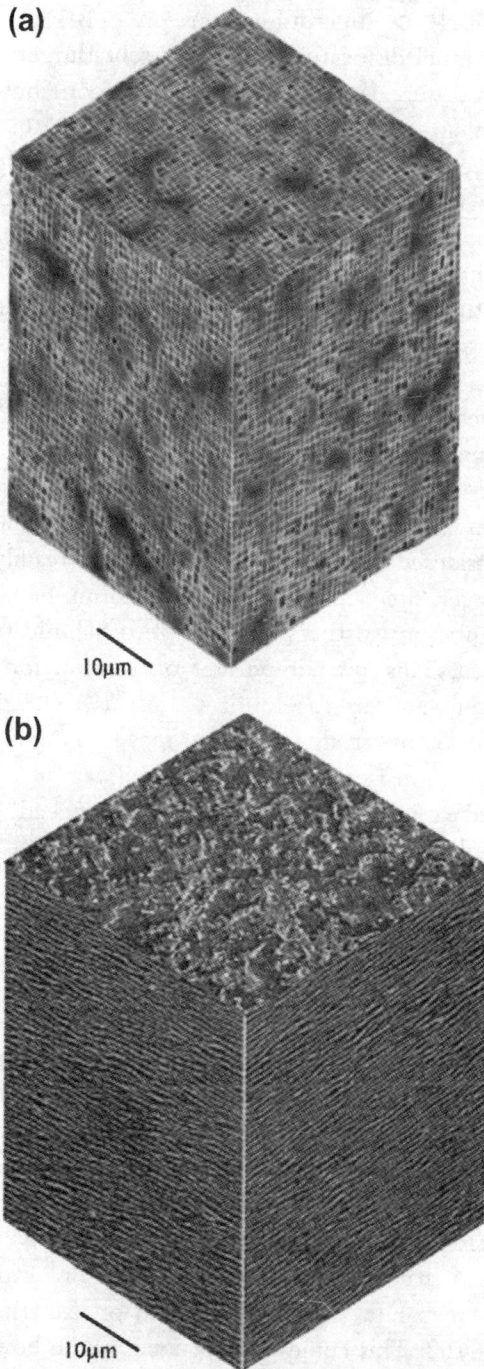

Figure 136 The three-dimensional microphotographs of CMSX-10 (variant) before (a) and after (b) rafting. The tensile stress is perpendicular to the plates and δ is negative. *Courtesy of S. Tin, IIT.*

precipitates by $a{<}100{>}$ superdislocations on $\{110\}$ planes. Here, two $a/2{<}110{>}$ γ-channel dislocations with different Burgers' vectors jointly shear the γ phase (just as with Pollock and Argon) but form a super-dislocation with an overall Burgers' vector of $a[010]$. Two different configurations were observed associated with the pure edge $a{<}010{>}$ and the $45°$ $a{<}001{>}$ dislocations by Srinivasan et al. They observed that the cores of these superdislocations are not compact but rather are composed of two different $a/2{<}110{>}$ dislocations. Movement of the super-dislocations in the γ' phase occurs by two superpartials moving by a combined process of glide and climb, which requires diffusional exchange of atoms/vacancies between the leading and the trailing superpartial. The process is "self-fed," and the overall vacancy equilibrium concentration is maintained. Minimum creep rates can be rationalized on the basis of the fluxes associated with the movement of superdislocations in the γ' phase. This mechanism is attractive in that realistic slip planes and Burgers vectors are considered and thermal activation is readily explained by diffusion. However, slip cannot, by this mechanism, be plainly explained for ${<}100{>}$ orientations to the stress axis (zero Schmid factor) in which creep is observed. This is an important orientation for turbine blades. Others, however, appear to confirm this $a{<}100{>}$ Burgers vector [995,1004-1006]. Sarosi et al. and, more recently, Agudo Jácome [833] address this orientation issue [984].

As mentioned earlier, two phenomena occur very early in creep at high temperatures: rafting and the formation of interfacial dislocations, and these will be discussed somewhat further.

4.1 Dislocations Networks

As mentioned earlier, dislocation networks form at the γ/γ' interface, quickly leading to rafting [600]. There appears to be the necessity of plastic deformation for rafting [989]. The interfacial dislocation networks are formed from $a/2{<}110{>}$ dislocations of the channels. The interfacial dis-locations have an edge character and reduce the interfacial strain energy. These reactions are described in particular detail elsewhere [600]. A variety of reactions between the matrix dislocations are suggested to give rise to the interfacial dislocation meshes. Reed [978], Zhang et al. [990], and Koizumi et al. [991] found that the smaller the interfacial dislocation spacing, the lower was the creep rate. These were based on experiments on TMS-75,−138, and −162. This could be explained by the bowing mechanism suggested by Pollock and Argon.

4.2 Rafting

Under external stress–free annealing, large magnitudes of δ lead to substantial multidimensional coarsening [987]. However, with the application of a uniaxial stress, the coarsening is directional, in one or two of the <100> directions, which is referred to as rafting. Parallel rods or plates are formed, as first reported by Tien et al. [988]. For negative δ, such as CMSX-4, the tensile loads are perpendicular to the plate rafts. When δ is positive, rod-shaped rafts are observed parallel to the applied tensile stress. Compressive stresses change the nature of the rafting but are similar to the tensile case for an accompanying change in the sign of δ. A rafted microstructure is illustrated in Figure 136.

Partially crept superalloy single crystals continue rafting even after the external stress is removed [978,1008] and the rafting "rate" appears unchanged after external stress removal. As Pollock and Field suggest, plasticity affects both the driving force and kinetics of the rafting process [600]. Matan et al. [989] suggest that there may even be a threshold strain in order for rafting to occur. All this appears to suggest that dislocations may relax the coherency strains in some orientations and a stress gradient may be created. Sarosi [984] and others suggest that a stress gradient leads to a vacancy flux and subsequent rafting.

It has been suggested that one consequence of rafting is that the γ' become the continuous phase rather than the γ [992]. Rafting may then increase the flow stress of the dislocations in the γ phase leading to reduced creep rates. It has also been suggested that lower stacking fault energies retard rafting by increasing the difficulty in forming interfacial networks that benefit from cross-slip.

CHAPTER 12

Creep in Amorphous Metals

M.E. Kassner, K.K. Smith

Contents

1. INTRODUCTION

Amorphous metals are a relatively new class of alloy, originating in about 1960 with the discovery of thin metallic ribbons by splat cooling [1015]. These are always alloys, and pure metal glasses have not yet been produced. Because these alloys are noncrystalline, they have no dislocations, at least in the sense normally described in crystalline materials. Thus, amorphous metals have yield stresses that are higher than crystalline alloys. High fracture stress, low elastic moduli, and sometimes-favorable fracture toughness are observed. Often, favorable corrosion properties were observed, as well, partly due to an absence of grain boundaries. Toward 1990, alloys with deep eutectics were developed that allowed liquid structures to be retained in thicker sections in the amorphous state on cooling to ambient temperature [1016–1027]. With this development, there has been fairly intensive study of bulk metallic glasses (BMGs) for possible structural applications. Most of the alloys in this chapter are relevant to BMGs. Table 12 lists some of the short-term mechanical properties of some BMGs taken from [1026], and some of the impressive properties are given.

Figure 137, based on an illustration [1049], is a time-temperature-transformation (T-T-T) diagram that illustrates some of the important temperatures for metallic glasses. First, there is the equilibrium liquid-to-solid transition at the melting temperature T_m, where, of course, multiple

Fundamentals of Creep in Metals and Alloys
ISBN 978-0-08-099427-7
http://dx.doi.org/10.1016/B978-0-08-099427-7.00012-8
275

Table 12 Mechanical properties of some glassy alloys

Material	E (GPa)	σ_y (MPa)	σ_f (MPa)	ε_y (%)	ε_p (%)	Reference
$Co_{43}Fe_{20}Ta_{5.5}B_{31.5}$	268		5185	2		[1028]
$Cu_{60}Hf_{25}Ti_{15}$	124	2024	2088		1.6	[1029]
$(Cu_{60}Hf_{25}Ti_{15})_{96}Nb_4$	130		2405		2.8	[1030]
$Cu_{47}Ti_{33}Zr_{11}Ni_6Sn_2Si_1$	84	1930	2250	1.7		[1031]
$Cu_{50}Zr_{50}$	92.3	1272	1794	2.2	6.2	[1032]
$Cu_{64}Zr_{36}$			2000			[1033]
$(Fe_{0.9}Co_{0.1})_{64.5}Mo_{14}C_{15}B_6Er_{0.5}$	192	3700	4100	2.1	0.55	[1034]
$Fe_{71}Nb_6B_{23}$			4850		1.6	[1035]
$Fe_{72}Si_4B_{20}Nb_4$	200		4200		1.9	[1036]
$Fe_{74}Mo_6P_{10}C_{7.5}B_{2.5}$		3330	3400		2.2	[1037]
$[(Fe_{0.6}Co_{0.4})_{0.75}B_{0.2}Si_{0.05}]_{96}Nb_4$	210	4100	4250	2	2.25	[1038]
$Fe_{49}Cr_{15}Mo_{14}C_{15}B_6Er_1$	220	3750	4140		0.25	[1039]
$Gd_{60}Co_{15}Al_{25}$	70		1380		1.97	[1040]
$Ni_{61}Zr_{22}Nb_7Al_4Ta_6$			3080		5	[1041]
$Pd_{77.5}Cu_6Si_{16.5}$		1476	1600		11.4	[1042]
$Pd_{79}Cu_6Si_{10}P_5$	82	1475	1575		3.5	[1043]
$Pt_{57.5}Cu_{14.7}Ni_{5.3}P_{22.5}$		1400	1470	2	20	[1044]
$Ti_{41.5}Zr_{2.5}Hf_5Cu_{42.5}Ni_{7.5}Si_1$	95		2040		0	[1045]
$Zr_{55}Cu_{30}Al_{10}Ni_5$		1410	1420			[1046]
$Zr_{41.25}Ti_{13.75}Cu_{12.5}Ni_{10}Be_{22.5}$	96	1900	1900	2		[1047]
$Zr_{57}Nb_5Al_{10}Cu_{15.4}Ni_{12.6}$	86.7	1800	1800	2		[1048]

Note. ε_y, elongation at yielding; ε_p, plastic elongation. All the tests were conducted under compression, generally at strain rates from $1-5 \times 10^{-4} \text{ s}^{-1}$.
From Ref. [1027].

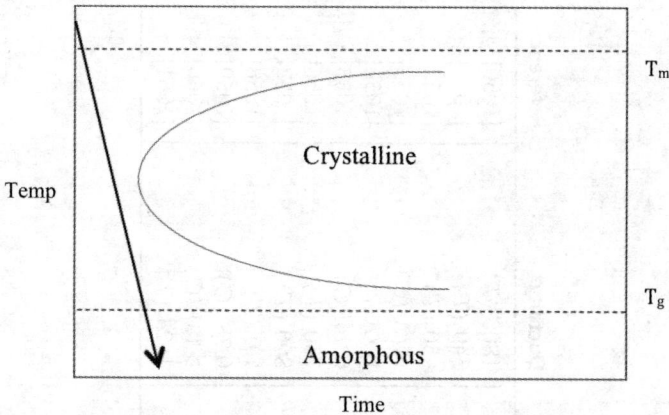

Figure 137 A time-temperature-transformation diagram that illustrates the important temperature regions of BMGs. *From Ref. [1049].*

solid crystalline phases form on cooling. Below this temperature, a T-T-T curve is illustrated. Cooling below T_m must be sufficiently rapid to avoid intersecting the "nose" of the curve. Also illustrated is the glass transition temperature, T_g. This is generally assigned to that temperature where there is a discontinuity in the change of a property (e.g., heat capacity, thermal expansion coefficient, etc.) with temperature. The region between T_m and T_g is generally referred to as the super-cooled liquid regime. Some values for various BMGs are listed in Table 13.

The discussions in this chapter will be largely confined to temperatures above $0.7\,T_g$. As will be discussed subsequently, this is the regime in which homogeneous deformation is observed. This review refers to this regime as a "creep regime" of amorphous metals. A practical importance of this regime is that this is where forming of a metallic glass is frequently performed. This regime is contrasted by the regime of lower temperatures where heterogeneous deformation or shear banding is often (but not always) observed.

2. MECHANISMS OF DEFORMATION

2.1 Overview

The suggested mechanisms have generally fallen into three categories: (1) dislocation-like defects [1062–1064], (2) diffusion-type deformation [1065], and (3) shear transformation zones (STZs)

Table 13 Deformation data of some BMGs in the super-cooled liquid region Alloys (in atomic percent)

	T_g (K)	T_x (K)	m value	Ductility[a]	Reference
$La_{55}Al_{25}Ni_{20}$	480	520	1.0	1800 (T)	[1050]
$Zr_{65}Al_{10}Ni_{10}Cu_{15}$	652	757	0.8–1.0	340 (T)	[1051]
$Zr_{52.5}Al_{10}Ti_5Cu_{17.9}Ni_{14.6}$	358	456	0.45–0.55	650 (T)	[1052]
$Zr_{55}Cu_{30}Al_{10}Ni_5$	683	763	0.5–1.0	N/A (C)	[1053]
$La_{60}Al_{20}Ni_{10}Co_5Cu_5$	451	523	1.0	N/A	[1054]
$Pd_{40}Ni_{40}P_{20}$	589	670	0.5–1.0	0.94 (C)	[1055]
$Zr_{65}Al_{10}Ni_{10}Cu_{15}$	652	757	0.83	750 (T)	[1056]
$Zr_{55}Al_{10}Cu_{30}Ni_5$	670	768	0.5–0.9	800 (T)	[1057]
$Ti_{45}Zr_{24}Ni_7Cu_8Be_{16}$	601	648	N/A	1.0 (T)	[1058]
$Cu_{60}Zr_{20}Hf_{10}Ti_{10}$	721	766	0.3–0.61	0.78 (C)	[1059]
$Zr_{52.5}Al_{10}Cu_{22}Ti_{2.5}Ni_{13}$	659	761	0.5–1.0	>1.0 (C)	[1060]
$Zr_{41.25}Ti_{13.75}Ni_{10}Cu_{12.5}Be_{22.5}$	614	698	0.4–1.0	1624 (T)	[1061]

[a] "T" and "C" indicate tension and compression, respectively.
From Ref. [1071].

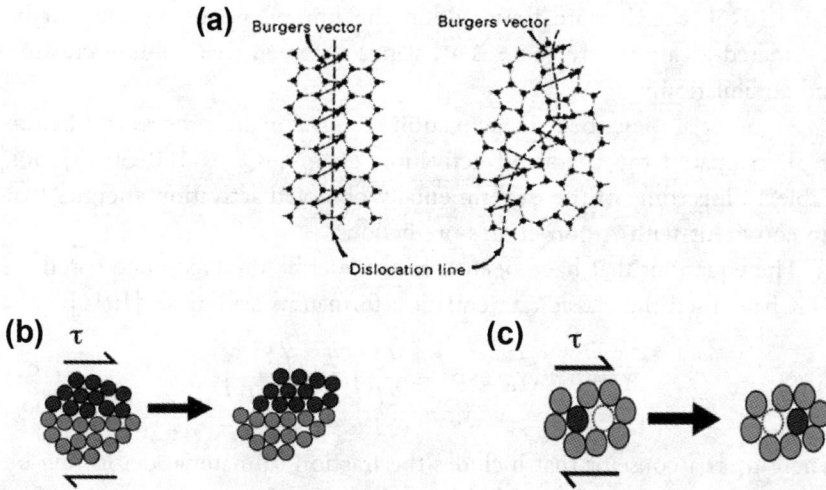

Figure 138 (a) Two-dimensional representation of a dislocation line in crystalline (*left*) and amorphous (*right*) solids; taken from [1026]; Atomistic deformation of amorphous metals in the form of (b) shear transformation zones (STZ); and (c) local atomic jump. *Adapted from [1016].*

[1066,1067]. These are illustrated in Figure 138 from Lu et al. [1068], as well as Schuh et al. [1069], and are all early explanations for plasticity, but it appears that the amorphous metals community has generally embraced the third, STZs [1026,1027,1070].

The essence of this latter mechanism is that there is a "free volume" in amorphous metals. The exact form and shape of these free volumes are not known. Increasing free volume would be associated with decreased density. Estimates for free volume for $Zr_{41.2}Ti_{13.8}Cu_{12.5}Ni_{10}Be_{22.5}$ (Vitreloy 1) is about 3% [1072]. Decreases in free volume (tighter packing) appear to increase ductility in homogeneous deformation at ambient temperature [1073].

With an applied stress, groups of atoms (e.g., a few to 100 [1016,1070,1074]), under an applied shear stress, τ, move and perform work. This constitutes an STZ. Argon et al. [1066,1067] considered that the STZ operation takes place within the elastic confinement of a surrounding glass matrix, and the shear distortion leads to stress and strain redistribution around the STZ region [1016,1066,1067]. When the STZs exist throughout the alloy, we have homogeneous deformation. STZs also occur in shear bands leading to heterogeneous deformation. They also have been observed to create free volume during homogeneous deformation

[1075,1076]. Steady-state flow within the homogeneous regime can be considered a case where there is a balance between free volume creation and annihilation.

Argon et al. described the activation energy for this process and Schuh et al. estimated the predicted activation energy, Q, as 100–500 kJ mol. Table 14 lists some of the experimentally observed activation energies that are consistent with Argon et al.'s predictions.

The equations that have been used to describe the creep rate based on STZ have used the classic rate equation formalism leading to [1016]

$$\dot{\gamma} = \alpha_o \nu_o \gamma_o \cdot \exp\left(-\frac{Q}{kT}\right) \sinh\left(\frac{\tau V}{kT}\right), \tag{191}$$

where α_o is a constant that includes the fraction of material deforming by activation, ν_o is an attempt frequency, γ_o is the characteristic strain of an STZ, and V is the activation volume. The hyperbolic sine function arises, as the there can be both a forward and a reverse "reaction."

At low stresses $(\tau \ll kT/V)$, this equation reduces to the Newtonian

$$\dot{\gamma} = \frac{\alpha_o \nu_o \gamma_o V}{kT} \cdot \exp\left(-\frac{Q}{kT}\right) \tau, \tag{192}$$

because "reverse" deformation is irrelevant.

Conversely, at stresses, $\tau \gg kT/V$

$$\dot{\gamma} = \frac{1}{2} \alpha_o \nu_o \gamma_o \cdot \exp\left(-\frac{Q - \tau V}{kT}\right), \tag{193}$$

Table 14 Activation energies for creep of selected metallic glasses

Composition	T_g (K)	T_{test} (K)	ΔQ (kJ mol)
$Al_{20}Cu_{25}Zr_{55}$	740	573	230.12
$Cu_{40}Zr_{60}$	677	543	218.82
$Cu_{56}Zr_{44}$	727	573	217.57
$Cu_{60}Zr_{40}$	750	573	228.45
$Pd_{80}Si_{20}$	673	546	191.63
$Zr_{55}Cu_{30}Al_{10}Ni_5$			410
$Au_{49}Ag_{5.5}Pd_{2.3}Cu_{26.9}Si_{16.3}$			103
$Zr_{44}Ti_{11}Cu_{10}Ni_{10}Be_{25}$	625/632		366

From [1067,1074,1077,1078].

Schuh et al. [1016] point out that Eqn (191) suggests a Newtonian region followed by, with increasing stress, continual increase in stress exponent. Examples of BMGs that have evinced Eqns (191)–(193) behaviors are illustrated in Figures 139 and 140, which plot the steady-state creep behavior of several BMGs [1016,1071].

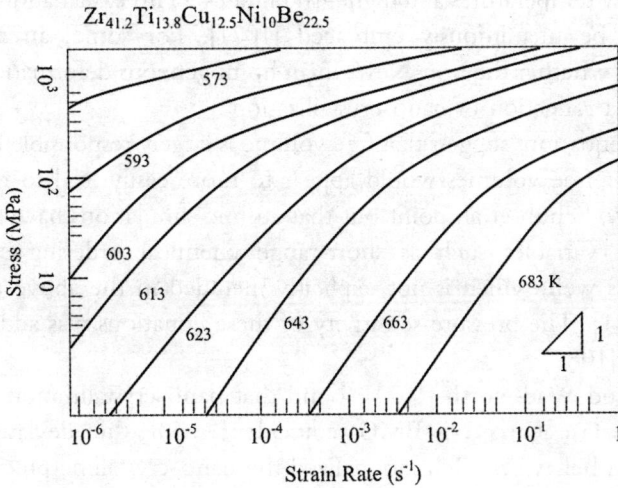

Figure 139 Steady-state homogeneous flow data for $Zr_{41.2}Ti_{13.8}Cu_{12.5}Ni_{10}Be_{22.5}$ metallic glass at elevated temperatures. *From the work of Lu et al. [1068]. Figure based on [1016].*

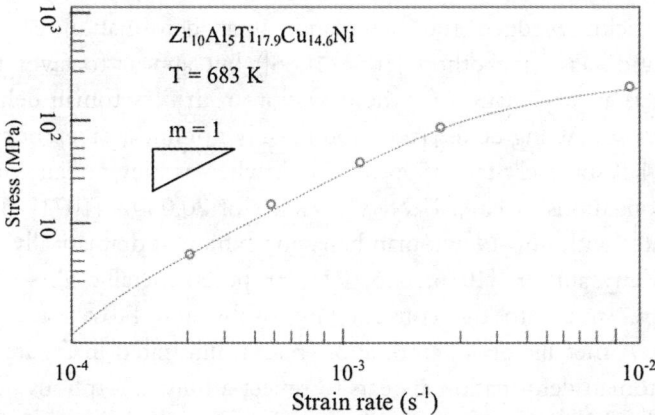

Figure 140 Stress–strain rate curve for a $Zr_{10}Al_5Ti_{17.9}Cu_{14.6}Ni$ glassy alloy shows Newtonian flow at low strain rates but non-newtonian at high strain rates. *Data from Ref. [1051]. Figure based on [1071].*

The figures illustrate steady-state behavior such that, with increasing strain rate and/or decreasing temperature, there is a breakdown in Newtonian behavior and the apparent stress-exponent increases. Generally, this has been regarded as a natural consequence of Eqn (192), the rate equation that predicts Newtonian behavior at low stresses (higher temperature and lower strain rates) but increased exponents with higher stresses (low temperatures and higher strain rates). This explanation does not appear to be unanimously embraced [1071]. For some, an important question is whether the non-Newtonian homogeneous deformation region is actually a reflection of nano-crystallization.

These equations suggest that free volume is largely responsible for plastic flow; larger free volumes would appear to more easily lead to regions of plastic flow. Schuh et al. point out that atomic simulations have suggested that other variables such as short-range chemical ordering can affect plasticity as well, which is not explicitly included in the above equations [1079–1081]. The pressure sensitivity of these equations was addressed by Sun et al. [1082].

Nieh and Wadsworth [1071] found that nano-crystallization occurred in $Zr_{10}Al_5Ti_{19.9}Cu_{14.6}Ni$ BMG coincident with the deviation from Newtonian behavior. Nieh rationalized the nano-crystalline precipitates as akin to dispersion strengthening. Suryanarayana and Inoue [1083] appear to suggest that the stress exponent increases due to second phase strengthening of the nanoparticles by a straightforward rule of mixtures for the flow strength. Schuh et al. [1016] referenced the Nieh and Wadsworth work and certainly acknowledged the observation that deformation can induce crystallization (as have others [1083–1086]) but appear to favor the rate equation as an explanation for the deviation from Newtonian behavior at higher stresses. Wang et al. [1087] found only nonlinear creep behavior in Vitreloy 1 if some crystallization occurred, whereas Newtonian conditions led to elongations in $La_{55}Al_{25}Ni_{20}$ in excess of 20,000% [1071]. Those at higher rates with non-Newtonian behavior exhibited dramatically reduced values. Many authors [1016,1026,1027] proposed metallic glass deformation maps, similar to the construct by Ashby and Frost for crystalline materials. A metallic glass deformation–map is illustrated in Figure 141.

Newtonian deformation appears to reflect a fully amorphous alloy, but at least in other regions, including heterogeneous deformation, nano-crystallization may be occurring [1083–1086]. Furthermore, as will be discussed in a subsequent section, homogeneous deformation may extend to low temperatures, in at least some cases.

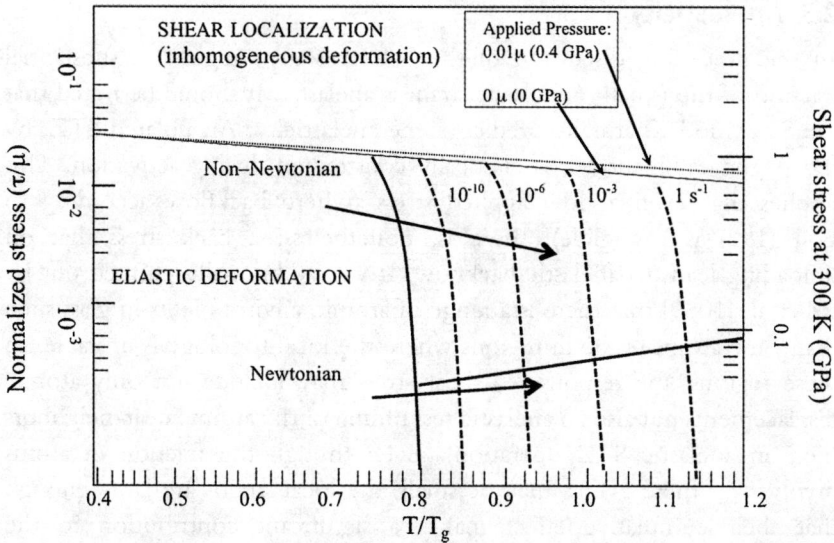

Figure 141 Deformation mechanism maps for metallic glass plotted in normalized stress versus normalized temperature. The absolute stress values indicated in the figure are for the $Zr_{41.2}Ti_{13.8}Cu_{12.5}Ni_{10}Be_{22.5}$ metallic glass. *Adapted from [1016,1026].*

2.2 Homogeneous Flow at Very Low Temperatures

Recent work [1075,1088–1090] shows that, given sufficient time, homogeneous deformation can be detected under "electrostatic" (i.e., at a stress less than the yield stress, σ_y) loading at room temperature (RT). The stress exponent was not assessed, so it was unclear whether Newtonian flow was observed. Alloys include $Zr_{46.75}Ti_{8.25}Cu_{7.5}Ni_{10}Be_{27.5}$, $Ni_{62}Nb_{38}$, $Cu_{50}Zr_{50}$, $Cu_{57}Zr_{43}$, and $Cu_{65}Zr_{35}$. Of course, some BMGs, such as $Zn_{20}Cu_{20}Tb_{20}(Li_{0.55}Mg_{0.45})_{20}$, may have a low T_g (323 K) allowing homogeneous deformation at RT [1091]. Alloys with higher packing densities exhibit greater plastic strain during homogeneous deformation at room temperature but show less global plasticity during inhomogeneous deformation in a typical compression test [1073]. Park et al. [1075] suggest deformation induced structural disordering by molecular dynamics (MD) simulations, although others [1090] imply STZ as the mechanism. Compression tests on $Pd_{77}Si_{23}$ showed that as the sample size decreased to the submicron range, homogeneous deformation occurs and was suggested to occur due to the necessity of a critical size volume for shear bands [1092]. Similar results were noted by others [1093].

2.3 Anelasticity

In the preceding discussion of the "electrostatic" regime, a substantial fraction of the (small) nonelastic strain is anelastic. It should be noted that the STZ model naturally predicts some anelasticity. An isolated STZ, by the Argon et al. model, is elastically constrained during activation. This implies that even at low applied stress (where backflow according to Eqn (192) is negligible), there is, nonetheless, a back stress that on unloading leads to anelastic backflow. It was additionally pointed out by Ke et al. [1089] that there is a range of atomic environments in glass such that some atoms reside in regions where the local topology is unstable. In these regions, the response to shear stress may include not only atomic displacements but also an anelastic reshuffling of the atomic near-neighbors (i.e., an anelastic STZ operation). Even though the fraction of atoms involved in these events may be small, the local strains are large enough that their cumulative effect makes a significant contribution to the macroscopic strain [1089].

2.4 Primary and Transient Creep (Non–Steady-State Flow)

Steady-state flow has principally been discussed, so far. It has been presumed that STZs create free volume (leading to softening) and that recovery processes promote the annihilation of free volume (leading to hardening). Therefore, steady state has been regarded as a balance between free volume creation and annihilation. Other hardening effects such as

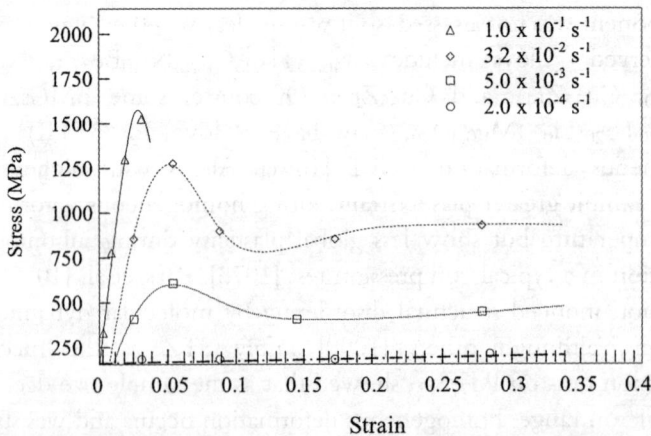

Figure 142 Effect of strain rate on the uniaxial stress-strain behavior of Vitreloy 1 at 643 K and strain rates of 1.0×10^{-1}, 3.2×10^{-2}, 5.0×10^{-3} and $2.0 \times 10^{-4}\,s^{-1}$ [1068].

chemical ordering have not been explicitly considered for steady state. It has been suggested that there can be a net free volume increase or decrease during deformation that precedes a steady state. Figure 142 from Lu et al. [1068] shows hardening at the onset of deformation that continues beyond the eventual steady state. The interpretation of this peak stress followed by softening to a steady state is unclear.

CHAPTER 13

Low-Temperature Creep Plasticity

M.E. Kassner, K.K. Smith

Contents

1. INTRODUCTION

Temperature ranges for creep can be subdivided into three categories: (1) high-temperature creep ($T > 0.6\ T_m$), (2) intermediate-temperature creep ($0.3\ T_m < T < 0.6\ T_m$), and (3) low-temperature creep ($T < 0.3\ T_m$). Generally, creep studies investigate high-temperature deformation; however, this chapter reviews the latter category. Less attention has been paid to low-temperature creep due to the fact that materials generally neither fail nor experience significant plasticity at lower (especially ambient and cryogenic) temperatures.

Creep at low temperature can be understood as time-dependent plasticity that occurs at $T < 0.3\ T_m$ and at stresses often below the macroscopic yield stress ($\sigma_y^{0.002}$). This is where creep is often not expected. Still, even with the lesser attention paid to this area of creep, many materials do experience very noticeable plasticity at lower temperatures. This has

Fundamentals of Creep in Metals and Alloys
ISBN 978-0-08-099427-7
http://dx.doi.org/10.1016/B978-0-08-099427-7.00013-X

some commercial importance. These materials include Ti alloys and steels [1094–1103], Al–Mg [1104], α-brass [1105], ionic solids [1106], pure Au, Cd, Cu, Al, Ti, Hg, Ta, Pb, and Zn [1107–1121], precipitation-hardened alloys [1122], and glass and rubber [1121].

Low-temperature creep has generally been investigated for two reasons: (1) materials may undergo plasticity that affects its intended performance; this category includes structural alloys and creep of Cu at cryogenic temperatures, and (2) there has been theoretical curiosity regarding low-temperature deformation and the mechanism of plasticity, particularly at cryogenic temperatures. This includes the validity of the dislocation intersection mechanism proposed by Seeger et al., [1123,1124] as investigated by others [1112]. Also, there have been investigations of the proposition of quantum mechanical tunneling of dislocations at very low temperature [1106,1113,1115,1117,1119,1120].

1.1 Phenomenology

Generally, but not always, low-temperature creep is a discussion of primary creep without the observation of a genuine mechanical steady state. One study has suggested steady state at 4.2 K, but there were problems with the data analysis [1125]. At high temperatures, primary creep is described by the equations

$$\varepsilon = bt^{1/3} + c_1 \tag{194}$$

as suggested long ago by Andrade [1126] and Orowan [1127]. Evans and Wilshire [1128] reviewed the high-temperature primary creep equations and suggested a refinement. This refinement led to an equation of the form

$$\varepsilon = at^{1/3} + ct + dt^{4/3} \tag{195}$$

This is now the common phenomenological equation used to describe primary creep at elevated temperatures. Variations to this equation include [1129]

$$\varepsilon = at^{1/3} + ct \tag{196}$$

and [1130]

$$\varepsilon = at^{1/3}bt^{2/3} + ct \tag{197}$$

or

$$\varepsilon = at^b + c^t \tag{198}$$

where [1131]

$$0 < b < 1$$

or [1094]

$$\varepsilon = at^b \qquad\qquad (199)$$

where

$$0 < b < 1$$

It is suggested that Eqns 194–199 are all of a similar (power–law) form. Another form of equations was suggested by Phillips [1121], Laurent and Eudier [1132], and Chévenard [1133]

$$\varepsilon_p = \alpha \ln t + c_2 \qquad\qquad (200)$$

Wyatt [1111], long ago, suggested for pure metals, such as Al, Cd, and Cu, that at higher temperatures, Eqn (194) was the proper descriptive equation, but at lower temperatures, he then suggested Eqn (200) was often the proper form.

1.2 Objectives

The following discussion will describe the phenomenological trends in greater detail. The data appear to best be presented/described by material category (e.g., alloy, metal, or ceramic). In particular, the low-temperature creep behavior of both alloys and pure metals will be described in separate sections. It will be shown that, generally, the descriptive equations generally fall within the forms of Eqn (194) or Eqn (195). Distinctions will be made for cases where the applied stress is above and below the conventional yield stress (at an ordinary strain rate; e.g., $10^{-4}\,\mathrm{s}^{-1}$), as well as at shorter times than a few hours and much longer times.

2. CREEP BEHAVIOR OF VARIOUS METALS AND ALLOYS

2.1 Titanium Alloys

A useful discussion of low-temperature creep in Ti alloys was presented by Neeraj et al. [1]. Figure 143 plots ambient-temperature creep data [1096,1104,1134] for Ti alloys with different microstructures and different compositions at 0.8 σ_y. Other, more recent, data are also available [1094].

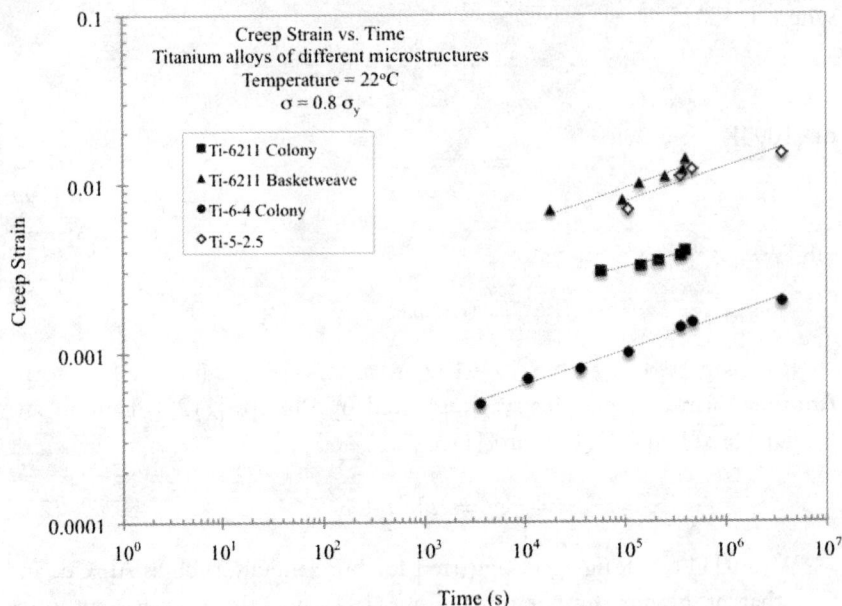

Figure 143 Ambient-temperature creep for various Ti-alloys of different microstructures and different compositions. From Neeraj et al., [1]. Power-law behavior is observed (Eqn (199)). Applied stresses are 0.8 σ_y.

A particular value of these data is the extension of creep to longer times (over a month). The data show that Eqn (194), the power–law relationship, reasonably describes the data. In contradiction to some of the earliest phenomenological equations, the exponent of 1/3 does not appear to be unique for the b-value. Figure 143 suggests a b-value closer to 0.2. Also, the applied stresses in this case are all below the macroscopic yield stress. As Neeraj et al., point out, other literature confirms that $0.03 < b < 1$. The relatively recent effort to theoretically justify a value for b = 1/3 by Cottrell [1135–1137] and Nabarro [1138] may have been misspent.

2.2 Steels

2.2.1 AISI 4340 Steel

Oehlert and Atrens [1098] performed creep studies at ambient temperature, where for all creep conditions, the applied stress is below the yield stress. One of the tests was conducted at an applied stress of just half the yield stress. The durations of the tests were relatively short, and the times are only up to 20 min. These creep curves were described by a modification to Eqn (200)

$$\varepsilon_p = \varepsilon_{py} + \beta(\ln t) \tag{201}$$

where ε_{py} is the plastic strain on loading that is a function of stress. One clumsiness with Eqn (201) is that, at $t = 0$, infinite creep rate is predicted. This was eliminated by modifying the equation to:

$$\varepsilon_p = \varepsilon_{py} + \alpha \ln(1 + \phi t) \tag{202}$$

where α and ϕ are constants. Oehlert and Atrens found that over the range of stresses, the constant α varied by a factor of nearly 10 and β by a factor of 2. The authors also examined 3.5 NiCr MoV and AeroMet100 with similar results.

2.2.2 304 Stainless Steel

Figure 144 shows the creep curves for annealed 304 stainless steel [2], which evinces some typical features of low-temperature creep. This was discussed in the previous section. Figure 144(a) is a creep plot of the total plastic strain, ε_p, versus time, t, for annealed 304. Figure 144(b) is a semi-log plot that illustrates the full time range of data [2]. Tests are conducted both above and below the yield stress. Tests above the yield stress were generally for a shorter term and strain was measured using an extensometer. However, below the yield stress, strain was measured for longer times using an optical comparator. It appears that the creep data for this material, at ambient temperature, follow a logarithmic behavior at a fixed stress, σ, of Eqn (201)

$$\varepsilon_p = \varepsilon_{py} + \beta(\ln t) \tag{203}$$

where,

$$\varepsilon_{py} = a + b\sigma \tag{204}$$

and a and b are constants and a basically reflects the strain on loading. Also, β is the slope in Figure 144(b) and appears to decrease with decreasing stress, approximated by

$$\beta = -k\sigma + C_o \tag{205}$$

where k and C_o are constants.

Regression analysis suggests that for Eqn (204):

Above yield stress: $a = -8.12$ and $b = 0.037$

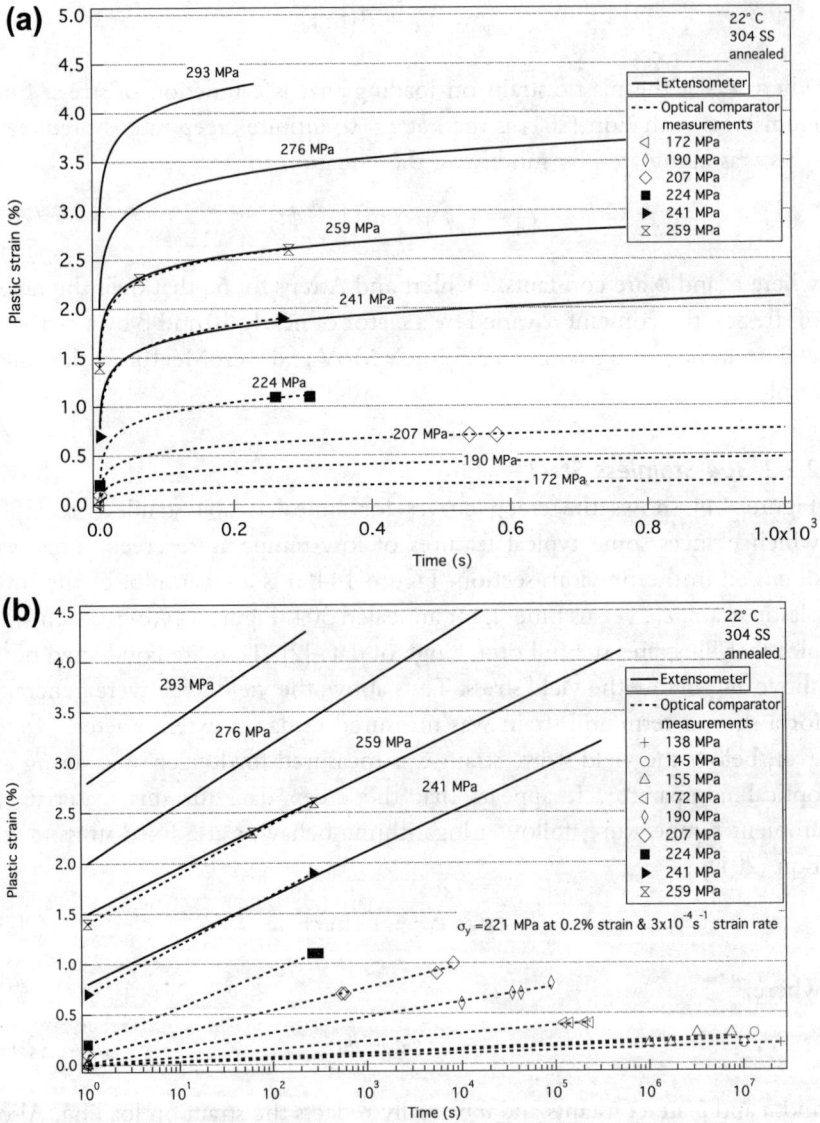

Figure 144 (a) The plastic strain vs time behavior of annealed 304 stainless steel under different stresses. (b) Plastic strain vs log time behavior of annealed 304 stainless steel [2]. Logarithmic behavior is observed (Eqn (200)).

Below yield stress: $a = -0.47$ and $b = 0.0027$

And for Eqn (205):

Above yield stress: $k = -0.0040$ and $C_o = -0.52$

Below yield stress: $k = -0.0028$ and $C_o = -0.38$

2.3 Pure Metals

2.3.1 Copper

Copper behavior is illustrated in Figure 145 with data from Wyatt for shorter-duration tests. Copper data for longer-term creep is also shown from Yen et al. [1125]. The two sets of data are not in agreement and the explanation is unclear. The strains for the Yen et al. strain data appear low for the magnitude of stress greater than the yield stress, σ_y. Eqn (200) appears reasonable, except, possibly, for the lower two stresses of the Yen data, which follow neither Eqn (194) nor Eqn (200).

2.3.2 Aluminum

The ambient-temperature $(T < T_m)$ creep behavior of annealed high-purity Al is illustrated in Figure 146. Here, it is unclear whether, power-law behavior (Eqn (199)) or logarithmic (Eqn (200)) behavior dominate.

2.3.3 Cadmium

Figure 147 illustrates the creep behavior of pure annealed Cd at 77 K and at ambient temperature. It was concluded by Wyatt that Eqn (200) better describes the data at $T < 0.3\,T_m$.

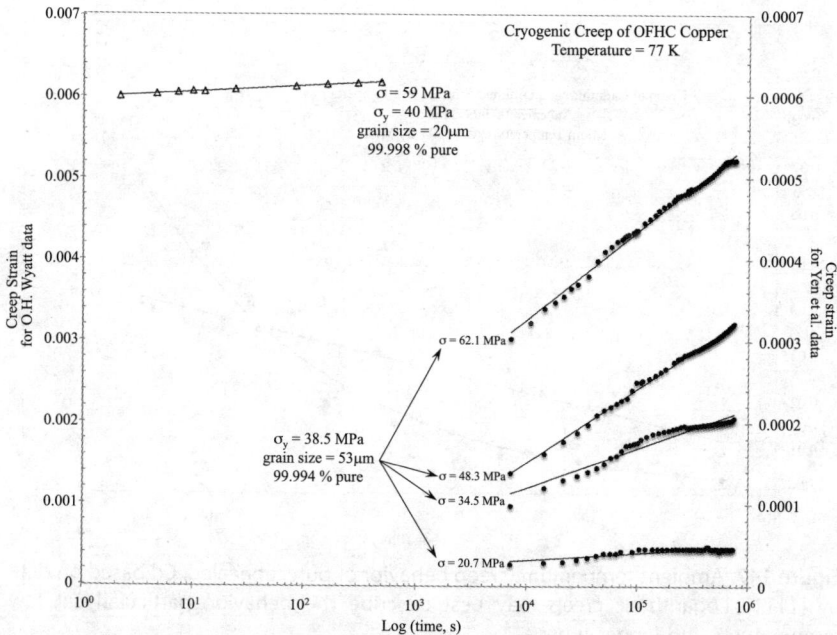

Figure 145 Creep strain vs log time data for annealed copper at various low temperatures and stresses [1111,1125]. Logarithmic behavior appears to dominate (Eqn (200)).

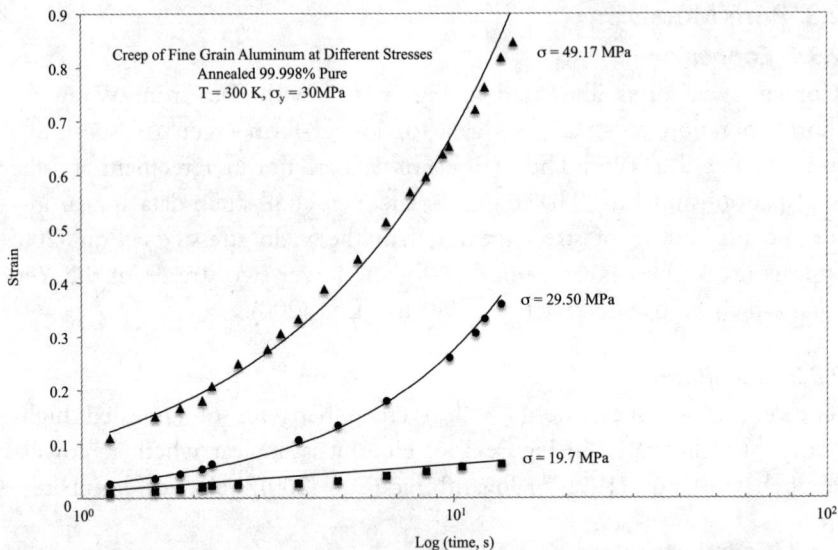

Figure 146 Ambient-temperature creep behavior pure, annealed, Al based on data by [1111]. Power-law creep is occurring but logarithmic behavior may be observed as well.

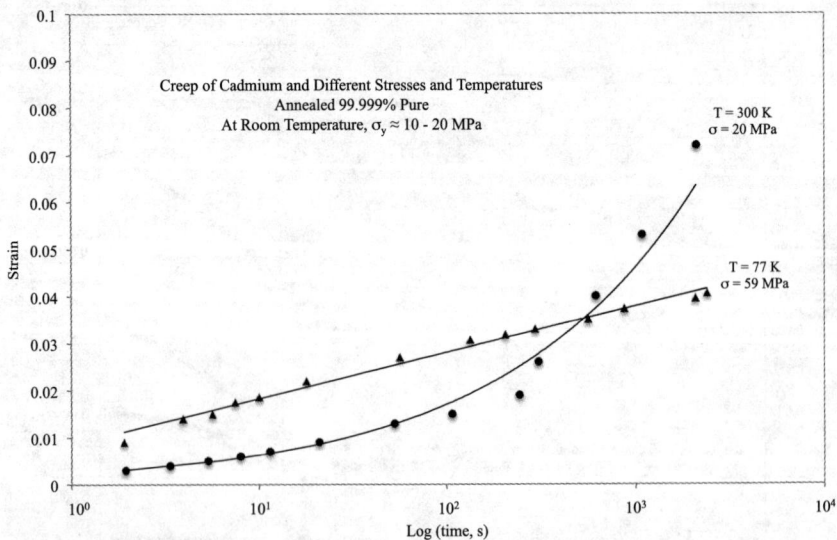

Figure 147 Ambient-temperature creep behavior of pure, annealed, Cd based on data by [1111]. Logarithmic creep may best describe the behavior, particularly at low temperatures and longer times.

3. MECHANISMS

3.1 Logarithmic Creep

The earliest discussion of low-temperature creep was based on dislocation glide and dislocation exhaustion by Mott and Nabarro [1139]. There has also been substantial discussion of low-temperature creep, especially low-temperature plasticity, with published work on the dislocation intersection mechanism by Seeger et al., [1123,1124].

Here, the concept of activation volume and area were probably first used. The product of the activation area, A, Burger's vector, b, the difference between the applied stress, τ, and the back stress or long-range internal stress (LRIS), τ_G, due to other dislocations, is the energy supplied but is also the applied stress to allow a dislocation to surmount an obstacle. The activation area is usually defined as the product of the width of the obstacle, d, and the obstacle spacing, ℓ. Alternatively, the activation volume, v, equals $Ab = \ell db$. This leads to the classic rate equation [1112,1123]

$$\dot{\gamma} = NAbv_{o}\,\exp\left\{\frac{-\Delta H_{o} - v(\tau - \tau_{G})}{kT}\right\} \qquad (206)$$

where $\dot{\gamma}$ = strain rate; N = number of dislocation segments per unit volume held up at the intersection points of mean spacing, ℓ. v_{o} is an atomic frequency of the order of the Debye frequency; and ΔH_{o} = energy required for the intersection process (i.e., the energy for jog formation $\approx Gb^{2}/10$) [1140].

$$\tau_{G} = \tau_{G}^{o} + \int_{0}^{\gamma} h d\gamma \qquad (207)$$

where τ_{G}^{o} = stress due to the dislocations initially in the crystal and h = strain-hardening coefficient, which is defined as $d\tau/d\gamma$. It appears that $\tau_G \ll \tau$ for small strains [1141].

Eqns (206) and (207) suggest

$$\dot{\gamma} = NAbv_{o}\,\exp\left\{\frac{-\Delta H_{o} - v(\tau - \tau_{G}^{o} - h\gamma)}{kT}\right\} \qquad (208)$$

Integrating

$$\exp\left\{\frac{\Delta H_{\mathrm{o}} - v\left(\tau - \tau_{\mathrm{G}}^{\mathrm{o}} - h\gamma\right)}{kT}\right\}d\gamma = (NAbv_{\mathrm{o}})dt \qquad (209)$$

$$\left(\frac{kT}{vh}\right)\exp\left\{\frac{\Delta H_{\mathrm{o}} - v\left(\tau - \tau_{\mathrm{G}}^{\mathrm{o}} - h\gamma\right)}{kT}\right\} = NAbv_{\mathrm{o}}t + D \qquad (210)$$

$$\exp\left\{\frac{\Delta H_{\mathrm{o}} - v\left(\tau - \tau_{\mathrm{G}}^{\mathrm{o}} - h\gamma\right)}{kT}\right\} = (NAbv_{\mathrm{o}})\frac{vht}{kT} + D' \qquad (211)$$

$$\frac{\Delta H_{\mathrm{o}} - v\left(\tau - \tau_{\mathrm{G}}^{\mathrm{o}} - h\gamma\right)}{kT} = \ln(v't + D') \qquad (212)$$

where

$$v' = \left(\frac{NAbv_{\mathrm{o}}vh}{kT}\right)$$

$$\left(\frac{hv\gamma}{kT}\right) = \ln(v't + D') - \frac{\Delta H_{\mathrm{o}} + v\left(\tau - \tau_{\mathrm{G}}^{\mathrm{o}}\right)}{kT} \qquad (213)$$

$$\gamma = \left(\frac{kT}{hv}\right)\ln(v't + D') - E \qquad (214)$$

Choose a value for D' so that $\gamma(0) = 0$.

$$\gamma = \left(\frac{kT}{hv}\right)\ln(v't + 1) \qquad (215)$$

The last equation is in the form of Eqn (200). One difficulty with this analysis is Eqn (207), which τ_{G} may equal nearly zero, in actuality. Other attempts to justify this equation were made by Wyatt [1111], but the methodology was unclear. Welch et al. appeared to attempt a similar approach to that of Wyatt.

Other mechanisms considered include quantum mechanical tunneling, which predicts athermal creep behavior at low temperatures. Early proponents include [1113,1117,1119,1120]. Conversely, subsequent work by [1106, 1115,1116] suggests that, even to 4 K, creep is time dependent and is a result of the thermal activation of dislocations. Dislocation kink mechanisms were suggested for body-centered cubic metals [1116]. However, it is unclear how this mechanism, by itself, explains the observed creep behavior. In materials with solutes, it is suggested that thermal activation past pinning solutes [1104] in Al-Mg.

Increasing creep resistance at low temperatures appears to be accomplished in similar ways as at elevated temperatures. Where cold work increased the creep resistance in 304 stainless steel at elevated temperatures, it also increases creep resistance at low temperatures [1142]. Others have suggested the role of other features [1096] such as twin boundaries. Of course, solute strengthening is a variable, such as in the 304 stainless steel. The precise mechanism by which the strengthening variables superimpose is unclear.

3.2 Power-Law Analysis

In an empirical analysis, Neeraj et al. [1094] assume the Holloman flow equation

$$\sigma = K\varepsilon^n \dot{\varepsilon}^m \tag{216}$$

where K is the strength parameter, n is the strain-hardening exponent, and m is the strain-rate sensitivity exponent. This equation was suggested to reasonably describe some Ti-alloy behavior. Thus, they show

$$\dot{\varepsilon} = \frac{d\varepsilon}{dt} = \left(\frac{\sigma}{K}\right)^{1/m} \varepsilon^{-n/m} \tag{217}$$

$$\int \varepsilon^{n/m} = \int \left(\frac{\sigma}{K}\right)^{1/m} dt \tag{218}$$

under constant stress

$$\varepsilon = \left(\frac{\sigma}{K}\right)^{1/m+n} \left(\frac{m+n}{m}\right)^{m/m+n} t^{m/m+n}$$

This is similar to Eqn 199

$$\varepsilon = at^b$$

with

$$b = \left(\frac{m}{m+n}\right) \tag{219}$$

$$a = \left(\frac{\sigma}{K}\right)^{1/m+n} \left(\frac{m+n}{m}\right)^{m/m+n} \tag{220}$$

These equations relate constant strain-rate behavior to constant stress behavior. These investigators show interesting predictions of the low-temperature creep behavior in Ti-alloys from a limited set of constant strain-rate data. They rationalized low-temperature creep of Ti-alloys as being due to low *n* and moderate *m* values. This rationale may be extended to steels. Certainly, as also pointed out by [1143], low *n* and high *m* certainly predisposes a material to significant creep plasticity at low temperatures.

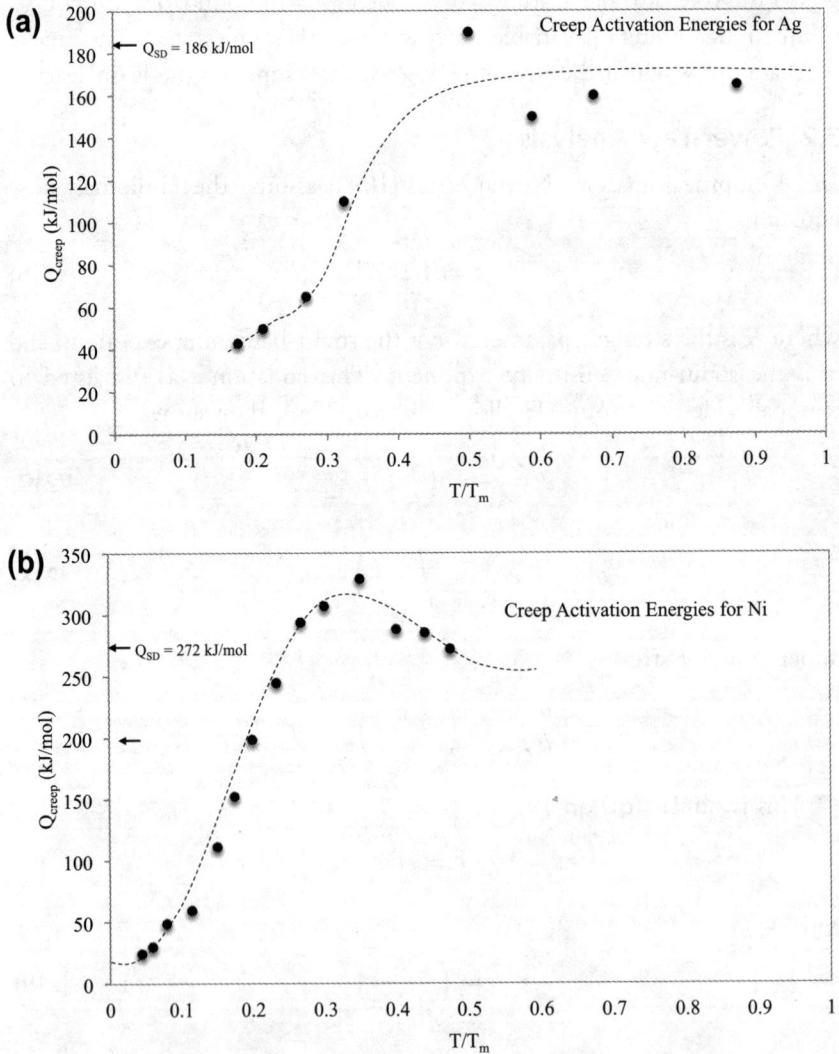

Figure 148 Activation energy for (steady-state) creep of (a) Ag, (b) Ni, (c) Cu, and (d) Al as a function of temperature. *Adapted from Ref. [1141].*

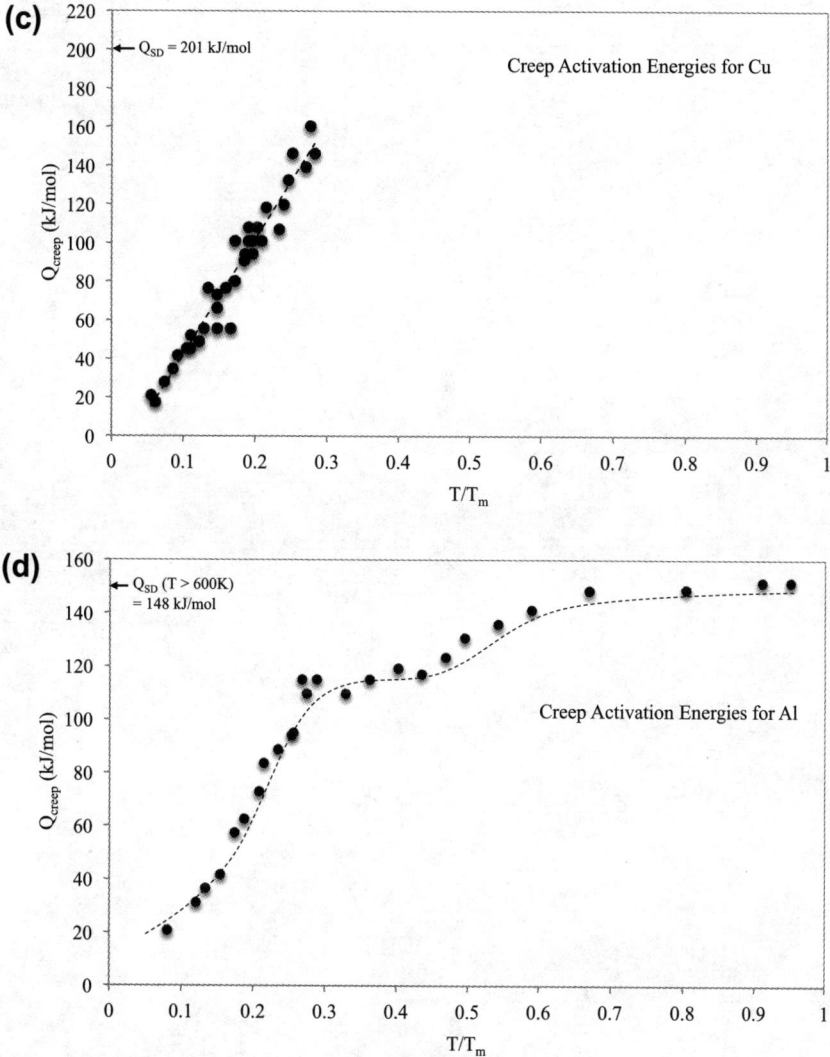

(c)

Creep Activation Energies for Cu

$Q_{SD} = 201$ kJ/mol

Q_{creep} (kJ/mol)

T/T_m

(d)

Q_{SD} (T > 600K) = 148 kJ/mol

Creep Activation Energies for Al

Q_{creep} (kJ/mol)

T/T_m

Figure 148 cont'd

3.3 Activation Energies

Figure 148 [1144,1145] shows that the activation energy decreases with temperature below 0.3 T_m and is much lower than at higher temperatures where Q is associated with lattice self-diffusion.

REFERENCES

[1] S. Suri, G.B. Viswanathan, T. Neeraj, D.-H. Hou, M.J. Mills, Acta Mater. 47 (1999) 1019.
[2] M.E. Kassner, P. Geantil, R.S. Rosen, J. Eng. Mater. Technol. 133 (2011) 5083.
[3] Z.S. Basinski, Philos. Mag. 4 (1959) 393.
[4] C.M. Young, S.L. Robinson, O.D. Sherby, Acta Metall. 23 (1975) 633.
[5] O.D. Sherby, J. Weertman, Acta Metall. 27 (1979) 387.
[6] O.D. Sherby, R.J. Klundt, A.K. Miller, Metall. Trans. A 8A (1977) 843.
[7] M.E. Kassner, Res Mech. 18 (1986) 179.
[8] M.E. Kassner, Scr. Metall. 16 (1982) 265.
[9] R. Logan, A.K. Mukherjee, R.G. Castro, Scr. Metall. 17 (1983) 741.
[10] M. Biberger, J.C. Gibeling, Acta Metall. Mater. 43 (1995) 3247.
[11] M.E. Kassner, J. Pollard, E. Cerri, E. Evangelista, Acta Metall. Mater. 42 (1994) 3223.
[12] M.E. Kassner, Metall. Trans. 20A (1989) 2001.
[13] D.A. Hughes, W.D. Nix, Metall. Trans. 19A (1988) 3013.
[14] H.K. Adenstedt, Met. Prog. 65 (1949) 658.
[15] T.A. Tozera, O.D. Sherby, J.E. Dorn, Trans. ASM 49 (1957) 173.
[16] O.D. Sherby, P.M. Burke, Prog. Mater. Sci. 13 (1967) 325.
[17] S. Takeuchi, A.S. Argon, J. Mater. Sci. 11 (1975) 1542.
[18] A.S. Argon, in: R.W. Cahn, P. Haasen (Eds.), Physical Metallurgy, Elsevier, 1996, p. 1957.
[19] A. Orlova, J. Cadek, Mater. Sci. Eng. 77 (1986) 1.
[20] J. Cadek, Creep in Metallic Materials, Elsevier, Amsterdam, 1988.
[21] A.K. Mukherjee, in: R.J. Arsenault (Ed.), Treatise on Materials Science and Technology, vol. 6, Academic, New York, 1975, p. 163.
[22] W. Blum, Plastic deformation and fracture, in: R.W. Cahn, P. Haasen, E.J. Kramer (Eds.), in: H. Mughrabi (Ed.), Materials Science and Technology, vol. 6, VCH Publishers, Wienheim, 1993, p. 339.
[23] F.R.N. Nabarro, H.L. de Villers, The Physics of Creep, Taylor and Francis, London, 1995.
[24] J. Weertman, in: R.W. Cahn, P. Haasen (Eds.), Physical Metallurgy, third ed., Elsevier, Amsterdam, 1983.
[25] J. Weertman, in: Mechanics and Materials Interlinkage, John Wiley, 1999.
[26] W.D. Nix, B. Ilschner, in: P. Haasen, V. Gerold, G. Kostorz (Eds.), Strength of Metals and Alloys, Pergamon, Oxford, 1980, p. 1503.
[27] M.E. Kassner, P. Geantil, L.E. Levine, Int. J. Plast. 45 (2013) 44.
[28] R.W. Evans, B. Wilshire, Creep of metals and alloys, Inst. Met. (1985). London, 313 pp.
[29] M.E. Kassner, M.-T. Pérez-Prado, Prog. Mater. Sci. 45 (2000) 1.
[30] H.E. Evans, Mechanisms of Creep Fracture, Elsevier App. Science, London, 1984.
[31] J.J. Shrzypek, Plasticity and Creep, CRC Press, Boca Raton, 1993.
[32] J. Gittus, Creep, Viscoelasticity and Creep Fracture in Solids, Applied Science Pub., London, 1975.
[33] H.J. Frost, M.E. Ashby, Deformation Mechanisms Maps, the Plasticity and Creep of Metals and Ceramics, Pergamon Press, 1982.

[34] E.H. Norton, Creep of Steel at High Temperatures, McGraw Hill, New York, 1929.

[35] F.C. Monkman, N.J. Grant, Proc. ASTM 56 (1956) 593.

[36] P. Yavari, T.G. Langdon, Acta Metall. 30 (1982) 2181.

[37] H.J. McQueen, J.K. Solberg, N. Ryum, E. Nes, Philos. Mag. A 60 (1989) 473.

[38] S.C. Shrivastava, J.J. Jonas, G.R. Canova, J. Mech. Phys. Solids 30 (1982) 75.

[39] M.E. Kassner, N.Q. Nguyen, G.A. Henshall, H.J. McQueen, Mater. Sci. Eng. A132 (1991) 97.

[40] A.N. Campbell, S.S. Tao, D. Turnbull, Acta Metall. 35 (1987) 2453.

[41] J.P. Porrier, Acta Metall. 26 (1978) 629.

[42] M.A. Morris, J.L. Martin, Acta Metall. 32 (1984) 1609.

[43] M.A. Morris, J.L. Martin, Acta Metall. 32 (1984) 549.

[44] H. Luthy, A.K. Miller, O.D. Sherby, Acta Metall. 28 (1980) 169.

[45] S.H. Goods, W.D. Nix, Acta Metall. 26 (1978) 753.

[46] D.A. Hughes, Q. Liu, D.C. Chrzan, N. Hansen, Acta Mater. 45 (1997) 105.

[47] J. Gil Sevillano, R.V. Van Houtte, E. Aernoudt, Prog. Mater. Sci. 25 (1980) 69.

[48] F. Garofalo, Fundamentals of Creep Rupture in Metals, Macmillan, New York, 1965.

[49] J.E. Bird, A.K. Mukherjee, J.E. Dorn, in: D.G. Brandon, A. Rosen (Eds.), Quantitative Relations between Properties and Microstructure, Israel Univ. Press, Jerusalem, 1969, p. 255.

[50] J. Harper, J.E. Dorn, Acta Metall. 5 (1957) 654.

[51] C. Herring, J. Appl. Phys. 21 (1950) 437.

[52] R.L. Coble, J. Appl. Phys. 34 (1963) 1679.

[53] M.Y. Wu, O.D. Sherby, Acta Metall. 32 (1984) 1561.

[54] A.J. Ardell, Acta Mater. 45 (1997) 2971.

[55] W. Blum, W. Maier, Phys. Status Solidi A 171 (1999) 467.

[56] D.M. Owen, T.G. Langdon, Mater. Sci. Eng. A216 (1996) 20.

[57] G.W. Greenwood, Scr. Metall. Mater. 30 (1994) 1527.

[58] J.B. Bilde-Sorenson, D.A. Smith, Scr. Metall. Mater. 30 (1994) 1527.

[59] J. Wolfenstine, O.A. Ruano, J. Wadsworth, O.D. Sherby, Scr. Metall. Mater. 29 (1993) 515.

[60] B. Burton, G.L. Reynolds, Mater. Sci. Eng. A191 (1995) 135.

[61] O.A. Ruano, J. Wadsworth, J. Wolfenstine, O.D. Sherby, Mater. Sci. Eng. A165 (1993) 133.

[62] G. Konig, W. Blum, in: P. Haasen, V. Gerold, G. Kostorz (Eds.), Strength of Metals and Alloys, Pergamon, Oxford, 1980, p. 363.

[63] P. Shewmon, Diffusion in Solids, second ed., TMS, Warrendale, PA, 1989.

[64] O.D. Sherby, J.L. Lytton, J.E. Dorn, Acta Metall. 5 (1957) 219.

[65] P.R. Landon, J.L. Lytton, L.A. Shepard, J.E. Dorn, Trans. ASM 51 (1959) 900.

[66] L.J. Cuddy, Metall. Trans. 1A (1970) 395.

[67] A.W. Thompson, B. Odegard, Metall. Trans. 4 (1973) 899.

[68] O.D. Sherby, A.K. Miller, J. Eng. Mater. Technol. 101 (1979) 387.

[69] M.E. Kassner, R.S. Rosen, G.A. Henshall, Metall. Trans. 21A (1990) 3085.

[70] S.V. Raj, T.G. Langdon, Acta Metall. 37 (1989) 843.

[71] H. Mecking, Y. Estrin, Scr. Metall. 14 (1980) 815.

[72] A.K. Mukherjee, J.E. Bird, J.E. Dorn, ASM Trans. Q. 62 (1969) 155.

[73] F.A. Mohamed, T.G. Langdon, Acta Metall. 22 (1974) 779.

[74] A.J. Ardell, O.D. Sherby, Trans. AIME 239 (1961) 1547.

[75] G.M. Hood, H. Zou, R.J. Schultz, E.H. Bromley, J.A. Jackman, J. Nucl. Mater. 217 (1994) 229.

[76] S. Straub, W. Blum, Scr. Metall. Mater. 24 (1990) 1837.

[77] W. Blum, in: T.G. Langdon, H.D. Merchant, J.G. Morris, M.A. Zaidi (Eds.), Hot Deformation of Aluminum Alloys, TMS, Warrendale, PA, 1991, p. 181.

[78] M.E. Kassner, M.-T. Perez-Prado, M. Long, K.S. Vecchio, Metall. Mater. Trans. 33A (2002) 311.

[79] G.V. Viswanathan, S. Karthikeyan, R.W. Hayes, H.J. Mills, Metall. Mater. Trans. 33 (2002) 329.

[80] T.A. Hayes, M.E. Kassner, R.S. Rosen, Metall. Mater. Trans. 334 (2002) 337.

[81] W. Blum, Private Communication, 1999.

[82] B. Wilshire, Metall. Mater. Trans. 334 (2002) 241.

[83] R.W. Evans, B. Wilshire, in: B. Wilshire, R.W. Evans (Eds.), Creep and Fracture of Structural Materials, Inst. Metals, London, 1987, p. 59.

[84] D. McLean, Trans. AIME 22 (1968) 1193.

[85] P. Ostrom, R. Lagneborg, Res Mech. 1 (1980) 59.

[86] A.J. Ardell, M.A. Przystupa, Mech. Mater. 3 (1984) 319.

[87] J. Friedel, Dislocations, Pergamon Press, Oxford, 1964.

[88] D. Hull, D.J. Bacon, Introduction to Dislocations, fifth ed., Butterworth, Heinemann, Oxford, 2011.

[89] T.J. Ginter, F.A. Mohamed, J. Mater. Sci. 17 (1982) 2007.

[90] S. Straub, W. Blum, in: H.J. McQueen, E.V. Konopleva, N.D. Ryan (Eds.), Hot Workability of Steels and Light Alloys, Canadian Inst. Mining, Metallurgy and Petroleum, Montreal, Quebec, 1996, p. 1889.

[91] P.B. Hirsch, A. Howie, P.B. Nicholson, D.W. Pashley, M.J. Whelan, Electron Microscopy of Thin Crystals, Kreiger, New York, 1977.

[92] W. Blum, A. Absenger, R. Feilhauer, in: P. Haasen, V. Gerold, G. Kostorz (Eds.), Strength of Metals and Alloys, Pergamon Press, Oxford, 1980, p. 265.

[93] J. Hausselt, W. Blum, Acta Metall. 24 (1976) 1027.

[94] V. Randle, The Measurement of Grain Boundary Geometry, IOP, London, 1993.

[95] D. Calliard, J.L. Martin, Acta Metall. 30 (1982) 791.

[96] D. Cailliard, J.L. Martin, Acta Metall. 30 (1982) 437.

[97] D. Cailliard, J.L. Martin, Acta Metall. 31 (1983) 813.

[98] M.E. Kassner, J.W. Elmer, C.J. Echer, Metall. Trans. A 17 (1986) 2093.

[99] M.E. Kassner, Mater. Lett. 5B (1984) 451.

[100] S.V. Raj, G.M. Pharr, Mater. Sci. Eng. 81 (1986) 217.

[101] A.J. Ardell, S.S. Lee, Acta Metall. 34 (1986) 2411.

[102] P. Lin, S.S. Lee, A.J. Ardell, Acta Metall. 37 (1989) 739.

[103] K.D. Challenger, J. Moteff, Metall. Trans. 4 (1973) 749.

[104] J.D. Parker, B. Wilshire, Philos. Mag. 41A (1980) 665.

[105] E. Evans, G. Knowles, Acta Metall. 25 (1977) 963.

[106] O. Ajaja, Scr. Metall. 24 (1990) 1435.

[107] M.E. Kassner, J. Mater. Sci. 25 (1990) 1997.

[108] L.Q. Shi, D.O. Northwood, Phys. Status Solidi A 149 (1995) 213.

[109] B. Burton, Philos. Mag. 45 (1982) 657.

[110] M.E. Kassner, A.K. Miller, O.D. Sherby, Metall. Trans. A 13 (1982) 1977.

[111] M.E. Kassner, A.A. Ziaai-Moayyed, A.K. Miller, Metall. Trans. A 16 (1985) 1069.

[112] S. Karashima, T. Iikubo, T. Watanabe, H. Oikawa, Trans. Jpn. Inst. Met. 12 (1971) 369.

[113] U. Hofmann, W. Blum (Eds.), 7th Inter. Symp. on Aspects of High Temperature Deformation and Fracture of Crystalline Materials, Jap. Inst. Metals, Sendai, 1993, p. 625.

[114] G.B. Gibbs, Philos. Mag. 13 (1966) 317.

[115] L.E. Levine, P. Geantil, B.C. Larson, J.Z. Tischler, M.E. Kassner, W. Liu, M.R. Stoudt, F. Tavazza, Acta Mater. 59 (2011) 5803.

[116] I.F. Lee, T.Q. Phan, L.E. Levine, J.Z. Tischler, P. Geantil, Y. Huang, T.G. Langdon, M.E. Kassner, Acta Mater. 61 (2013) 7741.

[117] J.C. Gibeling, W.D. Nix, Acta Metall. 28 (1980) 1743.

[118] M.J. Mills, J.C. Gibeling, W.D. Nix, Acta Metall. 33 (1985) 1503.

[119] M. Legros, O. Ferry, F. Houdellier, A. Jacques, A. George, Mater. Sci. Eng. A 483-4 (2008) 353.

[120] G.S. Nakayama, J.C. Gibeling, Acta Metall. 38 (1990) 2023.

[121] E. Weckert, W. Blum, in: H.J. McQueen, J.-P. Bailon, J.I. Dickson, J.J. Jonas, M.G. Akben (Eds.), Strength of Metals and Alloys, Pergamon Press, Oxford, 1979, p. 773.

[122] W. Blum, A. Cegielska, A. Rosen, J.L. Martin, Acta Metall. 37 (1989) 2439.

[123] W. Blum, A. Finkel, Acta Metall. 30 (1982) 1705.

[124] W. Blum, E. Weckert, Mater. Sci. Eng. 86 (1987) 147.

[125] W. Muller, M. Biberger, W. Blum, Philos. Mag. A66 (1992) 717.

[126] R.W. Evans, W.J.F. Roach, B. Wilshire, Mater. Sci. Eng. 73 (1985) L5.

[127] P.I. Ferreira, R.G. Stang, Acta Metall. 31 (1983) 585.

[128] F.A. Mohamed, M.S. Soliman, M.S. Mostofa, Philos. Mag. 51 (1985) 1837.

[129] Y. Huang, F.J. Humphreys, Acta Mater. 45 (1997) 4491.

[130] M. Carrard, J.L. Martin, in: H.J. McQueen, J.-P. Bailon, J.I. Dickson, J.J. Jonas, M.G. Akben (Eds.), Strength of Metals and Alloy, Pergamon Press, Oxford, 1985, p. 665.

[131] D.O. Northwood, I.O. Smith, Phys. Status Solidi A 115 (1989) 1495.

[132] M.E. Kassner, A.K. Mukherjee, Scr. Metall. 17 (1983) 741.

[133] U.F. Kocks, A.S. Argon, M.F. Ashby, Prog. Mater. Sci. 19 (1975) 1.

[134] J.D. Parker, B. Wilshire, Philos. Mag. 34 (1976) 485.

[135] W. Blum, J. Hausselt, G. König, Acta Metall. 24 (1976) 239.

[136] S. Straub, W. Blum, H.J. Maier, T. Ungar, A. Borberly, H. Renner, Acta Mater. 44 (1996) 4337.

[137] A.S. Argon, S. Takeuchi, Acta Metall. 29 (1981) 1877.

[138] H. Mughrabi, Acta Metall. 31 (1983) 1367.

[139] P. Kumar, M.E. Kassner, W. Blum, P. Eisenlohr, T.G. Langdon, Mater. Sci. Eng. 510-1 (2009) 20.

[140] P. Peralta, P. Llanes, J. Bassani, C. Laird, Philos. Mag. 70A (1994) 219.

[141] D.A. Hughes, A. Godfrey, in: T.A. Bieler, L.A. Lalli, S.R. MacEwen (Eds.), Hot Deformation of Aluminum Alloys, TMS, Warrendale, PA, 1998, p. 23.

[142] M.E. Kassner, Metall. Trans. 20A (1989) 2182.

[143] K.E. Thiesen, M.E. Kassner, J. Pollard, D.R. Hiatt, B. Bristow, in: F.H. Froes, I.L. Caplan (Eds.), Titanium '92, TMS, Warrendale, PA, 1993, p. 1717.

[144] H. Widersich, J. Met. 16 (1964) 423.

[145] U.F. Kocks, in: P. Haasen, V. Gerold, G. Kostorz (Eds.), Strength of Metals and Alloys, Pergamon Press, Oxford, 1980, p. 1661.

[146] M.E. Kassner, M.E. McMahon, Metall. Trans. 18A (1987) 835.

[147] M.E. Kassner, X. Li, Scr. Metall. Mater. 25 (1991) 2833.

[148] M.E. Kassner, Mater. Sci. Eng. 166 (1993) 81.

[149] H.J. Levinstein, W.H. Robinson (Eds.), Symp. at the National Phys. Lab., January 1963, p. 180.

[150] J.E. Bailey, P.B. Hirsch, Philos. Mag. 5 (1960) 485.

[151] R.L. Jones, H. Conrad, TMS, AIME 245 (1969) 779.

[152] O. Ajaja, A.J. Ardell, Scr. Metall. 11 (1977) 1089.

[153] O. Ajaja, A.J. Ardell, Philos. Mag. 39 (1979) 65.

[154] L. Shi, D.O. Northwood, Phys. Status Solidi A 137 (1993) 75.

[155] L. Shi, D.O. Northwood, Phys. Status Solidi A 140 (1993) 87.

[156] G.A. Henshall, M.E. Kassner, H.J. McQueen, Metall. Trans. 23A (1992) 881.

[157] M.J. Mills, unpublished research, presented at the Conference for Creep and Fracture of Structural Materials, Swansea, UK, April 1987.

[158] E.N. da C. Andrade, Proc. R. Soc. A84 (1910) 1.

[159] A.H. Cottrell, V. Aytekin, Nature 160 (1947) 328.

[160] J.B. Conway, Numerical Methods for Creep and Rupture Analyses, Gordon and Breach, New York, 1967.

[161] F. Garofalo, C. Richmond, W.F. Domis, F. von Gemmingen, in: Proc. Joint. Int. Conf. on Creep, Inst. Mech. Eng., London, 1963, p. 1.

[162] C.R. Barrett, N.D. Nix, O.D. Sherby, Trans. ASM 59 (1966) 3.

[163] V.K. Sikka, H. Nahm, J. Moteff, Mater. Sci. Eng. 20 (1975) 55.

[164] A. Orlova, M. Pahutova, J. Cadek, Philos. Mag. 25 (1972) 865.

[165] S. Daily, C.N. Ahlquist, Scr. Metall. 6 (1972) 95.

[166] A. Orlova, Z. Tobolova, J. Cadek, Philos. Mag. 26A (1972) 1263.

[167] S.H. Suh, J.B. Cohen, J. Weertman, Metall. Trans. 12 (1981) 361.

[168] F. Petry, F.P. Pchenitzka, Mater. Sci. Eng. 68 (1984) L7.

[169] M.E. Kassner, Acta Mater. 52 (2004) 1.

[170] T.G. Langdon, R.D. Vastava, P. Yavari, in: P. Haasen, V. Gerold, G. Kostorz (Eds.), Strength of Metals and Alloys, Pergamon, Oxford, 1980, p. 271.

[171] B.W. Evans, W.J.F. Roach, B. Wilshire, Scr. Metall. 19 (1985) 999.

[172] M. Delos-Reyes, M.S. Thesis, Department of Mechanical Engineering, Oregon State University, Corvallis, OR, 1996.

[173] A.W. Sleeswyk, M.R. James, D.H. Plantinga, W.S.T. Maathuis, Acta Metall. 126 (1978) 1265.

[174] E. Orowan, Internal Stress and Fatigue in Metals, in: General Motors Symposium, Elsevier, Amsterdam, 1959, p. 59.

[175] T. Hasegawa, Y. Ikeuchi, S. Karashima, Met. Sci. 6 (1972) 78.

[176] H. Mughrabi, Mater. Sci. Eng. A 85 (1981) 15.

[177] B. Derby, M.F. Ashby, Acta Metall. 35 (1987) 1349.

[178] S. Vogler, W. Blum, in: B. Wilshire, R.W. Evans (Eds.), Creep and Fracture of Engineering Materials and Structures, Inst. Metals, London, 1990, p. 65.

[179] J. Lepinoux, L.P. Kubin, Philos. Mag. A 57 (1985) 675.

[180] M.E. Kassner, M.A. Delos-Reyes, M.A. Wall, Metall. Mater. Trans. 28A (1997) 595.

[181] A. Borbély, G. Hoffmann, E. Aernoudt, T. Ungar, Acta Mater. 45 (1997) 89.

[182] M.E. Kassner, F.J. Weber, J. Koike, R.S. Rosen, J. Mater. Sci. 31 (1996) 2291.

[183] L.E. Levine, Private Communication, March 1998.

[184] I. Gaal, in: N. Hersel Andersen, M. Eldrup, N. Hansen, D. Juul Jensen, T. Leffers, H. Lilkolt, O.B. Pedersen, B.N. Singh (Eds.), Proc. 5th Int. Riso Symposium, 1984, p. 249. Roskilde, Denmark.

[185] M.E. Kassner, M.T. Perez-Prado, K.S. Vecchio, M.A. Wall, Acta Mater. 48 (2000) 4247.

[186] M.E. Kassner, M.-T. Pérez-Prado, K.S. Vecchio, Mater. Sci. Eng. A 319–321 (2001) 730.

[187] J. Weertman, in: J.C.M. Li, A.K. Mukherjee (Eds.), Rate Process in Plastic Deformation of Metals, ASM, Materials Park, OH, 1975, p. 315.

[188] J. Weertman, Trans. Q. ASM 61 (1968) 680.

[189] J. Weertman, J. Appl. Phys. 26 (1955) 1213.

[190] C.R. Barrett, W.D. Nix, Acta Metall. 12 (1965) 1247.

[191] T. Watanabe, S. Karashima, Trans. JIM 11 (1970) 159.

[192] L.I. Ivanov, V.A. Yanushkevich, Fiz. Met. Metall. 17 (1964) 112.

[193] W. Blum, Phys. Status Solidi 45 (1971) 561.

[194] J. Weertman, in: B. Wilshire, et al. (Eds.), Creep and Fracture of Engineering Materials, 1984, p. 1. Pineridge, Swansea.

[195] J. Weertman, J.R. Weertman, Constitutive relations and their physical basis, in: S.E. Andersen, J.B. Bilde-Sorensen, N. Hansen, T. Leffers, H. Lilholt, O.B. Pedersen, B. Ralph (Eds.), Proc. of 8th Risø Int. Symp. on Materials Science, Risø, Denmark, 1987, p. 191.

[196] S.K. Mitra, D. McLean, Met. Sci. 1 (1967) 192.

[197] R. Lagneborg, Met. Sci. J. 6 (1972) 127.

[198] N.F. Mott (Ed.), Conference on Creep and Fracture of Metals at High Temperatures, H.M. Stationery Office, London, 1956.

[199] L.E. Levine, B. Larson, J.Z. Tischler, M.E. Kassner, P. Geantil, W. Liu, M.R. Stoudt, F. Tavazza, Acta Mater, 59 (2011) 5083–5091.

[200] J. Gittus, Acta Metall. 22 (1974) 789.

[201] E. Nes, W. Blum, P. Eisenlohr, Metall. Mater. Trans. A 33 (2002) 305.

[202] K. Maruyama, S. Karashima, H. Oikawa, Res Mech. 7 (1983) 21.

[203] E. Orowan, Z. Phys. 89 (1934) 614.

[204] R.W. Bailey, Inst. Met. 35 (1926) 27.

[205] G.S. Daehn, H. Brehm, H. Lee, B.S. Lim, Mater. Sci. Eng. 387A (2004) 576.

[206] H. Brehm, G.S. Daehn, Metall. Mater. Trans. 33A (2002) 363.

[207] L. Bendersky, A. Rosen, A.K. Mukherjee, Int. Met. Rev. 30 (1985) 1.

[208] J.D. Parker, B. Wilshire, Met. Sci. 12 (1978) 453.

[209] J.C.M. Li, Acta Metall. 8 (1960) 296.

[210] M. Henderson-Brown, K.F. Hale, in: P.R. Swann, C.J. Humphreys, M.J. Goringe (Eds.), HVEM, Academic Press, London, 1974, p. 206.

[211] R.J. Klundt, (Ph.D. thesis), Department of Materials Science and Engineering, Stanford University, Stanford, CA, 1978.

[212] J.D. Parker, B. Wilshire, Mater. Sci. Eng. 43 (1980) 271.

[213] H.J. Levinstein, W.H. Robinson, The Relations between Structure and the Mechanical Properties of Metal, in: Symp. at the National Physical Lab., January 1963 (Her Majesty's Stationery Office), 1983, p. 180. From J. Weertman, J.L. Weertman, Physical Metallurgy, R.W. Cahn and P. Hassen, eds., Elsevier, p. 1259.

[214] W. Blum, Private Communication, 2002.

[215] E. Nes, Prog. Mater. Sci. 41 (1998) 129.

[216] J.A. Gorman, D.S. Wood, T. Vreeland, J. Appl. Phys. 40 (1969) 833.

[217] U.F. Kocks, M.G. Stout, A.D. Rollett, in: P.O. Kettunen, T.K. Lepistö, M.E. Lehtoness (Eds.), Strength of Metals and Alloys, vol. 1, Pergamon, Oxford, 1989, p. 25.

[218] M.E. Kassner, M.M. Myshlyaev, H.J. McQueen, Mater. Sci. Eng. 108A (1989) 45.

[219] H.J. McQueen, O. Knustad, N. Ryum, J.K. Solberg, Scr. Metall. 19 (1985) 73.

[220] M.M. Myshlyaev, O.N. Senkov, V.A. Likhachev, in: H.J. McQueen, J.-P. Bailon, J.I. Dickson, J.J. Jonas, M.G. Akben (Eds.), Strength of Metals and Alloys, Pergamon, Oxford, 1985, p. 841.

[221] C.G. Schmidt, C.M. Young, B. Walser, R.H. Klundt, O.D. Sherby, Metall. Trans. 13A (1982) 447.

[222] R.D. Doherty, D.A. Hughes, F.J. Humphreys, J.J. Jonas, D. Juul-Jensen, M.E. Kassner, W.E. King, T.R. McNelley, H.J. McQueen, A.D. Rollett, Mater. Sci. Eng. A238 (1997) 219.

[223] H.J. McQueen, S. Spigarelli, M.E. Kassner, E. Evangelista, Hot Deformation and Processing of Aluminum Alloys, CRC Press, Boca Raton, 2011.

[224] R.A. Petkovic, M.J. Luton, J.J. Jonas, Met. Sci. (1979) 569.

[225] H. Conrad, Acta Metall. 11 (1963) 75.

[226] D. Hansen, M.A. Wheeler, J. Inst. Met. 45 (1931) 229.

[227] F.A. Weinberg, Trans. AIME 212 (1958) 808.

[228] C.R. Barrett, J.L. Lytton, O.D. Sherby, Trans. TMS-AIME 239 (1967) 170.

[229] F. Garofalo, W.F. Domis, R. von Gemmingen, Trans. TMS-AIME 240 (1968) 1460.

[230] S. Kikuchi, A. Yamaguchi, in: H.J. McQueen, J.-P. Bailon, J.I. Dickson, J.J. Jonas, M.G. Akben (Eds.), Strength of Metals and Alloys, Pergammon Press, Oxford, 1985, p. 899.

[231] M.E. Kassner, in: M.A. Otooni, R.W. Armstrong, N.J. Grant, K. Rshizaki (Eds.), Grain Size and Mechanical Properties: Fundamentals and Applications, MRS, 1995, p. 157.

[232] J.T. Al-Haidary, M.J. Petch, E.R. Delos-Rios, Philos. Mag. 47A (1983) 863.

[233] C. Perdrix, Y.M. Perrin, F. Montheillet, Mem. Sci. Rev. Metall. 78 (1981) 309.

[234] W.J. Evans, B. Wilshire, Scr. Metall. 8 (1974) 497.

[235] W.J. Evans, B. Wilshire, Metall. Trans. 1 (1970) 2133.

[236] R.D. Warda, V. Fidleris, E. Teghtsoonian, Metall. Trans. 4 (1973) 1201.

[237] F.R.N. Nabarro, in: Rept. of Conf. on the Solids, The Physical Society, London, 1948, p. 75.

[238] G.W. Greenwood, Proc. R. Soc. Lond. A 436 (1992) 187.

[239] J. Fiala, J. Cadek, Mater. Sci. Eng. 75 (1985) 117.

[240] J. Fiala, J. Novotny, J. Cadek, Mater. Sci. Eng. 60 (1983) 195.

[241] I.G. Crossland, R.B. Jones, Met. Sci. 11 (1977) 504.

[242] T. Sritharan, H. Jones, Acta Metall. 27 (1979) 1293.

[243] R.S. Mishra, H. Jones, G.W. Greenwood, Philos. Mag. A 60 (6) (1989) 581.

[244] B. Ya Pines, A.F. Sirenko, Fiz. Met. Metall. 7 (1959) 766.

[245] R.B. Jones, Nature 207 (1965) 70.

[246] B. Burton, G.W. Greenwood, Acta Metall. 18 (1970) 1237.

[247] K.E. Harris, A.H. King, Acta Mater. 46 (17) (1998) 6195.

[248] R.L. Coble, High Strength Materials, John Wiley, New York, 1965.

[249] B. Burton, G.L. Reynolds, Physical Metallurgy of Reactor Fuel Elements, Metals Soc., London, 1975, p. 87.

[250] R. Folweiler, J. Appl. Phys. 21 (1950) 437.

[251] S.I. Warshau, F.H. Norton, J. Am. Ceram. Soc. 45 (1962) 479.

[252] R. Chang, J. Nucl. Mater. 2 (1959) 174.

[253] I.M. Bernstein, Trans. AIME 239 (1967) 1518.

[254] B. Burton, G.W. Greenwood, Met. Sci. J. 4 (1970) 215.

[255] I.G. Crossland, Physical Metallurgy of Reactor Fuel Elements, Metals Soc., London, 1975, p. 66.

[256] I.G. Crossland, B. Burton, B.D. Bastow, Met. Sci. 9 (1975) 327.

[257] D.J. Towle, H. Jones, Acta Metall. 24 (1975) 399.

[258] E.M. Passmore, R.H. Duff, T.S. Vasilos, J. Am. Ceram. Soc. 49 (1966) 594.

[259] T.G. Langdon, J.A. Pask, Acta Metall. 18 (1970) 505.

[260] D.B. Knorr, R.M. Cannon, R.L. Coble, Acta Metall. 37 (8) (1989) 2103.

[261] T.G. Langdon, Scr. Mater. 35 (6) (1996) 733.

[262] J.E. Harris, Met. Sci. J. 7 (1973) 1.

[263] R.L. Squires, R.T. Weiner, M. Phillips, J. Nucl. Mater. 8 (1) (1963) 77.

[264] M.E. Kassner, M.T. Perez-Prado, K.S. Vecchio, Cyclic Deformation Dislocation Microstructures, Advanced Materials for the 21st Century, 1999, in: J.R. Weertman Symposium, Y.-W. Chung, D.C. Dunand, P.K. Liaw, G.B. Olsen (Eds.), TMS, Warrendal, PA, 1999, pp. 3–13.

[265] O.A. Ruano, O.D. Sherby, J. Wadsworth, J. Wolfenstine, Scr. Mater. 38 (8) (1998) 1307.

[266] J. Wadsworth, O.A. Ruano, O.D. Sherby, Metall. Mater. Trans. 33A (2002) 219.

[267] O.A. Ruano, J. Wadsworth, O.D. Sherby, Scr. Metall. 22 (1988) 1907.

[268] O.A. Ruano, O.D. Sherby, J. Wadsworth, J. Wolfenstine, Mater. Sci. Eng. A 211 (1996) 66.

[269] C.R. Barrett, E.C. Muehleisen, W.D. Nix, Mater. Sci. Eng. A 10 (1972) 33.

[270] J.N. Wang, J. Mater. Sci. 29 (1994) 6139.

[271] L. Kloc, Scr. Mater. 35 (6) (1996) 733.

[272] J. Fiala, T.G. Langdon, Mater. Sci. Eng. A 151 (1992) 147.

[273] P. Greenfield, C.C. Smith, A.M. Taylor, Trans. AIME 221 (1961) 1065.

[274] K.R. McNee, G.W. Greenwood, H. Jones, Scr. Mater. 46 (2002) 437.

[275] K.R. McNee, G.W. Greenwood, H. Jones, Scr. Mater. 44 (2001) 351.

[276] F.R.N. Nabarro, Metall. Mater. Trans. A 33 (2002) 213.

[277] F.R.N. Nabarro, Private Communication, 1999. San Diego, CA.

[278] I.M. Lifshitz, Sov. Phys. (JETP) 17 (1963) 909.

[279] T. Mori, S. Onaka, K. Wakashima, J. Appl. Phys. 83 (12) (1998) 7547.

[280] S. Onaka, J.H. Huang, K. Wakashima, T. Mori, Acta Mater. 46 (11) (1998) 3821.

[281] S. Onaka, A. Madgwick, T. Mori, Acta Mater. 49 (2001) 2161.

[282] S.S. Sahay, G.S. Murty, Scr. Mater. 44 (2001) 841.

[283] B. Burton, Diffusional Creep of Polycrystalline Materials, Trans. Tech. Publications, Zurich, Switzerland, 1977, p. 61.

[284] R.N. Stevens, Philos. Mag. 23 (1971) 265.

[285] A.E. Aigeltinger, R.C. Gifkins, J. Mater. Sci. 10 (1975) 1889.

[286] W.R. Cannon, Philos. Mag. 25 (1972) 1489.

[287] R.S. Gates, Philos. Mag. 31 (1975) 367.

[288] A. Arieli, G. Gurewitz, A.K. Mukherjee, Met. Forum 4 (1981) 24.

[289] G.B. Gibbs, Mater. Sci. Eng. A 2 (1980) 262.

[290] R. Raj, M.F. Ashby, Metall. Trans. 2 (1971) 1113.

[291] W.B. Beere, Met. Sci. 10 (1976) 133.

[292] M.V. Speight, Acta Metall. 23 (1975) 779.

[293] N.F. Mott, Proc. R. Soc. A 220 (1953) 1.

[294] F.A. Mohamed, K.L. Murty, J.W. Morris, Metall. Trans. 4 (1973) 935.

[295] P. Yavari, D.A. Miller, T.G. Langdon, Acta Metall. 30 (1982) 871.

[296] F.A. Mohamed, T.J. Ginter, Acta Metall. 30 (1982) 1869.

[297] G. Malakondaiah, P. Rama Rao, Acta Metall. 29 (1981) 1263.

[298] J. Novotny, J. Fiala, J. Cadek, Acta Metall. 33 (1985) 905.

[299] G. Malakondaiah, P. Rama Rao, Mater. Sci. Eng. 52 (1982) 207.

[300] P.G. Dixon-Stubbs, B. Wilshire, Philos. Mag. A 45 (1982) 519.

[301] T.G. Langdon, Philos. Mag. A 47 (1983) L29.

[302] O.A. Ruano, J. Wolfenstine, J. Wadsworth, O.D. Sherby, Acta Metall. Mater. 39 (1991) 661.

[303] O.A. Ruano, J. Wolfenstine, J. Wadsworth, O.D. Sherby, J. Am. Ceram. Soc. 75 (7) (1992) 1737.

[304] K.S. Ramesh, E.Y. Yasuda, S. Kimura, J. Mater. Sci. 21 (1986) 3147.

[305] J.N. Wang, Scr. Metall. Mater. 30 (1994) 859.

[306] J.N. Wang, Philos. Mag. Lett. 70 (2) (1994) 81.

[307] J.N. Wang, T. Shimamoto, M. Toriumi, J. Mater. Sci. Lett. 13 (1994) 1451.

[308] J.N. Wang, J. Am. Ceram. Soc. 77 (11) (1994) 3036.

[309] W.B. Banerdt, C.G. Sammis, Phys. Earth Planet. Int. 41 (1985) 108.

[310] J. Wolfenstine, O.A. Ruano, J. Wadsworth, O.D. Sherby, Scr. Metall. Mater. 25 (1991) 2065.

[311] J.P. Poirier, J. Peyronneau, J.K. Gesland, G. Brebec, Phys. Earth Planet. Int. 32 (1983) 273.

[312] S. Beauchesne, J.P. Poirier, Phys. Earth Planet. Int. 61 (1990) 1982.

[313] J.N. Wang, B.E. Hobbs, A. Ord, T. Shimamoto, M. Toriumi, Science 265 (1994) 1204.

[314] J.N. Wang, Mater. Sci. Eng. A 183 (1994) 267.

[315] J.N. Wang, Mater. Sci. Eng. A 187 (1994) 97.

[316] M.Z. Berbon, T.G. Langdon, J. Mater. Sci. Lett. 15 (1996) 1664.

[317] F.A. Mohamed, Mater. Sci. Eng. A 32 (1978) 37.

[318] T.G. Langdon, P. Yavari, Acta Metall. 30 (1982) 881.

[319] J.N. Wang, T.G. Langdon, Acta Metall. Mater. 42 (7) (1994) 2487.

[320] J. Weertman, J. Blacic, Geophys. Res. Lett. 11 (1984) 117.

[321] F.R.N. Nabarro, Acta Metall. 37 (8) (1989) 2217.

[322] O.A. Ruano, J. Wadsworth, O.D. Sherby, Acta Metall. 36 (4) (1988) 1117.

[323] J.N. Wang, Scr. Metall. Mater. 29 (1993) 1267.

[324] J.N. Wang, T.G. Nieh, Acta Metall. 43 (4) (1995) 1415.

[325] J.N. Wang, Philos. Mag. 71A (1995) 105.

[326] J.N. Wang, J.S. Wu, D.Y. Ding, Mater. Sci. Eng. A 334 (2002) 275.

[327] J.N. Wang, Philos. Mag. A 71 (1) (1995) 105.

[328] J.N. Wang, Acta Metall. 44 (3) (1996) 855.

[329] M.A. Przystupa, A.J. Ardell, Metall. Mater. Trans. A 33 (2002) 231.

[330] F.R. Nabarro, Phys. Status Solidi 182 (2000) 627.

[331] W. Blum, P. Eisenlohr, F. Breutinger, Metall. Mater. Trans. A 33 (2002) 291.

[332] E. Nes, T. Pettersen, K. Marthinsen, Scr. Mater. 43 (2000) 55.

[333] K. Marthinsen, E. Nes, Mater. Sci. Technol. 17 (2001) 376.

[334] F.A. Mohamed, Metall. Mater. Trans. A 33 (2002) 261.

[335] T.J. Ginter, P.K. Chaudhury, F.A. Mohamed, Acta Mater. 49 (2001) 263.

[336] T.J. Ginter, F.A. Mohamed, Mater. Sci. Eng. A 322 (2002) 148.

[337] T.G. Langdon, Metall. Mater. Trans. A 33 (2002) 249.

[338] P. Yavari, F.A. Mohamed, T.G. Langdon, Acta Metall. 29 (1981) 1495.

[339] F.A. Mohamed, Mater. Sci. Eng. A 245 (1998) 242.

[340] T.G. Langdon, Mater. Trans. JIM 37 (3) (1996) 359.

[341] H. Oikawa, K. Sugawara, S. Karashima, Mater. Trans. JIM 19 (1978) 611.

[342] T. Endo, T. Shimada, T.G. Langdon, Acta Metall. 32 (11) (1984) 1991.

[343] O.D. Sherby, R.A. Anderson, J.E. Dorn, Trans. AIME 191 (1951) 643.

[344] J. Weertman, Trans. AIME 218 (1960) 207.

[345] J. Weertman, J. Appl. Phys. 28 (10) (1957) 1185.

[346] J. Lothe, J. Appl. Phys. 33 (6) (1962) 2116.

[347] J.P. Hirth, J. Lothe, Theory of Dislocations, McGraw-Hill, NY, 1968, p.584.

[348] H. Oikawa, N. Matsuno, S. Karashima, Met. Sci. 9 (1975) 209.

[349] R. Horiuchi, M. Otsuka, Trans. JIM 13 (1972) 284.

[350] A.H. Cottrell, M.A. Jaswon, Proc. R. Soc. Lond. A 199 (1949) 104.

[351] J.C. Fisher, Acta Metall. 2 (1954) 9.

[352] H. Suzuki, Sci. Rep. Res. Inst. Tohoku Univ. A 4 (1957) 455.

[353] G. Shoeck, in: J.E. Dorn (Ed.), Mechanical Behavior of Materials at Elevated Temperature, McGraw-Hill, New York, 1961, p. 77.

[354] J. Snoek, Physica. 9 (1942) 862.

[355] S. Takeuchi, A.S. Argon, Acta Metall. 24 (1976) 883.

[356] H. Oikawa, K. Honda, S. Ito, Mater. Sci. Eng. 64 (1984) 237.

[357] L.S. Darken, Trans. Am. Inst. Min. Engrs. 175 (1948) 184.

[358] F.A. Mohamed, Mater. Sci. Eng. 61 (1983) 149.

[359] M.S. Soliman, F.A. Mohamed, Metall. Trans. A 15 (1984) 1893.

[360] B. Liu, P. Eisenlohr, F. Roters, D. Raabe, Acta Mater. 60 (2012) 5380.

[361] W.R. Cannon, O.D. Sherby, Metall. Trans. 1 (1970) 1030.

[362] H.W. King, J. Mater. Sci. 1 (1966) 79.

[363] M.E. Kassner, H.J. McQueen, E. Evangelista, Mater. Sci. Forum. 113-115 (1993) 151.

[364] H. Laks, C.D. Wiseman, O.D. Sherby, J.E. Dorn, J. Appl. Mech. 24 (1957) 207.

[365] O.D. Sherby, T.A. Trozera, J.E. Dorn, Proc. Am. Soc. Test. Mat. 56 (1956) 789.

[366] O.D. Sherby, E.M. Taleff, Mater. Sci. Eng. A 322 (2002) 89.

[367] H. Oikawa, J. Kariya, S. Karashima, Met. Sci. 8 (1974) 106.

[368] K.L. Murty, Scr. Metall. 7 (1973) 899.

[369] K.L. Murty, Philos. Mag. 29 (1974) 429.

[370] A. Orlova, J. Cadek, Z. Metall. 65 (1974) 200.

[371] F.A. Mohamed, Mater. Sci. Eng. 38 (1979) 73.

[372] W. Blum, Mater. Sci. Eng. A 319–321 (2001) 8.

[373] S.S. Vagarali, T.G. Langdon, Acta Metall. 30 (1982) 1157.

[374] B.L. Jones, C.M. Sellars, Met. Sci. J. 4 (1970) 96.

[375] H. Oikawa, M. Maeda, S. Karashima, J. JIM 37 (1973) 599.

[376] N. Matsuno, H. Oikawa, S. Karashima, J. JIM 38 (1974) 1071.

[377] T.R. McNelley, D.J. Michel, A. Salama, Scr. Metall. 23 (1989) 1657.

[378] H.J. McQueen, M.E. Kassner, in: T.R. McNelley, C. Heikkenen (Eds.), Superplasticity in Aerospace II, TMS, Warrendale, PA, 1990, p. 77.

[379] M.S. Mostafa, F.A. Mohamed, Met. Trans. 17A (1986) 365.

[380] E.W. Hart, Acta Metall. 15 (1967) 351.

[381] E.M. Taleff, G.A. Henshall, T.G. Nieh, D.R. Lesuer, J. Wadsworth, J. Metall. Trans. 29 (1998) 1081.

[382] E.M. Taleff, D.R. Lesuer, J. Wadsworth, J. Metall. Trans. 27A (1996) 343.

[383] E.M. Taleff, P.J. Nevland, JOM 51 (1999) 34.

[384] E.M. Taleff, P.J. Nevland, Metall. Trans. 32 (2001) 1119.

[385] H. Watanabe, H. Tsutsui, T. Mukai, M. Kohzu, S. Tanabe, K. Higashi, Int. J. Plast. 17 (2001) 387.

[386] K.L. Murty, F.A. Mohamed, J.E. Dorn, Acta Metall. 20 (1972) 1009.

[387] P.K. Chaudhury, F.A. Mohamed, Mater. Sci. Eng. A 101 (1988) 13.

[388] A. Nortman, H. Neuhäuser, Phys. Status. Solidi. A 168 (1998) 87.

[389] J.J. Blandin, D. Giunchi, M. Suéry, E. Evangelista, Mater. Sci. Technol. 18 (2002) 333.

[390] Y. Li, T.G. Langdon, Metall. Trans. 30 (1999) 2059.

[391] J.C. Tan, M.J. Tan, Mater. Sci. Eng. A 339 (2003) 81.

[392] J. Wadsworth, S.E. Dougherty, P.A. Kramer, T.G. Nieh, Scr. Metall. Mater. 27 (1992) 71.

[393] R.W. Hayes, W.O. Soboyejo, Mater. Sci. Eng. A 319–321 (2001) 827.

[394] J. Mukhopadhyay, G.C. Kaschner, A.K. Mukherjee, in: T.R. McNelley, C. Heikkenen (Eds.), Superplasticity in Aerospace II, TMS, Warrendale, PA, 1990, p. 33.

[395] T.G. Nieh, W.C. Oliver, Scr. Metall. 23 (1989) 851.

[396] L.M. Hsiung, T.G. Nieh, Intermetallics. 7 (1999) 821.

[397] J. Wolfenstine, J. Mater. Sci. Lett. 9 (1990) 1091.

[398] D. Lin, A. Shan, D. Li, Scr. Metall. Mater. 31 (11) (1994) 1455.

[399] R.S. Sundar, T.R.G. Kutty, D.H. Sastry, Intermetallics 8 (2000) 427.

[400] D. Li, A. Shan, Y. Liu, D. Lin, Scr. Metall. Mater. 33 (1995) 681.

[401] A.J. Ardell, Scr. Mater. 69 (2013) 541.

[402] H.S. Yang, W.B. Lee, A.K. Mukherjee, in: R. Darolia, J.J. Lewandowski, C.T. Liu, P.L. Martin, D.B. Miracle, M.V. Nathal (Eds.), Structural Intermetallics, TMS, Warrendale, PA, 1993, p. 69.

[403] Y. Li, T.G. Langdon, Metall. Trans. 30 (1999) 315.

[404] G. González-Doncel, O.D. Sherby, Acta Metall. 41 (1993) 2797.

[405] T.G. Langdon, J. Wadsworth, in: S. Hori, M. Tokizane, N. Furushiro (Eds.), Proc. Int. Conf. on Superplasticity in Advanced Materials, The Japan Society of Research on Superplasticity, Osaka, Japan, 1991, p. 847.

[406] G.D. Bengough, J. Inst. Met. 7 (1912) 123.

[407] O.D. Sherby, T.G. Nieh, J. Wadsworth, Mater. Sci. Forum. 13 (1994) 170.

[408] C.E. Pearson, J. Inst. Met. 54 (1934) 111.

[409] R.Z. Valiev, A.V. Korznikov, R.R. Mulyukov, Mater. Sci. Eng. A 141 (1993) 168.

[410] F. Wakai, S. Sakaguchi, Y. Matsuno, Adv. Ceram. Mater. 1 (1986) 259.

[411] T.G. Langdon, Ceram. Int. 19 (1993) 279.

[412] M.Y. Wu, J. Wadsworth, O.D. Sherby, Scr. Metall. 18 (1984) 773.

[413] V.K. Sikka, C.T. Liu, E.A. Loria, in: F.H. Froes, S.J. Savage (Eds.), Processing of Structural Metals by Rapid Solidification, ASM, Metals Park, Ohio, 1987, p. 417.

[414] O.D. Sherby, J. Wadsworth, Prog. Mater. Sci. 33 (1989) 169.

[415] O.A. Kaibyshev, Superplasticity of Alloys, Intermetallics and Ceramics, Springer-Verlag, Berlin, Germany, 1992.

[416] A.H. Chokshi, A.K. Mukherjee, T.G. Langdon, Mater. Sci. Eng. R 10 (1993) 237.

[417] T.G. Nieh, J. Wadsworth, O.D. Sherby, in: D.R. Clarke, S. Suresh, I.M. Ward (Eds.), Superplasticity in Metals and Ceramics, Cambridge University Press, Cambridge, UK, 1997.

[418] T.G. Nieh, C.A. Henshall, J. Wadsworth, Scr. Metall. 18 (1984) 1405.

[419] K. Higashi, M. Mabuchi, T.G. Langdon, ISIJ Int. 36 (12) (1996) 1423.

[420] R.Z. Valiev, D.A. Salimonenko, N.K. Tsenev, P.B. Berbon, T.G. Langdon, Scr. Mater. 37 (12) (1997) 1945.

[421] S. Komura, P.B. Berbon, M. Furukawa, Z. Horita, M. Nemoto, T.G. Langdon, Scr. Mater. 38 (12) (1998) 1851.

[422] P.B. Berbon, M. Furukawa, Z. Horita, M. Nemoto, N.K. Tsenev, R.Z. Valiev, T.G. Langdon, Philos. Mag. Lett. 78 (4) (1998) 313.

[423] T.G. Langdon, M. Furukawa, Z. Horita, M. Nemoto, J. Met. 50 (6) (1998) 41.

[424] H.E. Cline, T.H. Alden, Trans. TMS-AIME 239 (1967) 710.

[425] S.W. Zehr, W.A. Backofen, Trans. ASM 61 (1968) 300.

[426] T.H. Alden, Trans. ASM 61 (1968) 559.

[427] T.H. Alden, Acta Met. 15 (1967) 469.

[428] D.L. Holt, W.A. Backofen, Trans. ASM 59 (1966) 755.

[429] D. Lee, Acta Met. 17 (8) (1969) 1057.

[430] E.W. Hart, Acta Met. 15 (1967) 1545.

[431] T.H. Alden, Acta Met. 17 (12) (1969) 1435.

[432] T.G. Langdon, Acta Metall. Mater. 42 (7) (1994) 2437.

[433] M.G. Zelin, A.K. Mukherjee, Mater. Sci. Eng. A 208 (1996) 210.

[434] M.F. Ashby, R.A. Verral, Acta Metall. 21 (1973) 149.

[435] J.R. Spingarn, W.D. Nix, Acta Metall. 26 (1978) 1389.

[436] K.A. Padmanabhan, Mater. Sci. Eng. 40 (1979) 285.

[437] T.G. Langdon, Mater. Sci. Eng. 137 (1991) 1.

[438] D.M.R. Taplin, G.L. Dunlop, T.G. Langdon, Annu. Rev. Mater. Sci. 9 (1979) 151.

[439] J. Weertman, Philos. Trans. R. Soc. Lond. A 288 (1978) 9.

[440] O.A. Ruano, J. Wadsworth, O.D. Sherby, Mater. Sci. Eng. 84 (1986) L1.

[441] A. Ball, M.M. Hutchison, Met. Sci. J. 3 (1969) 1.

[442] T.G. Langdon, Philos. Mag. 26 (1970) 945.

[443] A.K. Mukherjee, Mater. Sci. Eng. A 8 (1971) 83.

[444] A. Arieli, A.K. Mukherjee, Mater. Sci. Eng. A 45 (1980) 61.

[445] R.C. Gifkins, Metall. Trans. A 7 (1976) 1225.

[446] H. Fukuyo, H.C. Tsai, T. Oyama, O.D. Sherby, ISIJ Int. 31 (1991) 76.

[447] R.Z. Valiev, T.G. Langdon, Mater. Sci. Eng. A 137 (1993) 949.

[448] T.G. Langdon, Mater. Sci. Eng. A 174 (1994) 225.

[449] M.T. Pérez-Prado, (Ph.D. dissertation), Universidad Complutense de Madrid, 1998.

[450] C.P. Cutler, J.W. Edington, J.S. Kallend, K.N. Melton, Acta Metall. 22 (1974) 665.

[451] K.N. Melton, J.W. Edington, Scr. Metall. 8 (1974) 1141.

[452] R.H. Bricknell, J.W. Edington, Acta Metall. 27 (1979) 1303.

[453] R.H. Bricknell, J.W. Edington, Acta Metall. 27 (1979) 1313.

[454] M.T. Pérez-Prado, M.C. Cristina, O.A. Ruano, G. González-Doncel, Mater. Sci. Eng. 244 (1998) 216.

[455] M.T. Pérez-Prado, T.R. McNelley, O.A. Ruano, G. González-Doncel, Metall. Trans. A 29 (1998) 485.

[456] K. Tsuzaki, H. Matsuyama, M. Nagao, T. Maki, Mater. Trans. JIM 31 (1990) 983.

[457] L. Qing, H. Xiaoxu, Y. Mei, Y. Jinfeng, Acta Mater. 40 (1992) 1753.

[458] J. Liu, D.J. Chakrabarti, Acta Mater. 44 (1996) 4647.

[459] P.L. Blackwell, P.S. Bate, Metall. Trans. A 24 (1993) 1085.

[460] P.S. Bate, Metall. Trans. A 23 (1992) 1467.

[461] P.L. Blackwell, P.S. Bate, in: N. Ridley (Ed.), Superplasticity: 60 years after Pearson, The Institute of Materials, Manchester, UK, 1994, p. 183.

[462] R.H. Johnson, C.M. Parker, L. Anderson, O.D. Sherby, Philos. Mag. 18 (1968) 1309.

[463] N. Naziri, R. Pearce, J. Inst. Met. 98 (1970) 71.

[464] D.S. McDarmaid, A.W. Bowen, P.G. Partridge, Mater. Sci. Eng. A 64 (1984) 105.

[465] O.A. Ruano, O.D. Sherby, Rev. Phys. Appl. 23 (1988) 625.

[466] O.D. Sherby, O.A. Ruano, in: N.E. Paton, C.H. Hamilton (Eds.), Proc. Int. Conf. Superplastic Forming of Structural Alloys, TMS, Warrendale, PA, 1982, p. 241.

[467] B.M. Watts, M.J. Stowell, B.L. Baike, D.G.E. Owen, Met. Sci. J. 10 (1976) 189.

[468] R. Grimes, Advances and Future Directions in Superplastic Materials, NATO-AGARD Lecture Series, No. 168, 1988, p. 8.1.

[469] J.A. Wert, N.E. Paton, C.H. Hamilton, M.W. Mahoney, Metall. Trans. A 12 (1981) 1267.

[470] M.T. Pérez-Prado, M.E. McMahon, T.R. McNelley, in: E.M. Taleff, R.K. Mahidhara (Eds.), Modeling the Mechanical Response of Structural Materials, TMS, Warrendale, PA, 1998, p. 181.

[471] K.N. Melton, J.W. Edington, J.S. Kallend, C.P. Cutler, Acta Metall. 22 (1974) 165.

[472] V. Randle, Microtexture determination and its applications, Inst. Met. (1992) 416 pp.

[473] U.F. Kocks, C.N. Tomé, H.R. Wenk, Texture and Anisotropy, Cambridge Univ. Press, 1998.

[474] W.F. Hosford, The Mechanics of Crystals and Textured Polycrystals, Oxford Univ. Press, 1993.

[475] M. Mabuchi, K. Higashi, JOM 50 (6) (1998) 34.

[476] H. Hosokawa and K. Higashi, Materials Science Research International, vol. 6, (no.3), Fourth International Symposium on Microstructure and Mechanical Properties of Engineering Materials (IMMM'99), Beijing, China, September 20–23, 1999, Soc. Mater. Sci. Japan, eds., 2000, 153.

[477] R.S. Mishra, J. Met. 53 (3) (2001) 23.

[478] M. Mabuchi, J. Koike, H. Iwasaki, K. Higashi, T.G. Langdon, Mater. Sci. Forum. 170–172 (1994) 503.

[479] H.Y. Kim, S.H. Hong, Scr. Metall. Mater. 30 (1994) 297.

[480] X. Huang, Q. Liu, C. Yao, M. Yao, J. Mater. Sci. Lett. 10 (1991) 964.

[481] K. Matsuki, H. Matsumoto, M. Tokizawa, N. Takatsuji, M. Isogai, S. Murakami, Y. Murakami, in: K. Hirano, H. Oikawa, K. Ikeda (Eds.), Science and Engineering of Light Metals, The Japan Institute of Light Metals, Tokyo, 1991.

[482] T.G. Nieh, T. Imai, J. Wadsworth, S. Kojima, Scr. Metall. Mater. 31 (1994) 1685.

[483] M. Mabuchi, T. Imai, J. Mater. Sci. Lett. 9 (1990) 763.

[484] T. Imai, M. Mabuchi, Y. Tozawa, M. Yamada, J. Mater. Sci. Lett. 9 (1990) 255.

[485] M. Mabuchi, K. Higashi, K. Inoue, S. Tanimura, Scr. Metall. Mater. 26 (1992) 1839.

[486] T. Imai, G. L'Esperance, B.D. Hong, Scr. Metall. Mater. 25 (1991) 2503.

[487] T.G. Nieh, J. Wadsworth, Mater. Sci. Eng. 147 (1991) 129.

[488] T.G. Nieh, J. Wadsworth, T. Imai, Scr. Metall. Mater. 26 (1992) 703.

[489] M. Mabuchi, K. Higashi, T.G. Langdon, Acta Metall. Mater. 42 (5) (1994) 1739.

[490] M. Mabuchi, K. Higashi, Philos. Mag. Lett. 70 (1) (1994) 1.

[491] J. Koike, M. Mabuchi, K. Higashi, Acta Metall. Mater. 43 (1) (1994) 199.

[492] M. Mabuchi, K. Higashi, Mater. Trans. JIM 35 (6) (1994) 399.

[493] K. Higashi, T.G. Nieh, M. Mabuchi, J. Wadsworth, Scr. Metall. Mater. 32 (7) (1995) 1079.

[494] R.S. Mishra, A.K. Mukherjee, Scr. Metall. Mater. 25 (1991) 271.

[495] R.S. Mishra, T.R. Bieler, A.K. Mukherjee, Acta Metall. Mater. 43 (3) (1995) 877.

[496] T.R. Bieler, R.S. Mishra, A.K. Mukherjee, JOM 48 (1996) 52.

[497] E. Artz, M.F. Ashby, R.A. Verrall, Acta Metall. 31 (1983) 1977.

[498] R.S. Mishra, T.R. Bieler, A.K. Mukherjee, Acta Mater. 45 (2) (1997) 561.

[499] M. Mabuchi, K. Higashi, Mater. Trans. JIM 36 (3) (1995) 420.

[500] M. Mabuchi, K. Higashi, Scr. Mater. 34 (12) (1996) 1893.

[501] Y. Li, T.G. Langdon, Acta Mater. 46 (11) (1998) 3937.

[502] T.G. Langdon, Mater. Sci. Forum 304–306 (1999) 13.

[503] Y. Li, T.G. Langdon, Acta Mater. 47 (12) (1999) 3395.

[504] M. Mabuchi, K. Higashi, Philos. Mag. A 74 (1996) 887.

[505] K.T. Park, F.A. Mohamed, Metall. Mater. Trans. 26 (1995) 3119.

[506] T.G. Nieh, P.S. Gilman, J. Wadsworth, Scr. Metall. 19 (1985) 1375.

[507] T.R. Bieler, T.G. Nieh, J. Wadsworth, A.K. Mukherjee, Scr. Metall. Mater. 22 (1988) 81.

[508] T.R. Bieler, A.K. Mukherjee, Mater. Sci. Eng. A 128 (1990) 171.

[509] K. Higashi, T. Okada, T. Mukai, S. Tanimura, Mater. Sci. Eng. 159 (1992) L1.

[510] K. Higashi, Mater. Sci. Eng. 166 (1993) 109.

[511] J.K. Gregory, J.C. Gibeling, W.D. Nix, Metall. Trans. A 16 (1985) 777.

[512] K. Higashi, T.G. Nieh, J. Wadsworth, Acta Metall. Mater. 43 (9) (1995) 3275.

[513] H. Watanabe, T. Mukai, M. Mabuchi, K. Higashi, Scr. Metall. 41 (2) (1999) 209.

[514] I.C. Hsiao, J.C. Huang, Scr. Mater. 40 (6) (1999) 697.

[515] H. Watanabe, T. Mukai, K. Ishikawa, M. Mabuchi, K. Higashi, Mater. Sci. Eng. 307 (2001) 119.

[516] A.V. Sergueeva, V.V. Stolyarov, R.Z. Valiev, A.K. Mukherjee, Scr. Mater. 43 (2000) 819.

[517] Z. Horita, M. Furukawa, M. Nemoto, A.J. Barnes, T.G. Langdon, Acta Mater. 48 (2000) 3633.

[518] S.S. Bhattacharya, U. Betz, H. Hahn, Scr. Mater. 44 (2001) 1553.

[519] S. Komura, Z. Horita, M. Furukawa, M. Nemoto, T.G. Langdon, Metall. Trans. A 32 (2001) 707.

[520] F.A. Mohamed, Y. Li, Mater. Sci. Eng. A 298 (2001) 1.

[521] R.S. Mishra, R.Z. Valiev, S.X. McFadden, R.K. Islamgaliev, A.K. Mukherjee, Mater. Sci. Eng. 81 (2001) 37.

[522] S.X. McFadden, R.S. Mishra, R.Z. Valiev, A.P. Zhilyaev, A.K. Mukherjee, Nature 398 (1999) 684.

[523] R.S. Mishra, S.X. McFadden, A.K. Mukherjee, Mater. Sci. Forum. 304–306 (1999) 31.

[524] R.K. Islamgaliev, R.Z. Valiev, R.S. Mishra, A.K. Mukherjee, Mater. Sci. Eng. 304–306 (2001) 206.

[525] R. Bohn, T. Klassen, R. Bormann, Intermetallics. 9 (2001) 559.

[526] R.S. Mishra, A.K. Mukherjee, in: A.K. Ghosh, T.R. Bieler (Eds.), Int. Conf. on Superplasticity and Superplastic Forming, The Minerals, Metals & Materials Society, 1998, p. 109.

[527] R.Z. Valiev, R.M. Gayanov, H.S. Yang, A.K. Mukherjee, Scr. Metall. Mater. 25 (1991) 1945.

[528] R.Z. Valiev, C. Song, S.X. McFadden, A.K. Mukherjee, R.S. Mishra, Philos. Mag. A 81 (1) (2001) 25.

[529] V. Yamakov, D. Wolf, M. Salazar, S.R. Phillpot, H. Gleiter, Acta Mater. 49 (2001) 2713.

[530] D. Turnbull, J.C. Fisher, J. Chem. Phys. 17 (1949) 71.

[531] J.W. Christian, The Theory of Transformations in Metals and Alloys, second ed., Pergamon Press, Oxford, 1975.

[532] F.J. Humphreys, M. Hatherly, Recrystallization and Related Annealing Phenomena, Pergamon Press, Oxford, 1995.

[533] R.W. Cahn, J. Inst. Met. 76 (1949) 121.

[534] J.J. Jonas, Mater. Sci. Eng. A184 (1994) 155.

[535] T. Sakai, J.J. Jonas, Acta Metall. 32 (1984) 189.

[536] M.-T. Pérez-Prado, S. Barrabes, M.E. Kassner, E. Evangelista, Acta Mater. 53 (2005) 581.

[537] E. Hornbogen, U. Koster, in: F. Haessner, Riederer (Eds.), Recrystallization of Metallic Materials, Verlag, Berlin, 1978, p. 159.

[538] W. Blum, B. Reppich, in: B. Wilshire, R.W. Evans (Eds.), Creep Behavior of Crystalline Solids, Pineridge Press, Swansea, UK, 1985, p. 83.

[539] B. Reppich, in: R.W. Cahn, P. Haasen, E.J. Kramer (Eds.), Materials Science and Technology, vol. 6, VCH, Weinheim, 1993, p. 312.

[540] B. Reppich, Z. Metall. 93 (2002) 605.

[541] E. Arzt, Res Mech. 31 (1991) 399.

[542] E. Arzt, G. Dehm, P. Gumbsch, O. Kraft, D. Weiss, Prog. Mater. Sci. 46 (2001) 283.

[543] G.S. Ansell, in: G.S. Ansell, T.D. Cooper, F.V. Lenel (Eds.), Oxide Dispersion Strengthening, Gordon and Breach, New York, 1968, p. 61.

[544] M.F. Ashby, Strength of Metals and Alloys, ASM, Materials Park, OH, 1970, p. 507.

[545] L.M. Brown, Fatigue and creep of composite materials, in: H. Linholt, R. Talreja (Eds.), Proc. 3rd. Riso Int. Symp. on Metallurgy and Materials Science, 1982, p. 1.

[546] J.B. Bilde-Sorensen, Deformation of multi-phase and particle containing materials, in: J.B. Bilde-Sorensen (Ed.), 4th Riso Int. Symp. on Metallurgy and Materials Science, 1983, p. 1.

[547] C.M. Sellars, R. Petkovic-Luton, Mater. Sci. Eng. 46 (1980) 75.

[548] J. Lin, O.D. Sherby, Res Mech. 2 (1981) 251.

[549] J.W. Martin, Michromechanics in Particle-Hardened Alloys, Cambridge, 1980.

[550] M.F. Ashby, Acta Metall. 14 (1966) 679.

[551] Q. Zhu, W. Blum, H.J. McQueen, Mater. Sci. Forum 217–222 (1996) 1169.

[552] R.W. Lund, W.D. Nix, Acta Metall. 24 (1976) 469.

[553] G.M. Pharr, W.D. Nix, Scr. Metall. 10 (1976) 1007.

[554] G.S. Ansell, J. Weertman, Trans. AIME 215 (1959) 838.

[555] E. Arzt, M.F. Ashby, Scr. Metall. 16 (1982) 1282.

[556] L.M. Brown, R.K. Ham, in: A. Kelly, R.B. Nicholson (Eds.), Strengthening Methods in Crystals, Applied Science, London, 1971, p. 9.

[557] E. Arzt, D.S. Wilkinson, Acta Metall. 34 (1986) 1893.

[558] J.H. Hausselt, W.D. Nix, Acta Metall. 25 (1977) 1491.

[559] R.A. Stevens, P.E.J. Flewett, Acta Metall. 29 (1981) 867.

[560] H.E. Evans, G. Knowles, Met. Sci. 14 (1980) 262.

[561] D.J. Bacon, U.F. Kocks, R.O. Scattergood, Philos. Mag. 8 (1973) 1241.

[562] R. Lagneborg, Scr. Metall. 7 (1973) 605.

[563] V.C. Nardone, J.K. Tien, Scr. Metall. 17 (1983) 467.

[564] J.H. Schroder, E. Arzt, Scr. Metall. 19 (1985) 1129.

[565] B. Reppich, Acta Mater. 46 (1997) 61.

[566] D. Srolovitz, M.J. Luton, R. Petkovic-Luton, D.M. Barnett, W.D. Nix, Acta Metall. 32 (1984) 1079.

[567] R. Behr, J. Mayer, E. Arzt, Intermetallics 17 (1999) 423.

[568] J. Rosler, E. Arzt, Acta Metall. 36 (1988) 1043.

[569] E. Arzt, J. Rosler, Acta Metall. 36 (1988) 1053.

[570] J. Rosler, E. Arzt, Acta Metall. 38 (1990) 671.

[571] E. Arzt, E. Gohring, Acta Metall. 46 (1998) 6584.

[572] B. Reppich, F. Brungs, G. Hummer, H. Schmidt, in: B. Wilshire, R.W. Evans (Eds.), Creep and Fracture of Eng. Mater. and Structures, Inst. of Metals, London, 1990, p. 141.

[573] O. Ajaja, T.E. Towson, S. Purushothaman, J.K. Tien, Mater. Sci. Eng. 44 (1980) 165.

[574] B. Reppich, H. Bugler, R. Leistner, M. Schutze, in: B. Wilshire, D.R.J. Owen (Eds.), Creep and Fracture of Engineering Materials and Structures, Pineridge, 1984, p. 279.

[575] R. Lagneborg, B. Bergman, Met. Sci. 10 (1976) 20.

[576] M. Heilmaier, B. Reppich, in: R.S. Mishra, A.K. Mukherjee, K. Linga Murty (Eds.), Creep Behavior of Advanced Materials for the 21st Century, TMS, Warrendale, 1999, p. 267.

[577] E. Arzt, J. Rosler, in: Y.-W. Kim, W.M. Griffith (Eds.), Dispersion Strengthened Aluminum Alloys, TMS, Warrendale, 1988, p. 31.

[578] K. Kucharová, S.J. Zhu, J. Čadek, Mater. Sci. Eng. A. 348 (1–2) (2003) 170–179.

[579] J. Cadek, H. Oikawa, V. Sustek, Mater. Sci. Eng. A190 (1995) 9.
[580] A.H. Clauer, N. Hansen, Acta Metall. 32 (1984) 269.
[581] R. Timmins, E. Arzt, Scr. Metall. 22 (1988) 1353.
[582] S. Spigarelli, Mater. Sci. Eng. A337 (2002) 306.
[583] J.J. Stephens, W.D. Nix, Metall. Trans. 16A (1985) 1307.
[584] D.C. Dunand, A.M. Jansen, Acta Mater. 45 (1997) 4569.
[585] A.M. Jansen, D.C. Dunand, Acta Mater. 45 (1997) 4583.
[586] W. Blum, P.D. Portella, in: J.B. Bilde-Sorenson, et al. (Eds.), Deformation of Multi-phase and Particle Containing Materials, Riso National Lab., Roskilde, 1983, p. 493.
[587] G. Sauthoff, J. Peterseim, Steel Res. 57 (1986) 19.
[588] R.F. Singer, E. Arzt, in: W. Betz, et al. (Eds.), High Temperature Alloys for Gas Turbines and Other Applications, 1986, p. 97.
[589] T.G. Langdon, J. Mater. Sci. 44 (2009) 5998.
[590] D.N. Seidman, E.A. Marquis, D.C. Dunand, Acta Mater. 50 (2002) 4021.
[591] E.A. Marquis, D.C. Dunand, Scr. Mater. 47 (2002) 503.
[592] G. Sauthoff, Intermetallics, VCH, New York, 1995.
[593] P. Sadrabadi, P. Eisenlohr, G. Wehrhan, J. Stablein, L. Parthier, W. Blum, Mater. Sci. Eng. A 510-1 (2009) 46.
[594] N.S. Stoloff, V.K. Sikka (Eds.), Physical Metallurgy and Processing of Intermetallic Compounds, Chapman & Hall, New York, 1996.
[595] Y.S. Choi, D.M. Dimiduk, M.D. Uchic, T.A. Parthasarathy, Philos. Mag. 87 (2007) 939.
[596] R.W. Cahn, Mater. Sci. Eng. A 324 (2002) 1.
[597] O. Izumi, Mater. Trans. JIM 30 (1989) 627.
[598] M. Yamaguchi, Y. Umakoshi, Prog. Mater. Sci. 34 (1990) 1.
[599] S.C. Deevi, V.K. Sikka, C.T. Liu, Prog. Mater. Sci. 42 (1997) 177.
[600] T.M. Pollock, R.D. Field, in: F.R.N. Nabarro, M.S. Duesbery (Eds.), Dislocations in Solids, vol. 11, Elsevier, 2002, p. 546 (Chapter 63).
[601] P. Veyssière, Mater. Sci. Eng. A 309–310 (2001) 44.
[602] M. Yamaguchi, H. Inui, K. Ito, Acta Mater. 48 (2000) 307.
[603] H. Oikawa, T.G. Langdon, in: B. Wilshire, R.W. Evans (Eds.), Creep Behavior of Crystalline Solids, Pineridge, Swansea, UK, 1985, p. 33.
[604] W. Blum, J. Dvorak, P. Kral, P. Eisenlohr, V. Sklenicka, Mater. Sci. Eng. A 590 (2014) 423.
[605] J.H. Schneibel, P.M. Hazzledine, Appl. Sci. 213 (1992) 565.
[606] R.E. Smallman, T.S. Rong, I.P. Jones, The Johannes Weertman Symposium, TMS, Warrendale, PA, 1966, p. 11.
[607] G. Sauthoff, Structural Intermetallics, TMS, Warrendale, PA, 1993, p. 845.
[608] G. Sauthoff, in: B. Fultz, R.W. Chan, D. Gupta (Eds.), Diffusion in Ordered Alloys, TMS, Warrendale, PA, 1993, p. 205.
[609] H. Oikawa, The Processing, Properties and Applications of Metallic and Ceramic Materials, Warley, UK, 1992, p. 383.
[610] J. Kumpfert, Adv. Eng. Mater. 3 (2001) 851.
[611] R.S. Mishra, T.K. Nandy, P.K. Sagar, A.K. Gogia, D. Banerjee, Trans. India Inst. Met. 49 (1996) 331.
[612] N.S. Stoloff, D.A. Alven, C.G. McKamey, Nickel and Iron Aluminides: Processing, Properties, and Applications, ASM International, Materials Park, OH, 1997, p. 65.

[613] Y.W. Kim, J. Met. 46 (1994) 31.

[614] Y.W. Kim, J. Met. 41 (1989) 24.

[615] J. Beddoes, W. Wallace, L. Zhao, Int. Mater. Rev. 40 (1995) 197.

[616] W.J. Zhang, S.C. Deevi, in: K.J. Hemker, D.M. Dimiduk, H. Clemens, R. Darolia, H. Inui, J.M. Larsen, V.K. Sikka, M. Thomas, J.D. Whittenberger (Eds.), Structural Intermetallics 2001, TMS, Warrendale, PA, 2001, p. 699.

[617] E.L. Hall, S.C. Huang, Acta Mater. 38 (1990) 539.

[618] S.C. Huang, E.L. Hall, Metall. Trans. A 22 (1991) 427.

[619] J. Beddoes, L. Zhao, P. Au, D. Dudzinsky, J. Triantafillou, in: M.V. Nathal, R. Darolia, C.T. Liu, P.L. Martin, D.B. Miracle, R. Wagner, M. Yamaguchi (Eds.), Structural Intermetallics 1997, TMS, Warrendale, PA, 1997, p. 109.

[620] M. Es-Souni, A. Bartels, R. Wagner, Mater. Sci. Eng. A 192/193 (1995) 698.

[621] D.B. Worth, J.W. Jones, J.E. Allison, Metall. Trans. A 26 (1995) 2947.

[622] W.J. Zhang, S.C. Deevi, Intermetallics 10 (2002) 603.

[623] W.J. Zhang, S.C. Deevi, Intermetallics 11 (2003) 177.

[624] D.Y. Seo, H. Saari, J. Beddoes, L. Zhao, in: K.J. Hemker, D.M. Dimiduk, H. Clemens, R. Darolia, H. Inui, J.M. Larsen, V.K. Sikka, M. Thomas, J.D. Whittenberger (Eds.), Structural Intermetallics 2001, TMS, Warrendale, PA, 2001, p. 653.

[625] M. Weller, A. Chatterjee, G. Haneczok, A. Wanner, F. Appel, H. Clemens, in: K.J. Hemker, D.M. Dimiduk, H. Clemens, R. Darolia, H. Inui, J.M. Larsen, V.K. Sikka, M. Thomas, J.D. Whittenberger (Eds.), Structural Intermetallics 2001, TMS, Warrendale, PA, 2001, p. 465.

[626] J. Wolfenstine, G. González-Doncel, Mater. Lett. 18 (1994) 286.

[627] S. Kroll, H. Mehrer, N. Stolwijk, C. Herzig, R. Rosenkranz, G. Frommeyer, Z. Metall. 83 (1992) 591.

[628] M. Lu, K. Hemker, Acta Mater. 45 (1997) 3573.

[629] M. Es-Souni, A. Bartels, R. Wagner, Acta Metall. Mater. 43 (1995) 153.

[630] G.B. Viswanathan, S. Kartikeyan, M.J. Mills, V.K. Vasudevan, Mater. Sci. Eng. A 319–321 (2001) 833.

[631] G.B. Viswanathan, V.K. Vasudevan, M.J. Mills, Acta Mater. 47 (1999) 1399.

[632] Y. Ishikawa, H. Oikawa, Mater. Trans. JIM 35 (1994) 336.

[633] P.B. Hirsch, D.H. Warrington, Philos. Mag. A 6 (1961) 715.

[634] J. Friedel, Philos. Mag. A 46 (1956) 1169.

[635] J.N. Wang, T.G. Nieh, Acta Mater. 46 (1998) 1887.

[636] P.I. Gouma, K. Subramanian, Y.W. Kim, M.J. Mills, Intermetallics 6 (1998) 689.

[637] W.R. Chen, L. Zhao, J. Beddoes, Scr. Mater. 41 (1999) 597.

[638] D.Y. Seo, J. Beddoes, L. Zhao, G.A. Botton, Mater. Sci. Eng. A 329–331 (2002) 810.

[639] W.R. Chen, J. Beddoes, L. Zhao, Mater. Sci. Eng. A 323 (2002) 306.

[640] W.R. Chen, J. Triantafillou, J. Beddoes, L. Zhao, Intermetallics 7 (1999) 171.

[641] T.A. Parthasarathy, P.R. Subramanian, M.G. Mendiratta, D.M. Dimiduk, Acta Mater. 48 (2000) 541.

[642] T.R. Bieler, D.Y. Seo, T.R. Everard, P.A. McQuay, in: R.S. Mishra, A.K. Mukherjee, K.L. Murty (Eds.), Creep Behavior of Materials for the 21st Century, TMS, Warrendale, PA, 1999, p. 181.

[643] J. Beddoes, D.Y. Seo, W.R. Chen, L. Zhao, Intermetallics 9 (2001) 915.

[644] C.T. Liu, J. Stringer, J.N. Mundy, L.L. Horton, P. Angelini, Intermetallics 5 (1997) 579.

[645] I. Baker, P.R. Munroe, Int. Mater. Rev. 42 (1997) 181.

[646] N.S. Stoloff, Mater. Sci. Eng. A 258 (1998) 1.

[647] T.B. Massalski, Binary Alloy Phase Diagrams, ASM, Metals Park, OH, 1986 p.112.

[648] D.G. Morris, M.A. Morris, Mater. Sci. Eng. A 239–240 (1997) 23.

[649] E.P. George, I. Baker, Intermetallics 6 (1998) 759.

[650] D.G. Morris, C.T. Liu, E.P. George, Intermetallics 7 (1999) 1059.

[651] D.G. Morris, P. Zhao, M.A. Morris-Muñoz, Mater. Sci. Eng. A 297 (2001) 256.

[652] J.W. Park, Intermetallics 10 (2002) 683.

[653] I. Baker, H. Xiao, O. Klein, C. Nelson, J.D. Whittenberger, Acta Metall. Mater. 30 (1995) 863.

[654] N.S. Stoloff, R.G. Davies, Prog. Mater. Sci. 13 (1966) 1.

[655] Y. Umakoshi, M. Yamaguchi, Y. Namba, M. Murakami, Acta Metall. 24 (1976) 89.

[656] S. Hanada, S. Watanabe, T. Sato, O. Izumi, Scr. Metall. 15 (1981) 1345.

[657] W. Schroer, C. Hartig, H. Mecking, Z. Metall. 84 (1993) 294.

[658] K. Yoshimi, S. Hanada, M.H. Yoo, Acta Metall. Mater. 44 (1995) 4141.

[659] D.G. Morris, M.A. Morris, Intermetallics 5 (1997) 245.

[660] D.G. Morris, in: C.T. Liu, R.W. Cahn, G. Sauthoff (Eds.), Ordered Intermetallics-Physical Metallurgy and Mechanical Behavior, Kluwer, Dordrecht, 1992, p. 123.

[661] D. Paris, P. Lesbats, J. Nucl. Mater. 69/70 (1978) 628.

[662] R. Wurschum, C. Grupp, H. Schaefer, Phys. Rev. Lett. 75 (1995) 97.

[663] J.L. Jordan, S.C. Deevi, Intermetallics 11 (2003) 507.

[664] R. Carleton, E.P. George, R.H. Zee, Intermetallics 3 (1995) 433.

[665] E.P. George, I. Baker, Philos. Mag. 77 (1998) 737.

[666] J. Pesicka, G. Schmitz, Intermetallics 10 (2002) 717.

[667] F. Stein, A. Schneider, G. Frommeyer, Intermetallics 11 (2003) 71.

[668] N. Ziegler, Trans. AIME 100 (1932) 267.

[669] R.G. Davies, Trans. AIME 227 (1963) 22.

[670] C.G. McKamey, P.J. Masiasz, J.W. Jones, J. Mater. Res. 7 (1992) 2089.

[671] V.K. Sikka, B.G. Gieseke, R.H. Baldwin, in: K. Natesan, D.J. Tillack (Eds.), Heat Resistant Materials, ASM, Materials Park, OH, 1991, p. 363.

[672] J. Phillips, G. Eggeler, B. Ilschner, E. Batawi, Scr. Mater. 36 (1997) 693.

[673] S.C. Deevi, R.W. Swindeman, Mater. Sci. Eng. 258 (1998) 203.

[674] B. Voyzelle, J.D. Boyd, Mater. Sci. Eng. A 258 (1998) 243.

[675] M.A. Morris-Muñoz, Intermetallics 7 (1999) 653.

[676] D.H. Sastry, Y.V.R.K. Prasad, S.C. Deevi, Mater. Sci. Eng. A 299 (2001) 157.

[677] P. Málek, O. Kratochvíl, J. Pešička, P. Hanus, I. Šedivá, Intermetallics 10 (2002) 895.

[678] J.D. Whittenberger, Mater. Sci. Eng. A 77 (1986) 103.

[679] J.D. Whittenberger, Mater. Sci. Eng. A 57 (1983) 77.

[680] D.G. Morris, Intermetallics 6 (1998) 753.

[681] D. Lin, D. Li, Y. Liu, Intermetallics 6 (1998) 243.

[682] J.P. Chu, J.H. Wu, H.Y. Yasuda, Y. Umakoshi, K. Inoue, Intermetallics 8 (2000) 39.

[683] J.P. Chu, W. Kai, H.Y. Yasuda, Y. Umakoshi, K. Inoue, Mater. Sci. Eng. A 329–331 (2002) 878.

[684] D. Lin, Y. Liu, Mater. Sci. Eng. A 329–331 (2002) 863.

[685] C. García-Oca, (Ph.D. dissertation), Universidad Complutense de Madrid-CENIM, Madrid, Spain, 2003.

[686] M.A. Muñoz-Morris, C. Garcia-Oca, D.G. Morris, Acta Mater. 50 (2002) 2825.

[687] D.V. Kolluru, R.G. Baligidad, Mater. Sci. Eng. A 328 (2002) 58.

[688] R.G. Baligidad, A. Radhakrishna, U. Prakash, Mater. Sci. Eng. A 257 (1998) 235.

[689] R.S. Sundar, S.C. Deevi, Mater. Sci. Eng. A 357 (2003) 124.

[690] K. Aoki, O. Izumi, J. Jpn. Inst. Met. 43 (1979) 1190.

[691] K.J. Hemker, M.J. Mills, W.D. Nix, Acta Metall. Mater. 39 (1991) 1901.

[692] W.H. Zhu, D. Fort, I.P. Jones, R.E. Smallman, Acta Mater. 46 (1998) 3873.

[693] C. Knobloch, K. Glock, U. Glatzel, High-temperature ordered intermetallic alloys VIII, MRS 556 (1999) 1.

[694] S. Miura, Z.-L. Peng, Y. Mishima, High-temperature ordered intermetallic alloys VII, MRS 460 (1997) 431.

[695] D.H. Shah, Scr. Metall. 17 (1983) 997.

[696] K.J. Hemker, W.D. Nix, Metall. Trans. A 24 (1993) 335.

[697] M.V. Nathal, J.O. Diaz, R.V. Miner, High temperature ordered intermetallic alloys III, MRS 133 (1989) 269.

[698] J. Wolfenstine, H.K. Kim, J.C. Earthman, Mater. Sci. Eng. A 192/193 (1994) 811.

[699] R.K. Ham, R.H. Cook, G.R. Purdy, G. Willoughby, Met. Sci. J. 6 (1972) 205.

[700] D.L. Anton, D.D. Pearson, D.B. Snow, High temperature ordered intermetallic alloys II, MRS 81 (1987) 287.

[701] M.D. Uchic, D.C. Chrzan, W.D. Nix, Intermetallics 9 (2001) 963.

[702] T.S. Rong, I.P. Jones, R.E. Smallman, Acta Mater. 45 (1997) 2139.

[703] S. Miura, T. Hayashi, M. Takekawa, Y. Mishima, T. Suzuki, High-temperature ordered intermetallic alloys IV, MRS 213 (1991) 623.

[704] S.E. Hsu, T.S. Lee, C.C. Yang, C.Y. Wang, C.H. Hong, Appl. Sci. 213 (1992) 597.

[705] S.E. Hsu, C.H. Tong, T.S. Lee, T.S. Liu, High-temperature ordered intermetallic alloys III, MRS 133 (1989) 275.

[706] K. Glock, C. Knobloch, U. Glatzel, Metall. Mater. Trans. A 31 (2000) 1733.

[707] J.H. Schneibel, J.A. Horton, J. Mater. Res. 3 (1988) 651.

[708] T.S. Rong, I.P. Jones, R.R. Smallman, Acta Metall. Mater. 43 (1995) 1385.

[709] R.E. Smallman, T.S. Rong, S.C.D. Lee, I.P. Jones, Mater. Sci. Eng. A 329–331 (2002) 852.

[710] J. Wolfenstine, H.K. Kim, J.C. Earthman, Scr. Metall. Mater. 26 (1992) 1823.

[711] P.A. Flinn, Trans. AIME 218 (1960) 145.

[712] J.R. Nicholls, R.D. Rawlings, J. Mater. Sci. 12 (1977) 2456.

[713] J.H. Schneibel, G.F. Petersen, C.T. Liu, J. Mater. Res. 1 (1986) 68.

[714] T.S. Rong, I.P. Jones, R.E. Smallman, Scr. Metall. Mater. 30 (1994) 19.

[715] T. Hayashi, T. Shinoda, Y. Mishima, T. Suzuki, High-temperature ordered inter-metallic alloys IV, MRS 213 (1991) 617.

[716] J.H. Schneibel, W.D. Porter, J. Mater. Res. 3 (1988) 403.

[717] H.S. Yang, P. Jin, A.K. Mukhejee, Mat. Trans. JIM 33 (1992) 38.

[718] M. Nemoto, H. Takesue, Z. Horita, Mater. Sci. Eng. A 234–236 (1997) 327.

[719] Y. Zhang, D.L. Lin, High-temperature ordered intermetallic alloys V, MRS 288 (1993) 611.

[720] R.D. Rawlings, A.E. Staton-Bevan, J. Mater. Sci. 10 (1975) 505.

[721] M. Cabibbo, E. Evangelista, M.E. Kassner, M.A. Meyers, W. Blum, Metall. Mater. Trans. 39 (2008) 181.

[722] P.H. Thornton, R.G. Davis, T.L. Johnston, Metall. Trans. 1 (1970) 207.

[723] M.J. Lunt, Y.Q. Sun, Mater. Sci. Eng. A 239–240 (1997) 445.

[724] P.M. Hazzeldine, J.H. Schneibel, Scr. Metall. 23 (1989) 1887.

[725] J. Lapin, Intermetallics 7 (1999) 599.

[726] F. Carreño, J.A. Jiménez, O.A. Ruano, Mater. Sci. Eng. A 278 (2000) 272.

[727] U.E. Klotz, R.P. Mason, E. Gohring, E. Artz, Mater. Sci. Eng. A 231 (1997) 198.

[728] R.P. Mason, N.J. Grant, Mater. Sci. Eng. A 192/193 (1995) 741.

[729] R.P. Mason, N.J. Grant, High-temperature ordered intermetallic alloys VI, MRS 364 (1995) 861.

[730] P. Veyssiere, D.L. Guan, J. Rabier, Philos. Mag. A 49 (1984) 45.

[731] N.S. Stoloff, Int. Mater. Rev. 34 (1989) 153.

[732] S.V. Raj, Mater. Sci. Eng. A 356 (2003) 283.

[733] C. Knobloch, V.N. Toloraia, U. Glatzel, Scr. Mater. 37 (1997) 1491.

[734] T. Link, C. Knobloch, U. Glatzel, Scr. Mater. 40 (1999) 85.

[735] Z.L. Peng, S. Miura, Y. Mishima, Mater. Trans. JIM 38 (1997) 653.

[736] P. Caron, T. Khan, P. Veyssiere, Philos. Mag. A 60 (1989) 267.

[737] H. Brehm, U. Glatzel, Int. J. Plast. 15 (1998) 285.

[738] S.E. Hsu, N.N. Hsu, C.H. Tong, C.Y. Ma, S.Y. Lee, High temperature ordered intermetallic alloys II, MRS 81 (1987), p. 507.

[739] A. Fujita, T. Matsmoto, M. Nakamura, Y. Takeda, High-temperature ordered intermetallic alloys III, MRS 133 (1989), p. 573.

[740] D.M. Shah, D.N. Duhl, High temperature ordered intermetallic alloys II, MRS 81 (1987), p. 411.

[741] M. Nazmy, M. Staubli, Scr. Metall. Mater. 25 (1991) 1305.

[742] T. Khan, P. Caron, S. Naka, in: S.H. Whang, C.T. Liu, D.P. Pope, J.O. Stiegler (Eds.), High Temperature Aluminides and Intermetallics, TMS, Warrendale, PA, 1990, p. 219.

[743] D.P. Pope, S.S. Ezz, Int. Met. Rev. 29 (1984) 136.

[744] J.D. Whittenberger, M.V. Nathal, P.O. Book, Scr. Metall. 28 (1993) 53.

[745] D.B. Miracle, Acta Metall. Mater. 41 (1993) 649.

[746] R.D. Noebe, R.R. Bowman, M.V. Nathal, Int. Mater. Rev. 38 (1993) 193.

[747] J.D. Whittenberger, S.K. Mannan, K.S. Kumar, Scr. Metall. 23 (1989) 2055.

[748] J.D. Whittenberger, L.J. Westfall, M.V. Nathal, Scr. Metall. 23 (1989) 2127.

[749] J.D. Whittenberger, R. Reviere, R.D. Noebe, B.F. Oliver, Scr. Metall. Mater. 26 (1992) 987.

[750] J.D. Whittenberger, R. Ray, S.C. Jha, S. Draper, Mater. Sci. Eng. A 138 (1991) 83.

[751] E. Artz, P. Grahle, Acta Mater. 46 (1998) 2717.

[752] K. Xu, R.J. Arsenault, Acta Mater. 47 (1999) 3023.

[753] J.D. Whittenberger, R. Ray, S.C. Jha, Mater. Sci. Eng. A 151 (1992) 137.

[754] J.D. Whittenberger, E. Artz, M.J. Luton, Scr. Metall. Mater. 26 (1992) 1925.

[755] J.D. Whittenberger, R.D. Noebe, D.R. Johnson, B.F. Oliver, Intermetallics 5 (1997) 173.

[756] W. Yang, R.A. Dodd, P.R. Strutt, Metall. Trans. 3 (1972) 2049.

[757] W.J. Yang, R.A. Dodd, Met. Sci. J. 7 (1973) 41.

[758] P.R. Strutt, R.S. Polvani, B.H. Kear, Scr. Metall. 7 (1973) 949.

[759] A. Prakash, R.A. Dodd, J. Mater. Sci. 16 (1981) 2495.

[760] M. Rudy, G. Sauthoff, Mater. Sci. Eng. A 81 (1986) 525.

[761] J.D. Whittenberger, J. Mater. Sci. 23 (1988) 235.

[762] J.D. Whittenberger, J. Mater. Sci. 22 (1987) 394.

[763] J.D. Whittenberger, R.K. Viswanadham, S.K. Mannan, B. Sprissler, J. Mater. Sci. 25 (1990) 35.

[764] M. Rudy, G. Sauthoff, High temperature ordered intermetallic alloys, MRS 39 (1985), p. 327.

[765] R.R. Vandervoort, A.K. Mukherjee, J.E. Dorn, Trans. ASM 59 (1966) 930.

[766] L.A. Hocking, P.R. Strutt, R.A. Dodd, J. Inst. Met. 99 (1971) 98.

[767] J. Bevk, R.A. Dodd, P.R. Strutt, Metall. Trans. 4 (1973) 159.

[768] K.R. Forbes, U. Glatzel, R. Darolia, W.D. Nix, High temperature ordered inter-metallic alloys V, MRS 288 (1993), p. 45.

[769] S.V. Raj, S.C. Farmer, High temperature ordered intermetallic alloys V, MRS 288 (1993), p. 647.

[770] I. Jung, M. Rudy, G. Sauthoff, High temperature ordered intermetallic alloys, MRS 81 (1987), p. 263.

[771] J.D. Whittenberger, R.D. Noebe, C.L. Cullers, K.S. Kumar, S.K. Mannan, Metall. Trans. A 22 (1991) 1595.

[772] J.S. Waddington, K.J. Lofthouse, J. Nucl. Mater. 22 (1967) 205.

[773] J.O. Stiegler, K. Farrell, B.T.M. Loh, H.E. McCoy, ASM Trans. Q. 60 (1967) 494.

[774] T.H. Courtney, Mechanical Behavior of Materials, McGraw-Hill, New York, 1990.

[775] I.W. Chen, A.S. Argon, Acta Metall. 29 (1981) 1321.

[776] A.C.F. Cocks, M.F. Ashby, Prog. Mater. Sci. 27 (1982) 189.

[777] W.D. Nix, Mater. Sci. Eng. A103 (1988) 103.

[778] A. Needleman, J.R. Rice, Acta Metall. 28 (1980) 1315.

[779] W. Beere, Scr. Metall. 17 (1983) 13.

[780] I.-W. Chen, Scr. Metall. 17 (1983) 17.

[781] A.S. Argon, Scr. Metall. 17 (1983) 5.

[782] T.Q. Phan, I.F. Lee, L.E. Levine, J.Z. Tischler, Y. Huang, A.G. Fox, T.G Langdon, M.E. Kassner, Scr. Mater., in press.

[783] J.Y. Zhang, Z.D. Sha, P.S. Branicio, Y.W. Zhang, V. Sorkin, Q.X. Pei, D.J. Srolovitz, Scr. Mater. 69 (2013) 525.

[784] B.F. Dyson, Scr. Metall. 17 (1983) 31.

[785] H. Riedel, Fracture at High Temperatures, Springer-Verlag, Berlin, 1987.

[786] J. Mackerle, Int. J. Pressure Vessels Piping. 77 (2000) 53.

[787] P. Feltham, J.D. Meakin, Acta Metall. 7 (1959) 614.

[788] D.C. Dunand, B.Q. Han, A.M. Jansen, Metall. Trans. 30A (1999) 829.

[789] F. Dobes, K. Miliska, Met. Sci. 10 (1976) 382.

[790] E. Molinie, R. Piques, A. Pineau, Fat. Fract. Eng. Mater. Struct. 14 (1991) 531.

[791] F.R. Larson, J. Miller, Trans. ASME 74 (1952) 765.

[792] K.L. Murty, Y. Zhou, B. Davarajan, in: R.S. Mishra, J.C. Earthman, S.V. Raj (Eds.), Creep Deformation: Fundamentals and Applications, TMS, Warrendale, PA, 2002, p. 3.

[793] A. Ayensu, T.G. Langdon, Metall. Trans. 27A (1996) 901.

[794] R.T. Chen, J.R. Weertman, Mater. Sci. Eng. 64 (1984) 15.

[795] M. Arai, T. Ogata, A. Nitta, JSME Int. J. 39 (1996) 382.

[796] H. Hosokawa, H. Iwasaki, T. Mori, M. Mabuchi, T. Tagata, K. Higashi, Acta Mater. 47 (1999) 1859.

[797] L.C. Lim, H.H. Lu, Scr. Metall. Mater. 31 (1994) 723.

[798] P. Yavari, T.G. Langdon, J. Mater. Sci. Lett. 2 (1983) 522.

[799] J. Weertman, J. Appl. Phys. 60 (1986) 1877.

[800] R. Raj, M.F. Ashby, Acta Metall. 23 (1975) 699.

[801] K. Davanas, A.A. Solomon, Acta Metall. Mater. 38 (1990) 1905.

[802] A.S. Argon, I.-W. Chen, C.W. Lau, in: R.M. Pelloux, N.S. Stoloff (Eds.), Proc. Symp. Creep Fatigue-Environment Interactions, TMS, Warrendale, 1980, p. 46.

[803] H. Riedel, Acta Metall. 32 (1984) 313.

[804] C.W. Chen, E.S. Machlin, Acta Metall. 4 (1956) 655.

[805] R.G. Fleck, D.M.R. Taplin, C.J. Beevers, Acta Metall. 23 (1975) 415.

[806] C. Gandhi, R. Raj, Acta Metall. 30 (1982) 505.

[807] D.K. Dewald, T.C. Lee, I.M. Robertson, H.K. Birnbaum, Metall. Trans. 21A (1990) 2411.

[808] I.-W. Chen, Metall. Trans. 14A (1983) 2289.

[809] M.H. Yoo, H. Trinkaus, Acta Metall. 34 (1986) 2381.

[810] H. Trinkaus, M.H. Yoo, Philos. Mag. 55 (1987) 269.

[811] T.G. Nieh, W.D. Nix, Scr. Metall. 14 (1980) 365.

[812] B.F. Dyson, M.S. Loveday, M.J. Rodgers, Proc. R. Soc. Lond. A 349 (1976) 245.

[813] T. Watanabe, P.W. Davies, Philos. Mag. 37 (1978) 649.

[814] J.N. Greenwood, D.R. Miller, J.W. Suiter, Acta Metall. 2 (1954) 250.

[815] B.M. Morrow, R.W. Kozar, K.R. Anderson, M.J. Mills, Acta Mater. 61 (2013) 4452.

[816] R. Lombard, H. Vehoff, Scr. Metall. Mater. 24 (1990) 581.

[817] B.J. Cane, Metal Sci. 13 (1979) 287.

[818] F.A. McClintock, J. Appl. Mech. 35 (1968) 363.

[819] Y.S. Lee, J. Yu, Metall. Trans. A 30A (1999) 2331.

[820] Y.K. Oh, G.S. Kim, J.E. Indacochea, Scr. Mater. 41 (1999) 7.

[821] E.P. George, R.L. Kennedy, D.P. Pope, Phys. Status. Solidi. A 167 (1998) 313.

[822] A. Yousefani, F.A. Mohamed, J.C. Earthman, Metall. Trans. 31A (2000) 2807.

[823] R.P. Wei, H. Liu, M. Gao, Acta Mater. 46 (1997) 313.

[824] J. Svoboda, V. Sklenicka, Acta Metall. Mater. 38 (1990) 1141.

[825] M.E. Kassner, T.C. Kennedy, K.K. Schrems, Acta Mater. 46 (1998) 6445.

[826] V. Randle, Mater. Sci. Forum 113–115 (1993) 189.

[827] M.S. Yang, J.R. Weertman, M. Roth, in: B. Wilshire, D.R.J. Owen (Eds.), Proc. Sec. Int. Conf. Creep and Fracture of Eng. Mater. and Structures, 1984, p. 149. Pineridge, Swansea.

[828] P.M. Andersen, R.G. Shewmen, Mech. Mater. 32 (2000) 175.

[829] D. Hull, D.E. Rimmer, Philos. Mag. 4 (1959) 673.

[830] M.V. Speight, W. Beere, Met. Sci. 9 (1975) 190.

[831] R. Raj, H.M. Shih, H.H. Johnson, Scr. Met. 11 (1977) 839.

[832] J. Weertman, Scr. Metall. 7 (1973) 1129.

[833] L. Agudo Jácome, P. Nörtershäuser, C. Somsen, A. Dlouhy, G. Eggeler, Acta Mater. 69 (2014) 246.

[834] T.G. Nieh, W.D. Nix, Acta Metall. 28 (1980) 557.

[835] D.A. Miller, T.G. Langdon, Metall. Trans. 11A (1980) 955.

[836] L.-E. Svensson, G.L. Dunlop, Met. Sci. 16 (1982) 57.

[837] M.D. Hanna, G.W. Greenwood, Acta Metall. 30 (1982) 719.

[838] H.C. Cho, J. Yu, I.S. Park, Metall. Trans. 23A (1992) 201.

[839] S.E. Broyles, K.R. Anderson, J. Groza, J.C. Gibeling, Metall. Trans. 27 (1996) 1217.

[840] B.J. Cane, Met. Sci. 15 (1981) 302.

[841] J.M. Mintz, A.K. Mukherjee, Metall. Trans. 19A (1988) 821.

[842] R. Raj, Acta Metall. 26 (1978) 341.

[843] N.G. Needham, T. Gladman, Met. Sci. 14 (1980) 64.

[844] N.G. Needham, T. Gladman, Met. Sci. Technol. 2 (1986) 368.

[845] S.E. Stanzl, A.S. Argon, E.K. Tschegg, Acta Metall. 34 (1986) 2381.

[846] T.-J. Chuang, J.R. Rice, Acta Metall. 21 (1973) 1625.

[847] W.D. Nix, K.S. Yu, J.S. Wang, Metall. Trans. 14A (1983) 563.

[848] H.E. Evans, Met. Sci. J. 3 (1969) 33.

[849] A. Chakraborty, J.C. Earthman, Acta Mater. 45 (1997) 4615.

[850] B.L. Adams, Mater. Sci. Eng. A166 (1993) 59.

[851] T. Watanabe, Mater. Sci. Eng. A166 (1993) 11.

[852] B.F. Dyson, Met. Sci. 10 (1976) 349.

[853] J.R. Rice, Acta Metall. 29 (1981) 675.

[854] H. Riedel, Z. Metall. 76 (1985) 669.

[855] P.M. Anderson, J.R. Rice, Acta Metall. 33 (1985) 409.

[856] S.J. Williams, M.R. Bache, B. Wilshire, Mater. Sci. Technol. 26 (2010) 1332.

[857] A. Yousefiani, A.A. El-Nasr, F.A. Mohamed, J.C. Earthman, in: J.C. Earthman, F.A. Mohamed (Eds.), Creep and Fracture of Engineering Materials and Structures, TMS, Warrendale, 1997, p. 439.

[858] E. van der Giessen, V. Tvergaard, Int. J. Fract. 48 (1991) 153.

[859] B.F. Dyson, in: R. Mishra, J.C. Earthman, S.V. Raj (Eds.), Creep Deformation: Fundamentals and Applications, TMS, Warrendale, PA, 2002, p. 309.

[860] J.W. Hancock, Met. Sci. 10 (1976) 319.

[861] W. Beere, M.V. Speight, Met. Sci. 12 (1978) 172.

[862] J.S. Wang, L. Martinez, W.D. Nix, Acta Metall. 31 (1983) 873.

[863] I.-W. Chen, A.S. Argon, Acta Metall. 29 (1981) 1759.

[864] Y.S. Lee, T.A. Kozlosky, T.J. Batt, Acta Metall. Mater. 41 (1993) 1841.

[865] J.M. Schneibel, L. Martinez, Scr. Metall. 21 (1987) 495.

[866] M. Lu, T.J. Delph, Scr. Metall. 29 (1993) 281.

[867] J. Cadek, Mater. Sci. Eng. A 117 (1989) L5.

[868] E. van der Giessen, M.W.D. van der Burg, A.I. Needleman, V. Tvergaard, J. Mech. Phys. Solids. 43 (1995) 123.

[869] T.J. Delph, Metall. Mater. Trans. 33A (2002) 383.

[870] A.C.F. Cocks, M.F. Ashby, Met. Sci. 16 (1982) 465.

[871] L.E. Forero, D.A. Koss, Scr. Met. Mater. 31 (1994) 419.

[872] B.Q. Han, D.C. Dunand, in: R.S. Mishra, J.C. Earthman, S.V. Raj (Eds.), Creep Deformation: Fundamentals and Applications, TMS, Warrendale, PA, 2002, p. 377.

[873] Y. Huang, J.W. Hutchinson, V. Tvergaard, J. Mech. Phys. Solids 39 (1991) 223.

[874] D.G. Harlow, T.G. Delph, J. Eng. Mater. Technol., Trans. ASME 122 (2000) 342.

[875] A.H. Sherry, R. Pilkington, Mater. Sci. Eng. A172 (1993) 51.

[876] S.H. Ai, V. Lupinc, M. Maldini, Scr. Metall. Mater. 26 (1992) 579.

[877] W.D. Nix, D.K. Matlock, R.J. Dimelfi, Acta Metall. 25 (1977) 495.

[878] J.D. Landes, J.A. Begley, Mechanics of Crack Growth, ASTM STP 590, ASTM, 1976, p. 128.
[879] C. Wiesner, J.C. Earthman, G. Eggeler, B. Ilschner, Acta Metall. 37 (1989) 2733.
[880] J.T. Staley, A. Saxena, Acta Metall. Mater. 38 (1990) 897.
[881] M. Tabuchi, K. Kubo, K. Yogi, in: B. Wilshire, R.W. Evans (Eds.), Creep and Fracture of Engineering Materials and Structures, Inst. Metals, London, 1993, p. 449.
[882] R. Raj, S. Baik, Met. Sci. 14 (1980) 385.
[883] W.E. Churley, J.C. Earthman, Metall. Trans. 28A (1997) 763.
[884] D.S. Wilkinson, V. Vitek, Acta Metall. 30 (1982) 1723.
[885] D.S. Wilkinson, Mater. Sci. Eng. 49 (1981) 31.
[886] D.A. Miller, R. Pilkington, Metall. Trans. 11A (1980) 177.
[887] J. Wolfenstine, Trans. Br. Ceram. Soc. 89 (1990) 175.
[888] J.N. Wang, M. Toriumi, Mater. Sci. Eng. A187 (1994) 97.
[889] D.M. Owen, T.G. Langdon, Mater. Sci. Eng. A216 (1996) 20.
[890] M. Pahutova, J. Cadek, P. Rys, Philos. Mag. 23 (1971) 509.
[891] V. Srivastava, K.R. McNee, H. Jones, G.W. Greenwood, Mater. Sci. Technol. 21 (2005) 701.
[892] B. Wilshire, C.J. Palmer, Scr. Mater. 46 (2002) 483.
[893] E.C. Muehleisen, C.R. Barrett, W.D. Nix, Scr. Metall. 4 (1970) 9.
[894] G. Gottstein, A.S. Argon, Acta Metall. 35 (1987) 1261.
[895] C.R. Barrett, O.D. Sherby, Trans. AIME 230 (1964) 1322.
[896] P. Geantil, B. Devincre, M.E. Kassner, Advanced Materials Modeling for Structures, Springer, Berlin, 2013, p. 177.
[897] S.V. Raj, T.G. Langdon, Acta Metall. 39 (1991) 1823.
[898] H. Duong, J. Wolfenstine, J. Am. Ceram. Soc. 74 (1991) 2697.
[899] D.S. Lloyd, J.D. Embury, Met. Sci. 4 (1970) 6.
[900] J.G. Harper, L.A. Shepard, J.E. Dorn, Acta Metall. 6 (1958) 509.
[901] M.E. Kassner, Acta Mater. 52 (2004) 1.
[902] N. Prasad, G. Malkondaiah, P. Rama Rao, Scr. Metall. 26 (1992) 541.
[903] M.G. Justice, E.K. Graham, R.E. Tressler, I.S.T. Tsong, Geophys. Res. Lett. 9 (1982) 1005.
[904] T.G. Langdon, Z. Metall. 96 (2005) 522.
[905] K.R. McNee, H. Jones, G.W. Greenwood, in: J.D. Parker (Ed.), Creep and Fracture of Engineering Materials and Structures, Inst. Metals, London, 2001, p. 3.
[906] M. Biberger, W. Blum, Scr. Metall. 23 (1989) 1419.
[907] W.R. Cannon, T.G. Langdon, J. Mater. Sci. 18 (1983) 1.
[908] W.R. Cannon, T.G. Langdon, J. Mater. Sci. 23 (1988) 1.
[909] M.D. Uchic, D.M. Dimiduk, J.N. Florando, W.D. Nix, Science 303 (2004) 986.
[910] J.A. Van Orman, Geophys. Res. Lett. 31 (2004) 1.
[911] J. Wolfenstine, D.L. Kohlstedt, J. Mater. Sci. 23 (1988) 3550.
[912] R.L. Cummerow, J. Appl. Phys. 34 (1963) 1724.
[913] W.S. Rothwell, A.S. Neiman, J. Appl. Phys. 36 (1965) 2309.
[914] T.G. Langdon, J.A. Pask, Acta Metall. 18 (1970) 505.
[915] [a] J.H. Hensler, G.V. Cullen, J. Am. Ceram. Soc. 51 (1968) 178;
      [b] R.T. Tremper, (Ph.D. dissertation), University of Utah, 1971.
[916] J.P. Porrier, Philos. Mag. 26 (1972) 701.
[917] W. Blum, B. Ilschner, Phys. Status. Solidi. 20 (1967) 629.

[918] P. Burke, (Ph.D. dissertation), Stanford University, 1968.

[919] C. Relandeau, Geophys. Res. Lett. 8 (1981) 733.

[920] M. Darot, Y. Guerguen, J. Geophys. Res. B86 (1981) 6219.

[921] D.L. Kohlstedt, C. Goetze, J. Geophys. Res. 79 (1974) 2045.

[922] M.B. Schwenn, C. Goetze, Techtonophysics 49 (1978) 41.

[923] S.V. Raj, Scr. Metall. 19 (1985) 1069.

[924] V. Raman, S.V. Raj, Scr. Metall. 19 (1985) 629.

[925] [a] J.L. Routbort, Acta Metall. 27 (1979) 649;
      [b] D.R. Cropper, T.G. Langdon, Philos. Mag. 18 (1968) 1181.

[926] G. Streb, B. Reppich, Phys. Status Solidi 16A (1972) 493.

[927] D.R. Cropper, J.A. Pask, Philos. Mag. 27 (1973) 1105.

[928] A.L. Ruoff, C.V.S. Narayan Rao, J. Am. Ceram. Soc. 58 (1975) 503.

[929] E.C. Yu, J.C.M. Li, Philos. Mag. 36 (1977) 811.

[930] S. Karthikeyan, G.B. Viswanathan, M.J. Mills, Acta Mater. 52 (2004) 2577–2589.

[931] S. Karthikeyan, G.B. Viswanathan, P.I. Gouma, V.K. Vasudevan, Y.W. Kim, M.J. Mills, Mater. Sci. Eng. A 329–331 (2002) 621.

[932] S. Karthikeyan, Mechanisms and Effect of Microstructure on High Temperature Deformation of γ-TiAl Based Alloys (Ph.D. dissertation), Ohio State University, 2003.

[933] M.A. Morris, T. Lipe, Intermetallics 5 (1997) 329.

[934] B. Viguier, K.J. Hemker, J. Bonneville, F. Louchet, J.L. Martin, Philos. Mag. A 71 (1995) 1295.

[935] S. Sriram, D.M. Dimiduk, P.M. Hazzledine, V.K. Vasudevan, Philos. Mag. A 76 (1997) 965.

[936] W.T. Marketz, F.D. Fisher, H. Clemens, Int. J. Plast. 19 (2003) 281–321.

[937] A. Chatterjee, H. Mecking, E. Arzt, H. Clemens, Mater. Sci. Eng. A 329–331 (2002) 840.

[938] P. Kumar, M.E. Kassner, Scr. Mater. 60 (2009) 60.

[939] F. Perdrix, M.F. Trichet, J.L. Bonnentien, M. Cornet, J. Bigot, Intermetallics 9 (2001) 807.

[940] P.I. Gouma, M.J. Mills, Y.K. Kim, Philos. Mag. Lett. 78 (1998) 59.

[941] H.Y. Kim, G. Wegmann, K. Maruyama, Mater. Sci. Eng. A 329 (2002) 795.

[942] G. Wegmann, T. Suda, K. Maruyama, Intermetallics 8 (2000) 165.

[943] K. Maruyama, H.Y. Kim, H. Zhu, Mater. Sci. Eng. A 387–389 (2004) 910.

[944] P.D. Crofts, P. Bowen, I.P. Jones, Scr. Mater. 35 (1996) 1391.

[945] R. Yamamoto, K. Mizoguchi, G. Wegmann, K. Maruyama, Intermetallics 6 (1998) 699.

[946] W. Blum, S. Straub, S. Vogler, in: D.B. Brandon, R. Chaim, A. Rosen (Eds.), Strength of Metals and Alloys, ICSMA, Haifa, 1991, p. 111.

[947] P. Weidinger, (Ph.D. thesis), University of Erlangen – Nuremberg, 1998.

[948] M.E. Kassner, P. Kumar, W. Blum, Int. J. Plast. 23 (2007) 980.

[949] H.M. Zbib, T.D. de la Rubia, M. Rhee, J.P. Hirth, J. Nucl. Mater. 276 (2000) 154.

[950] D. Rodney, R. Phillips, Phys. Rev. Lett. 82 (1998) 1704.

[951] J.C. Fisher, E.W. Hart, R.H. Pry, Phy. Rev. 87 (1952) 958.

[952] P. Kumar, M.E. Kassner, T.G. Langdon, J. Mater. Sci. 43 (2008) 4801.

[953] A. Borbély, W. Blum, T. Ungar, Mater. Sci. Eng. 276 (2000) 186.

[954] M.E. Kassner, M.A. Wall, Metall. Mater. Trans. 30A (1999) 777.

[955] H. Mughrabi, T. Ungar, Dislocations in Solids (Chapter 60), Elsevier, 2002, p. 345.

[956] T. Ungar, H. Mughrabi, D. Ronnpagel, M. Wilkens, Acta Met. 32 (1984) 332.

[957] H. Mughrabi, T. Ungar, W. Kienle, M. Wilkens, Philos. Mag. A 53 (1986) 793.

[958] B. Jakobsen, H.F. Poulsen, U. Lienert, J. Almer, S.D. Shastri, H.O. Sørensen, C. Gundlach, W. Panteleon, Science 32 (2006) 889.

[959] T. Ungar, H. Mughrabi, M. Wilkens, A. Hilscher, Philos. Mag. A 64 (1994) 495.

[960] M. Hecker, E. Thiele, C. Holste, Mater. Sci. Eng. A234-236 (1997) 806.

[961] H. Mughrabi, F. Pschenitzka, Philos. Mag. 85 (2005) 3029.

[962] H. Mughrabi, Private Communication, Los Angeles, CA, March 2006.

[963] B. Tippelt, J. Bretschneider, P. Hahner, Phys. Status Solidi A 163 (1997) 11.

[964] J. Bretschneider, C. Holste, B. Tippelt, Acta Mater. 45 (1997) 3775.

[965] M. Cabibbo, W. Blum, E. Evangelista, M.E. Kassner, M.A. Meyers, Metall. Mater. Trans. 39A (2008) 181.

[966] H. Mughrabi, Philos. Mag. A 86 (2006) 4037.

[967] H. Mughrabbi, Acta Mater. 54 (2006) 3417.

[968] A.S. Argon, W.C. Moffatt, Acta Metall. 29 (1981) 293.

[969] L.E. Levine, B.C. Larson, W. Yang, M.E. Kassner, J.Z. Tischler, M.A. Delos-Reyes, R.J. Fields, W. Liu, Nat. Mater. 5 (2006) 619.

[970] R. Sedlacek, W. Blum, J. Kratochvil, S. Forest, Metall. Mater. Trans. 33A (2002) 319.

[971] Z.S. Basinksi, S.J. Basinski, Prog. Mater. Sci. 36 (1992) 89–148.

[972] C. Laird, in: F.R.N. Nabarro (Ed.), Dislocations in Solids, vol. 6, North Holland, 1983, pp. 1–120.

[973] B.C. Larson, W. Yang, G.E. Ice, J.D. Budai, J.S. Tischler, Nature 415 (2002) 887.

[974] G.E. Ice, J.S. Chung, J.S. Tischler, A. Lunt, L. Assoufid, Rev. Sci. Instr. 71 (2000) 2635.

[975] Y. Li, Q.P. Kong, Phys. Status Solidi 113 (1989) 345.

[976] P. Veyssiere, G. Saada, in: F.R.N. Nabarro, M.S. Duesbery (Eds.), Dislocations in Solids, Elsevier, 1996, p. 254. Chapter 53.

[977] R.C. Reed, The Superalloys: Fundamentals and Applications, Cambridge Press, Cambridge, 2006.

[978] Y.-Q. Sun, Creep and Rafting of γ/γ' Superalloys, Internal Rept. AFML, Wright Patterson, AFB, OH, May 29, 2007.

[979] F.J. Bremer, M. Beyes, E. Karthaus, A. Hellwig, T. Schober, J.-M. Welter, H. Wenzl, J. Cryst. Growth 87 (1988) 185.

[980] T.M. Pollock, A.S. Argon, Acta Metall. Mater. 42 (1994) 1859.

[981] S. Karthikeyan, R.R. Unicic, P.M. Sarosi, G.B. Viswanathan, D.D. Whitis, M.J. Mills, Scr. Mater. 54 (2006) 1157.

[982] M. Kolbe, Mater. Sci. Eng. A 319–321 (2001) 383.

[983] P.M. Sarosi, R. Srinivasan, G.F. Eggeler, M.V. Nathal, M.J. Mills, Acta Mater. 55 (2007) 2509.

[984] G.R. Leverant, B.H. Kear, Metall. Trans. 1 (1970) 491.

[985] T.M. Pollock, A.S. Argon, Acta Metall. 40 (1992) 1.

[986] M. Fahrmann, P. Fratzel, O. Paris, E. Fahrmann, W.C. Johnson, Acta Metall. Mater. 43 (1995) 1007.

[987] J.K. Tien, S.M. Copley, Metall. Trans. 2 (1971) 543.

[988] N. Matan, D.C. Cox, C.F.M. Rae, R.C. Reed, Acta Mater. 47 (1999) 2031.

[989] J.X. Zhang, T. Murakumo, H. Harada, Y. Koizumni, T. Kobayashi, in: K.A. Green, T.M. Pollock, H. Harada (Eds.), Superalloys 2004, TMS, Warrendale, PA, 2004, p. 189.

[990] Y. Koizumi, T. Kobayashi, T. Yokokawa, et al., in: K.A. Green, T.M. Pollock, H. Harada (Eds.), Superalloys 2004, TMS, Warrendale, PA, 2004, p. 35.

[991] R.C. Reed, N. Matan, D.C. Cox, M.A. Rist, C.M.F. Rae, Acta Mater. 47 (1999) 3367.

[992] R. Srinivasan, G.F. Eggeler, M.J. Mills, Acta Mater. 48 (2000) 4867.

[993] G. Eggeler, A. Dlouhy, Acta Mater. 45 (1997) 4251.

[994] G.B. Viswanathan, P.M. Sarosi, M.F. Henry, D.D. Whitis, W.W. Milligan, M.J. Mills, Acta Mater. 53 (2005) 3041.

[995] B. Decamps, S. Raujol, A. Coujou, F. Pettinari-Sturmel, N. Clement, D. Locq, P. Caron, Philos. Mag. A 84 (2004) 91.

[996] D. Mukherji, F. Jiao, R.P. Wahi, Acta Metall. Mater. 39 (1991) 1515.

[997] L.J. Carroll, Q. Feng, T.M. Pollock, Metall. Mater. Trans. 29A (2008) 1290.

[998] M.V. Nathal, L.J. Ebert, Metall. Trans. 16A (1985) 427.

[999] W. Schneider, J. Hammer, H. Mughrabi, in: S.D. Antolovich, et al. (Eds.), Superalloys 1992, TMS, Warrendale, 1992, p. 589.

[1000] T.E. Strangma, G.S. Hoppin, C.M. Philips, K. Harris, R.W. Schwer, in: J.K. Tien, et al. (Eds.), Superalloys 1980, ASM, Materials Park, OH, 1980, p. 215.

[1001] K. Harris, G.L. Erickson, S.L. Skikenga, W.D. Brentnall, J.M. Aurrecoechea, K.G. Kubarych, in: S.D. Antolovich, et al. (Eds.), Superalloys 1992, TMS, Warrendale PA, 1992, p. 297.

[1002] J.X. Zhang, H. Harada, Y. Koizumi, J. Mater. Res. 21 (2006) 2006.

[1003] T. Lin, A. Epishin, M. Klaus, U. Bruckner, A. Reznicek, Mater. Sci. Eng. A 405 (2005) 254.

[1004] A. Epishin, T. Link, Philos. Mag. A 84 (2004) 1979.

[1005] T. Pollock, Private Communication, July 2008.

[1006] J.Y. Buffmire, M. Ignat, Acta Metall. Mater. 43 (1995) 1791.

[1007] T.M. Pollock, S. Tin, J. Propul. Power 22 (2006) 361.

[1008] N. Matan, D.C. Cox, P. Carter, M.A. Rist, C.M.F. Rae, R.C. Reed, Acta Mater. 47 (1999) 1549.

[1009] F.J. Bremer, M. Beyss, E. Karthaus, A. Hellwig, T. Schober, J.M. Welter, H. Wenzl, J. Cryst. Growth 87 (1988) 185.

[1010] H. Okamoto, J. Phase Equilib. Diffus. 25 (4) (2004) 394.

[1011] B. Wilson, G. Fuchs, J. Met. (2008) 43.

[1012] Y. Cheng, M. Chauhan, F. Mohamed, Metall. Mater. Trans. 30A (2009) 80.

[1013] W. Klement, R.H. Willens, P. Duwez, Nature 187 (1960) 869.

[1014] C.A. Schuh, T.C. Hufnagel, U. Ramamurty, Acta Mater. 55 (2007) 4067–4109.

[1015] H.W. Kui, A.L. Greer, D. Turnbull, Appl. Phys. Lett. 45 (1984) 615–616.

[1016] N. Nishiyama, A. Inoue, Mater. Trans. 37 (1996) 1531–1539.

[1017] I.R. Lu, G. Wilde, G.P. Görler, R. Willnecker, J. Non-Cryst. Solids 250-2 (1999) 577–581.

[1018] A. Inoue, H. Yamaguchi, T. Zhang, T. Masumoto, Mater. Trans. 31 (1990) 104–109.

[1019] A. Inoue, T. Zhang, T. Masumoto, Mater. Trans. 31 (1990) 177–183.

[1020] A. Peker, W.L. Johnson, Appl. Phys. Lett. 63 (1993) 2342–2344.

[1021] J. Schroers, W.J. Johnson, Appl. Phys. Lett. 84 (2004) 3666–3668.

[1022] V. Ponnambalam, S.J. Poon, G.J. Shiftlet, J. Mater. Res. 19 (2004) 1320–1323.

[1023] Z.P. Lu, C.T. Liu, J.R. Thompson, W.D. Porter, Phys. Rev. Lett. 92 (2004) 245503.

[1024] M.M. Trexler, N. Thandani, Prog. Mater. Sci. 55 (2010) 759–839.

[1025] C. Suryanarayana, A. Inoue, Bulk Metallic Glasses, CRC Press, 2011.

[1026] A. Inoue, B.L. Shen, H. Koshiba, H. Kato, A.R. Yavari, Nat. Mater. 2 (2003) 661–663.

[1027] C.L. Qin, W. Zhang, K. Asami, H.M. Kimura, X.M. Wang, A. Inoue, Acta Mater. 54 (2006) 3713–3719.

[1028] C.L. Qin, W. Zhang, K. Asami, N. Ohtsu, A. Inoue, Acta Mater. 53 (2005) 3903–3911.

[1029] H.M. Fu, H.F. Zhang, H. Wang, Q.S. Zhang, Z.Q. Hu, Scr. Mater. 52 (2005) 669–673.

[1030] J. Das, M.B. Tang, K.B. Kim, R. Theissmann, F. Baier, W.H. Wang, J. Eckert, Phys. Rev. Lett. 94 (2005) 205501-1–205501-4.

[1031] D.H. Xu, B. Lohwongwatana, G. Duan, W.L. Johnson, C. Garland, Acta Mater. 52 (2004) 2621–2624.

[1032] X.J. Gu, S.J. Poon, G.J. Shiflet, J. Mater. Res. 22 (2007) 344–351.

[1033] J.H. Yao, J.Q. Wang, Y. Li, Appl. Phys. Lett. 92 (2008) 251906-1–251906-3.

[1034] K. Amiya, A. Urata, N. Nishiyama, A. Inoue, Mater. Trans. 45 (2004) 1214–1218.

[1035] F.J. Liu, Q.W. Yang, S.J. Pang, C.L. Ma, T. Zhang, Mater. Trans. 49 (2008) 231–234.

[1036] A. Inoue, B.L. Shen, C.T. Chang, Acta Mater. 52 (2004) 4093–4099.

[1037] X.J. Gu, S.J. Poon, G.J. Shiflet, Scr. Mater. 57 (2007) 289–292.

[1038] D. Chen, A. Takeuchi, A. Inoue, Mater. Sci. Eng. A 457 (2007) 226–230.

[1039] J.H. Na, J.M. Park, K.H. Han, B.J. Park, W.T. Kim, D.H. Kim, Mater. Sci. Eng. A 431 (2006) 306–310.

[1040] K.F. Yao, Y.Q. Yang, B. Chen, Intermetallics 15 (2007) 639–643.

[1041] L. Liu, A. Inoue, T. Zhang, Mater. Trans. 46 (2005) 376–378.

[1042] J. Schroers, W.L. Johnson, Phys. Rev. Lett. 93 (2004) 255506-1–255506-4.

[1043] C.L. Ma, H. Soejima, S. Ishihara, K. Amiya, N. Nishiyama, A. Inoue, Mater. Trans. 45 (2004) 3223–3227.

[1044] G.Q. Xie, D.V. Louzguine-Luzgin, H.M. Kimura, A. Inoue, Mater. Trans. 48 (2007) 1600–1604.

[1045] R.D. Conner, R.B. Dandliker, W.L. Johnson, Acta Metall. 46 (1998) 6089–6102.

[1046] H. Choi-Yim, R.D. Conner, F. Szuecs, W.L. Johnson, Acta Mater. 50 (2002) 2737–2745.

[1047] J. Schroers, J. Met. (2005) 35–39.

[1048] Y. Kawamura, T. Nakamura, A. Inoue, T. Masumoto, Mater. Trans. 40 (1999) 794–803.

[1049] Y. Kawamura, T. Nakamura, A. Inoue, Scr. Mater. 39 (1998) 301–306.

[1050] T.G. Nieh, T. Mukai, C.T. Liu, J. Wadsworth, Scr. Mater. 40 (1999) 1021–1027.

[1051] A. Reger-Leonhard, M. Heilmaier, J. Eckert, Scr. Mater. 43 (2000) 459–464.

[1052] Y. Saotome, T. Hatori, T. Zhang, A. Inoue, Mater. Sci. Eng. A 304 (2001) 716–720.

[1053] J.P. Chu, C.L. Chiang, T.G. Nieh, Y. Kawamura, Intermetallics 10 (2002) 1191–1195.

[1054] W.J. Kim, D.S. Ma, H.G. Jeong, Scr. Mater. 49 (2003) 1067–1073.

[1055] J.P. Chu, C.L. Chiang, T. Mahalingam, T.G. Nieh, Scr. Mater. (2003) 435–440.

[1056] D.H. Bae, J.M. Park, J.H. Na, D.H. Kim, Y.C. Kim, J.K. Lee, J. Mater. Res. 19 (2004) 937–942.

[1057] C.L. Chiang, J.P. Chu, C.T. Lo, Z.X. Wang, W.H. Wang, J.G. Wang, et al., Intermetallics 12 (2004) 1057–1061.

[1058] M. Bletry, P. Guyot, Y. Brechet, J.J. Blandin, J.L. Soubeyroux, Mater. Sci. Eng. A 387-9 (2004) 1005–1011.

[1059] G. Wang, J. Shen, J.F. Sun, Y.J. Huang, J. Zou, Z.P. Lu, et al., J. Non-Cryst. Solids 351 (2005) 209–217.

[1060] J.C.M. Li, Mech. Prop. Amorphous Met. 2 (1982) 1335–1340.

[1061] J.J. Gilman, J. Appl. Phys. 44 (1973) 675–679.

[1062] J.C.M. Li, Metallic Glasses, American Society for Metals, 1978 (Chapter 9).

[1063] F. Spaepen, Acta Metall. 25 (1977) 407–415.

[1064] A.S. Argon, Acta Metall. 27 (1979) 47–58.

[1065] A.S. Argon, L.T. Shi, Acta Metall. 31 (1983) 499–507.

[1066] J. Lu, G. Ravichandran, W.L. Johnson, Acta Mater. 51 (2003) 3429–3443.

[1067] A. Lund, C. Schuh, Intermetallics 12 (2004) 1159–1165.

[1068] F. Spaepen, Scr. Mater. 54 (2006) 363–367.

[1069] T.G. Nieh, J. Wadsworth, Scr. Mater. 55 (2006) 387–392.

[1070] A. Masuhr, R. Busch, W.L. Johnson, J. Non-Cryst. Solids 250-2 (1) (1999) 566–571.

[1071] K.W. Park, C.M. Lee, E. Fleury, J.C. Lee, Scr. Mater. 61 (2009) 363–366.

[1072] S.X. Song, Y.H. Lai, J.C. Huang, T.G. Nieh, Appl. Phys. Lett. 94 (2009) 061911.

[1073] K.W. Park, C.M. Lee, M. Wakeda, Y. Shibutani, M.L. Falk, J.C. Lee, Acta Mater. 56 (2008) 5440–5450.

[1074] P. De Hey, J. Sietsma, A. Van Den Beukel, Acta Mater. 46 (1998) 5873–5882.

[1075] D. Fátay, J. Gubicza, J. Lendvai, J. Alloys Compd. 75 (2007) 434–435.

[1076] B.S.S. Daniel, M. Heilmaier, A. Reger-Leonhard, J. Eckert, L. Schultz, MRS Proc. 644 (2001) L10.7.

[1077] M.L. Falk, Phys. Rev. B 60 (1999) 7062.

[1078] Y. Shi, M.L. Falk, Scr. Mater. 54 (2006) 381–386.

[1079] Y. Shi, M.L. Falk, Phys. Rev. Lett. 95 (2005) 095502.

[1080] L. Sun, M.Q. Jiang, L.H. Dai, Scr. Mater. 63 (2010) 945–948.

[1081] M.W. Chen, A. Inoue, W. Zhang, T. Sakurai, Phys. Rev. Lett. 96 (2006) 245502–245504.

[1082] M. Heggen, F. Spaepen, M.J. Feuerbacher, J. Appl. Phys. 97 (2005) 033506 1–8.

[1083] A. Inoue, W. Zhang, T. Tsurui, A.R. Yavari, A.L. Greer, Philos. Mag. Lett. 85 (2005) 221–229.

[1084] J. Shen, G. Wang, J.F. Sun, Z.H. Stachurski, C. Yan, L. Ye, B.D. Zhou, Intermetallics 13 (2005) 79–85.

[1085] Q. Wang, J.J. Blandin, M. Suery, B. Van de Moortéle, J.M. Pelletier, J. Mater. Sci. Technol. 19 (2003) 557–560.

[1086] S.C. Lee, C.M. Lee, J.W. Yang, J.C. Lee, Scr. Mater. 58 (2008) 591–594.

[1087] H.B. Ke, P. Wen, H.L. Ping, W.H. Wang, A.L. Greer, Scr. Mater. 64 (2011) 966–969.

[1088] T. Fujita, Z. Wang, Y. Liu, H. Sheng, W. Wang, M. Chen, Acta Mater. 60 (2012) 3741–3747.

[1089] K. Zhao, X.X. Xia, H.Y. Bai, D.Q. Zhao, W.H. Wang, Appl. Phys. Lett. 98 (2011) 141913-3.

[1090] C.A. Volkert, A. Donohue, F. Spaepen, J. Appl. Phys. 103 (2008) 083539.

[1091] H. Guo, P.F. Yan, Y.B. Wang, J. Tan, Z.F. Zhang, M.L. Sui, E. Ma, Nat. Mater. 6 (2007) 735–739.

[1092] M.C. Brandes, K. Kinsel, M.J. Mills, Proceedings of the 12th International Creep and fracture of materials and structures, Kyoto, Japan, ISBN978-4-88903-407-3 C3057, Japan Institute of Metals, Sendai, Japan, (2012).

[1093] A.W. Thompson, B.C. Odegard, Metall. Trans. 4 (1973) 899–908.

[1094] T. Kameyama, T. Matsunaga, E. Sato, K. Kuribayashi, Mater. Sci. Eng. A, 510-1, 364–367.

[1095] C. Liu, P. Liu, Z. Zhao, D.O. Northwood, Mater. Des. 22 (2001) 325–328.

[1096] A. Oehlert, A. Atrens, Acta Mater. 42 (1994) 1493–1508.

[1097] E. Krempl, J. Mech. Phys. Solids 27 (1979) 363–375.

[1098] V. Sklenicka, J. Dvorak, P. Knal, M. Svoboda, M. Kvapilova, V.I. Kopylov, S.A. Nekulin, S.V. Dobatkin, Acta Physica Polonica A 122 (2012) 485.

[1099] R. Castro, R. Tricot, D. Rousseau, Steel Strengthening Mechanisms, Climax Molybdenum Co, 1969, pp. 117–134.

[1100] M.E. Kassner, Y. Kosaka, J. Hall, Metall. Trans. 30A (1999) 2383–2389.

[1101] A.W. Thompson, B.C. Odegard, Metall. Trans. 5 (1974) 1207–1213.

[1102] Y.V.R.K. Prasad, D.H. Sastry, K.I. Vasu, Mater. Sci. Eng. 6 (1970) 327–333.

[1103] Y.V.R.K. Prasad, T. Ramchandran, Scr. Metall. 5 (1971) 411–416.

[1104] J.R. Tesh, R.W. Whitworth, Phys. Status Solidi 39 (1970) 627–633.

[1105] T. Matsunaga, K. Takahashi, T. Kameyama, E. Sato, Mater. Sci. Eng. A 510-1 (2009) 356–358.

[1106] J.C.M. Li, Acta Metall. 11 (1963) 1269–1270.

[1107] P.R. Thornton, P.B. Hirsch, Philos. Mag. 31 (1958) 738–761.

[1108] N. Karanjgaokar, F. Stump, P. Geubelle, I. Chasiotis, Scr. Mater. 68 (2013) 551–554.

[1109] O.H. Wyatt, Proc. Phys. Soc. B 66 (1953) 459–480.

[1110] H. Conrad, Acta Metall. 6 (1958) 339–350.

[1111] N.F. Mott, Philos. Mag. 1 (1956) 568–572.

[1112] R. Zeyfang, R. Martin, H. Conrad, Mater. Sci. Eng. 8 (1971) 134–140.

[1113] A.C. Arko, J. Weertman, Acta Metall. 17 (1969) 687–699.

[1114] R.J. Arsenault, Acta Metall. 14 (1966) 831–838.

[1115] J.W. Glen, Philos. Mag. 1 (1956) 400–408.

[1116] D.H. Sastry, Y.V.R.K. Prasad, K.I. Vasu, J. Mater. Sci. 6 (1971) 1433–1440.

[1117] V.I. Startsev, V.P. Soldatov, V.D. Natsik, V.V. Abraimov, Phys. Status Solidi 59 (1980) 377–388.

[1118] A.I. Osetskii, V.P. Soldatov, V.I. Startsev, V.D. Natsik, Phys. Status Solidi 22 (1974) 739–748.

[1119] F.P. Phillips, Philos. Mag. 9 (1905) 513.

[1120] G.C.E. Olds, Proc. Phys. Soc. B 67 (1954) 832–842.

[1121] A. Seeger, Philos. Mag. 46 (1955) 1194–1217.

[1122] A. Seeger, J. Diehl, S. Mader, H. Rebstock, Philos. Mag. 2 (1957) 323–350.

[1123] C. Yen, T. Caulfield, L.D. Roth, J.M. Wells, J.K. Tien, Cryogenics 24 (1984) 371–377.

[1124] E.N. da C. Andrade, Proc. R. Soc. A 84 (1910) 1–12.

[1125] E. Orowan, The Creep of Metals, West Scotland Iron and Steel Institute, 1947.

[1126] H.E. Evans, B. Wilshire, Creep of metals and alloys, Inst. Met. (1985) 313 pp.

[1127] A.H. Cottrell, V. Aytekin, Nature 160 (1947) 328–329.

[1128] J.B. Conway, Numerical methods for creep and rupture analysis, Adv. Mater. Res. (1967) 212 pp.

[1129] F. Garofalo, C. Richmond, W.F. Domis, F. Von Gemmingen (Eds.), Joint International Conference on Creep, 1963, pp. 1–31.

[1130] P. Laurent, M. Eudier, Rev. Métall. 47 (1950) 582–587.

[1131] P. Chévenard, Rev. Métall. 31 (1934) 473–535.

[1132] W.H. Miller, R.T. Chen, E.A. Starke Jr, Metall. Trans. A 18A (1987) 1451–1468.

[1133] A.H. Cottrell, Philos. Mag. Lett. 73 (1996) 35–37.

[1134] A.H. Cottrell, Philos. Mag. Lett. 74 (1996) 375–379.

[1135] A.H. Cottrell, Philos. Mag. Lett. 75 (1997) 301–308.

[1136] F.R.N. Nabarro, Philos. Mag. Lett. 75 (1997) 227–233.

[1137] N.F. Mott, F.R.N. Nabarro, Dislocation theory and transient creep, in: Bristol Physical Society Conference, 1948, pp. 1–19.

[1138] D. Hull, D.J. Bacon, Introduction to Dislocations, fifth ed., Elsevier, 2011.

[1139] M.E. Kassner, P. Geantil, L.E. Levine, Int. J. Plast. 45 (2013) 44–60.

[1140] M.E. Kassner, R.S. Rasen, G.A. Henshall, Metall. Trans. 21A (1990) 3085–3100.

[1141] M.E. Kassner, Fundamentals of Creep in Metals and Alloys, Elsevier, 2009.

[1142] P.R. Landon, J.L. Lytton, L.A. Shepard, J.E. Dorn, Trans. ASM 51 (1959) 900–910.

[1143] M.E. Kassner, Metall. Trans. 20A (1989) 2001–2010.

INDEX

Note: Page numbers followed by "f" and "t" indicate figures and tables respectively.

图书在版编目（C I P）数据

金属与合金蠕变的基本原理：第 3 版：英文／（美）迈克尔·卡斯纳
（M. E. Kassner）著. --长沙：中南大学出版社，2017.9
ISBN 978 - 7 - 5487 - 2993 - 8

Ⅰ.①金… Ⅱ.①迈… Ⅲ.①金属材料－蠕变－研究－英文 ②合
金－蠕变－研究－英文 Ⅳ.①TG14 ②TG13

中国版本图书馆 CIP 数据核字（2017）第 230379 号

金属与合金蠕变的基本原理(第 3 版)
JINSHU YU HEJIN RUBIAN DE JIBEN YUANLI (DI 3 BAN)

M. E. Kassner   著

□责任编辑	胡 炜	
□责任印制	易红卫	
□出版发行	中南大学出版社	
	社址：长沙市麓山南路	邮编：410083
	发行科电话：0731 - 88876770	传真：0731 - 88710482
□印　装	长沙鸿和印务有限公司	

□开　本	720×1000　1/16	□印张 23.25	□字数 592 千字	
□版　次	2017 年 9 月第 1 版	□2017 年 9 月第 1 次印刷		
□书　号	ISBN 978 - 7 - 5487 - 2993 - 8			
□定　价	126.00 元			

图书出现印装问题，请与经销商调换